THE WAR ON BUGS

THE WAR ON BUGS

Will Allen

Chelsea Green Publishing
White River Junction,
Vermont

Developmental Editor: Ben Watson
Project Manager: Emily Foote
Copy Editor: Cannon Labrie
Concept and Form Editor: Kate Duesterberg
Senior Researcher: Ed De Anda
Proofreader: Nancy Ringer
Designer: Roberto Carra
Assistant Designer: Anna Pantaleo

Printed in the United States of America

First printing, January 2008

10 9 8 7 6 5 4 3 2 1   08 09 10 11

Library of Congress Cataloging-in-Publication Data

Allen, Will, 1936–
  The War on bugs / Will Allen.
      p. cm.
  Includes bibliographical references and index.
  ISBN 978-1-933392-46-2
  1. Pesticides—Environmental aspects—United States. 2. Agricultural
chemicals industry—United States—History.
I. Title.

  S584.75.U6A45 2007
  363.17'920973—dc22

2007034781

Our Commitment to Green Publishing
Chelsea Green sees publishing as a tool for cultural change and ecological stewardship. We strive to align our book manufacturing practices with our editorial
mission and to reduce the impact of our business enterprise in the environment.
We print our books and catalogs on chlorine-free recycled paper, using soy-based
inks whenever possible. This book may cost slightly more because we use recycled
paper, and we hope you'll agree that it's worth it. Chelsea Green is a member of
the Green Press Initiative (www.greenpressinitiative.org), a nonprofit coalition of
publishers, manufacturers, and authors working to protect the world's endangered
forests and conserve natural resources. *The War on Bugs* was printed on 78-lb.
Renew Matte, a 20-percent postconsumer-waste old-growth-forest-free recycled
paper supplied by RR Donnelley.

Chelsea Green Publishing Company
Post Office Box 428
White River Junction, VT 05001
(802) 295-6300
www.chelseagreen.com

Dedication

For my wife, Kate Duesterberg; my son, Coupe, and his wife, Shelby; my daughter, Zuri, and her husband, Mike; my late sons Cameron and Jeff; my grandsons Cole, Taylor, and Jordan and my granddaughter Zurael; and Jess, Jan, and Noah.

There is not a single instance in history in which the introduction of a major technological innovation has had only benign consequences for the natural world, or even for the social world. New technologies allowed corporations and nations to exploit and expropriate nature for short-term gain, before damages finally escalated to such a level that they became necessary to regulate. The actual gain from these new technologies has always been illusory and instead resulted in long-term pollution, depletion, or destabilization of some portion of the biosphere.

—Daniel Imhoff, August 2000

Acknowledgments

This project has taken more than ten years, and during that time an uncountable number of people have given me tips, criticism, or advertisements. Six people have been especially generous with their time, patience, and ideas. First is my wife, Kate Duesterberg, who read every word and edited every chapter several times. Her love, insight, inspiration, organizational skills, and support are irreplaceable. Second is Eddie De Anda, who helped me research advertisements, sales promotions, intellectual arguments, and bibliographic references. He painstakingly located and documented thousands of advertisements, editorials, articles, essays, and books. Third is Daniel Imhoff, whose early brainstorms with me developed the book's outline and title. He edited earlier versions of several chapters and helped write chapters 19, 21, and 22. His support and advocacy for the book have been unflagging. Fourth is Roberto Carra, who came up with his extraordinary vision of how to design and graphically present a book that has so many graphics. Fifth is Ben Watson, my editor on the book, to whom I owe the idea for writing it. After viewing our poster display of pesticide ads he encouraged me to write the book. He also edited an earlier version of the book and, in spite of its fractured nature, continued to encourage me. Sixth is my late son Cameron Allen, who researched, found, and purchased most of the rare books that I used as source material.

Special thanks go to Carolyn Mugar, who continuously prodded and encouraged me, and to her late husband, Johhny O'Connor, who was very supportive of the book until his untimely death.

Thanks go to Chelsea Green, whose publisher, Margo Baldwin, and chief editors, John Barstow and Shay Totten, and managing editor, Emily Foote, have been very cooperative and supportive of the book, as has the entire staff. Thanks also go to copyright attorney Ike Williams and his staff, who reviewed the book for possible copyright infringement.

I am very appreciative of supporters at the charitable foundations that gave me direct support for writing and research. These include Yvon and Melinda Chionaurd at the Patagonia Foundation; Doug Thompkins, Quincy Thompkins, and Chris McDivitt at the Foundation for Deep Ecology; and Peter Buckley at the Eco-Literacy Foundation. Other foundations funded our work at the Sustainable Cotton Project and indirectly contributed to the research and writing of the book. They include Carolyn Mugar, Glenda Yoder, and Harry Smith at Farm Aid; Susan Clark at the Columbia Foundation; Bruce Hirsch at the Heller Foundation; Jean Wallace Douglas at the Wallace Genetics Foundation; Diana Donlan at the Goldman Foundation; Donald Ross at Rockerfeller Brothers; Vic DeLuca at the Jessie Smith Noyes Foundation; and Oren Hesterman at the Kellogg Foundation.

We obtained our advertisements and conducted most of our research at the Shields Library at the University of California at Davis. We received considerable help from the periodicals department and invaluable help from the staff at the Special Collections division of the library, especially archivist John Skarstad. We also conducted extensive research at the Dartmouth College Baker Library in Hanover,

New Hampshire, and at the University of Vermont Library in Burlington, Vermont.

My thanks go to others who have been especially helpful, supportive, and tolerant of my obsession. These include my daughter Zuri Allen, who did library research and gave emotional support, her husband Michael Kanter, who researched pesticide use in California, and their daughter Zurael; my son Coupe Snyder, his wife Shelby, and their sons Jordan, Cole, and Taylor, who housed and entertained me on my frequent trips to the Davis Library; my stepdaughter Jessica Duesterberg Chavez and her husband Jan, who provided important European pesticide data and information on the E.U. REACH program; and my stepson Noah Duesterberg Chavez, who offered some important insights and lots of laughs. Eric Sotelo conducted research on the book in its early days, and he, Eddie De Anda, and JoAnne Baumgartner helped develop the first poster display that led to the writing of the book. To them I owe a great debt, as I do to Richard Reed, who helped fundraise for the book in its early days. An important breakthrough in my understanding came when several people clued me in to the importance of Dr. Seuss in the promotion of Standard Oil's pesticide Flit and of popularizing pesticides in general. Another important factor was understanding the chemical dominance in agriculture from the perspective of a crop duster. I owe Larry Landis great thanks for flying me over thousands of miles of farms to show me where the chemical damage was hidden.

Other friends who supported the project but prodded me to get it done include Ronnie Cummins, Chuck Benbrook, Ellen Hickey, Monica Moore, Teddy Goldsmith, John Stauber, Augie Feeder, Vandana Shiva, Debbie Barker, Andy Kimbrell, Jerry Mander, Mark Ritchie, Joe Mendelson, Brian Baker, Bob Bugg, Jamie Liebman, Amigo Cantisano, Rick Knoll, Christie Knoll, Vern Grubinger, Fred Magdoff, Luke Joanis, Justin Park, John Passacantando, and the late Al Krebs.

Thanks to the Bioneers, the Ecological Farming Association, NOFA-Vermont, and the Kellogg Foundation for letting me display posters of the ads at their conferences. These showings provided me with considerable feedback and ideas about where to search. Although I have received considerable help in preparing this book, any errors of misinterpretation are entirely my responsibility.

Contents

INSECTICIDE WARNING

Keep on the safe side with insecticides. The National Safety Council warns summer gardeners that plant sprays and poisons claim many child lives each year. Keep poisons locked up.

———◆———

"The agricultural population produces the bravest men, the most valiant soldiers, and a class of citizens the least given of all to evil designs."—Cato.

Preface

My interest in how farmers became comfortable with using dangerous chemicals began more than thirty years ago, as many of us converted our farms from chemical to organic production. Along with several close friends, I had come to the realization that we did not need to use so many dangerous poisons on our farms since we were getting good yields and high quality without them. This realization was an epiphany for those of us brought up believing that the chemicals were Necessary, Critical, Essential, Modern, Progressive, Profitable, Economical, Miraculous, even Heroic—all in capital letters.

At farm field days, meetings, potlucks, and Farm Aid concerts, we began recounting how each of us had become convinced that farm chemicals were indispensable. All of us recalled how farmers, extension agents, schoolteachers, feed store salesmen, and billboard ads claimed that the chemicals were miraculously effective and safe.

As farm kids, we knew that the chemicals were effective. We knew that arsenic, nicotine, and lead killed pests and that the chemical fertilizers produced good yields, even though most of our folks were small farmers who rarely used them. All of us knew, however, that the claims about safety were B.S., because we would get our butts whipped if we went near the chemical storehouse. At my grandma's farm in Hemet, California, she and my aunt repeatedly told us to keep away from the shed with the chemicals. At home, my mom would always warn us: "Remember Bobby Arbuckle? He played with arsenic, and he's dead." Then she would follow with, "And don't forget that boy Danny what's-his-name, who lived down the road—he got into that Black Leaf 40 tobacco poison and it burned him like a fire."

One time, a friend and I were smoking one of our first homemade cigarettes made from straw and a little bit of tobacco from a cigarette butt we found. We were smoking and coughing behind his father's fertilizer shed and the manager of the ranch caught us. He was hopping mad. He chased us away with a stick, yelling after us that if we got one spark on the fertilizer it could blow up the whole place. Threats and warnings such as these definitely had their impact. They convinced all of us that, while farm chemicals produced bumper crops, they were dangerous.

In contrast to our fears, and all the threats and warnings, we all had a story or two to tell about hearing local large-scale farmers who laughed off anyone's concerns about health and safety as being ridiculous. Instead, these farmers, and the chemical salesmen they hung out with, emphasized that

From *Pacific Rural Press*, 1935. From farmer collection.

BASIC TYPE
ARSENATE OF LEAD
DRY POWDERED

FLAG
BRAND

MANUFACTURED BY
National Chemical Co. of California
Home Office   Manufactured at
SAN FRANCISCO   PITTSBURG, CALIF.
Distributed in Southern California by
McKesson-Western Wholesale Drug Co.,   Los Angeles, Calif.

From *California Cultivator*, May 1930.

pesticides and fertilizers were not only safe to use, but also necessary to make a profit, to conduct the Second World War, and to feed all the hungry people in the world. "If you read the label and follow the instructions, you can't get hurt," they would say.

I remembered how my teachers used to praise the war chemicals, the scientific revolution, and the heroic effect of DDT against typhus and malaria during the war. In class we watched documentary newsreels praising the chemicals. Friends found old articles where similar praise appeared in newspapers and magazines. Others recalled hearing reports on the radio or seeing movie newsreels and shorts at Saturday afternoon matinees that also praised the virtues of the war chemicals.

Several of us recalled when the feedstore salesmen began selling DDT near the close of World War II. Our salesman and family friend, Arnold, came to the house and delivered a practiced speech about the safety and effectiveness of this scientific wonder. He told us that DDT was a war hero, deadly to insects, typhus, and malaria, but harmless to people. He opened a bottle of the stuff, and in a few minutes, flies on the table and floor began writhing around in their death throes. Within the hour, all the flies and mosquitoes in the house died. He claimed that just opening the bottle usually "knocked 'em dead." We all snickered because his pitch was so canned. But we didn't laugh at how effective DDT was. Its killing power amazed us, and my folks bought some that day.

We always had some cows, a few goats, chickens, turkeys, ducks, lots of pigs—and, as a result, too many flies. My least favorite chore as a kid was swatting flies in the house and around the outside of the doorways. Right after we got DDT, my fly-swatting chores all but

vanished. My mom began to spray the flies and all us kids with the hand-pump sprayer filled with DDT. "Arnold said it won't hurt you," she would say as she used her "Flit Gun" to zap the flies, and us. For a while after that, I swatted only the flies that survived to enter the house. I loved not having to kill as many flies. But I hated to spray or to be sprayed with the stuff; it stunk, and it left a sticky-sweet film on my skin.

At the time, in spite of its significant drawbacks, DDT seemed better than fly swatting to this twelve-year-old farm kid. However, despite all the advertisements, promotions, hopes, and promises my furlough from fly swatting was short-lived. Within a few years the flies no longer rolled over in their death throes when we sprayed DDT. So we increased the dosage. Still, after a bit, the stronger dosage of DDT also failed to kill them. Suddenly I seemed to be swatting more flies than ever. Arnold said we should mix other chemicals with the DDT to make it stronger. First we mixed it with chlordane, then lindane; later we used lindane alone.

Several other "miraculous" chemicals followed as chlordane and lindane, like DDT, rapidly proved completely useless for killing flies. After spraying a chemical for only a couple of years, each one seemed to lose its killing power, and the flies returned with a vengeance. Many people had stopped using the fly sprays Flit or Bif some years earlier because they too had become useless. In spite of these setbacks, a belief in the necessity and ease of using the chemicals had seeped into our minds, and gradually it came to dominate nearly everyone's pest-management decisions. We were becoming hooked on pesticides, just like the large-scale farmers.

In the mid-1950s I joined the Marine Corps, and my brother and sisters took over my fly-swatting and pesticide-spraying duties. In the Marines I was an electrical technician and an atomic, biological, and chemical warfare paramedic. After the service I went to the University of California and the University of

Cyanide gas ad, from 1933 *California Cultivator* and *Pacific Rural Press*. From farmer collection.

FIGHT the HOPPERS!

enlist with me

-against the grape leaf hopper

CYANOGAS, THE PROVEN CONTROL METHOD,
destroys both winged adults and crawling nymphs.

Use Cyanogas to destroy crawling nymphs and winged adult-hoppers, or they'll suck the life out of vineyards . . . reduce sugar content . . . lower tonnage and mean heavy loss to growers.

CYANOGAS

REG. U.S. PAT. OFF.

Gaviota-Pacific guano fertilizer ad, from *California Cultivator* February, 1938, p. 142.

LACCO DDT dust and duster ad. From *California Cultivator,* April 13, 1946.

Illinois and was fortunate to do research in the tropical forests of Peru and to live with forest farmers.

The ability of these farmers to produce surpluses without chemicals in an environment ravaged by pests started me thinking that maybe the miracle chemicals that the salesmen pushed were not so necessary after all. I had never seen so many animal and insect pests and yet they were getting bumper yields. Thereafter, everything that I saw or learned about farming was filtered through that experience. I left the university in 1970 and worked on more than a dozen farms as a laborer, fence builder, planter, picker, mechanic, tractor driver, cultivator, manager, plumber, carpenter, cowboy, and researcher. As I worked on all these farms, I began to realize that American farming practices had become much more poisonous and dangerous than when I was a kid.

In the early 1980s, I enrolled in a pesticide- and fertilizer-applicator's course at the local college to learn more about spray rates for foliar fertilizers. I also hoped that I might understand why most of my neighbors and all my bosses continued to feel so comfortable with farm chemicals, while I had become fearful. The course provided a wealth of practical information about spray rates and nutrient requirements that helped me feed my plants better and certified me as a licensed pesticide applicator.

The course also left me more alarmed than ever about the dangers of farm and home pesticides. I was shocked to find that most of the chemicals in common use on farms were modified versions of the nerve poisons and antipersonnel weapons that I learned about when studying chemical warfare in the Marine Corps.

After the course, I had several contentious discussions with neighbors, friends, relatives, and employers who were addicted to chemicals. Literally everyone I talked with argued that farm chemicals were not dangerous if properly used. No matter what I said or how much evidence I produced to the contrary, deep down, most of these people believed that the fear of farm chemicals was blown out of proportion. More importantly, folks honestly felt that without chemical fertilizers their crops wouldn't grow, and without toxic pesticides the insects and weeds would destroy their plants. Nearly all maintained that if they didn't have the chemicals, the little profit they now enjoyed would be wiped out.

Often, when I tried to discuss the dangers of chemicals with friends and neighbors, many appeared to feel they were being accused of poisoning their families and their land with the "tools" they thought they needed so badly. Instead of seeing chemicals as synthetically produced poisons, these people viewed them as "their tools," and so mentally they minimized the threat that they posed. Most farmers too felt an ownership of the chemical "tools" as much as they felt for other pieces of farm equipment. They ignored the risks of using chemicals because they believed they needed them to make a profit, just as they needed tractors or rototillers or combines, which were also very dangerous if used incorrectly. In the minds of these yield- and price-dependent farmers, chemicals had become a necessary means of survival.

Part of the problem is that the toxicological analysis of farm chemicals is not required to be on the labels or in the advertisements for the products. Consequently, most farmers actually know very little about the dangers of the chemicals they use. Many find it hard to believe that the most heavily used poisons can cause a wide variety of cancers or birth defects or are incredibly damaging nerve poisons.

Farmers I spoke with wondered why they should bother to know all the chemistry or toxicology of each product. Several explained that they were more concerned with the killing power of the pesticide than its chemistry or toxicity. They were farmers, not chemists, they said. Many felt that understanding the chemical part was the job of the pest-control advisor and the university extension specialist at the agricultural experiment stations. They argued that if the government regulators and their banker allowed

the use of these chemicals, then they must be safe. Many times local chemical salesmen or bankers badgered my neighbors and friends about the necessity of using chemicals when I was present. I would laugh at them and argue that their poisons were unnecessary and dangerous. They in turn argued that my fears were exaggerated and proceeded to "guarantee" the safety of the chemicals.

Clearly, the job of the chemical sales staff is to convince farmers that they can't get along without their products, so no one can fault them for being aggressive—they're salespeople, after all. Many chemical salespeople get paid a commission on the basis of quantity of material sold. As a result, for them,

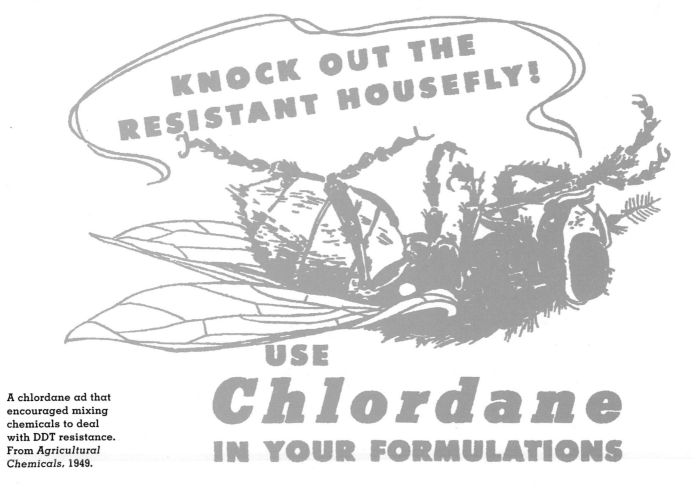

A chlordane ad that encouraged mixing chemicals to deal with DDT resistance. From *Agricultural Chemicals*, 1949.

selling chemicals and convincing farmers to buy more, whether the farmer needs them or not, has become a survival thing. For many, their survival and salary depend on the volume of pesticides they sell.

The other and often most important influence on farmers is their banker. Many farmers borrow money to farm, and if they don't pay out or don't get as high a yield as the bank expects, it can be difficult getting next year's operating loan. Some bankers merely prodded the farmers to use chemicals, while others required them to use chemicals to protect their loans. All this pressure has driven up the use of highly toxic pesticides, as table 1 illustrates.

By the late 1990s, tests required by the California Environmental Protection Agency finally proved how hazardous pesticides were to farmers and their families. However, in spite of the proven hazards that these tests revealed, the chemical corporations continued to block the cancellation of deadly birth-defect- and cancer-causing chemicals. My organic-farming friends and I were amazed at just how dangerous and deadly the chemicals proved to be,

but we were not surprised that the chemical corporations blocked the cancellation of these dangerous products. After all, we're talking about billions of dollars a year in profits.

Our neighbors also seemed amazed at the results of the government tests, but for different reasons. Many still couldn't see how the tests applied to them or their families. They too were not surprised that the chemical corporations defended their products and stonewalled their registration terminations. Many of them argued, "Oh, the state (CALEPA) and federal government (EPA and FDA) make the chemical corporations do tests on rats and dogs and ferrets, not people. What do you expect? They feed these little rats massive doses, of course they are going to get sick or die." Then they would finish with something like: "You would have to eat a roomful of apples to get the same amount of poison that the rats ate."

Dumbfounded, I would laugh and remind them that many of these pesticides were not just deadly to rats. They were the same as or similar to chemicals that Saddam Hussein used on the Kurds in Iraq or that

TABLE 1

Pesticide use in the 1990s was double what it was in the 1980s

Source: California EPA, Department of Pesticide Regulation

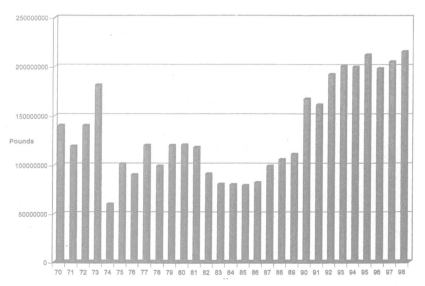

Pesticides Used in California 1970–98

Osama bin Laden threatened us with from Afghanistan, that we had used in the war in Vietnam, and that the Nazis had used in concentration camps. Somehow, they continued to have a comfort level with these terrible poisons, but not with the testing procedures or the test results. They didn't trust the tests or the government agencies that were requiring and evaluating them. Yes, they admitted that chemicals caused cancer and birth defects and even killed the lab rats, but they never failed to remind me that they were not lab rats or ferrets or dogs.

I told them about one test in which dogs inhaled only 268 and 283 parts per million of methyl bromide to determine a concentration that could be used in a twenty-eight-day test and a one-year test on dogs as required by California's 1984 Birth Defect Prevention Act.

The dogs were supposed to be exposed for four days, but the study had to be terminated after two days due to the observation of the following: severe neurotoxicity (delirium, thrashing and vocalization, tremors, traumatizing behavior [defined as slamming the head and body into cage walls]), depression, ataxia (irregular gait), rales (abnormal sounds when breathing), and a cachectic appearance (general wasting and malnutrition, associated with chronic disease).[1]

In spite of such damning evidence about their pesticide tools and the growing consumer alarm about the increased use of the most toxic poisons, my friends and relatives continued to argue that most pesticides were still registered, and that neither the FDA nor the EPA seemed anxious to prohibit their use. Therefore, they had concluded they must be safe enough to use. I argued that the FDA and the EPA determined that their most important constituencies and concerns were big chemical and corporate farming interests, not the taxpaying consumer or the health of the farmers. Consequently, neither agency felt obliged to eliminate chemicals that the powerful corporations developed and protected, even though both agencies and the chemical corporations knew that pesticides were dangerous and often deadly—not just to rats, but to people as well.

I began to ask elderly folks when and why farmers had started using these poisons. No one had a very complete picture of the origin and history of chemicals on U.S. farms. All of them assured me that, even though they didn't know the whole story, they believed that the use of chemicals had started long before they began to farm. Many felt that the wealthy farmers had always used pesticides and chemical fertilizers.

I finally realized that the deep-seated acceptance of chemicals was both economic and historic, because the grandparents and great-grandparents of the large-scale farmers had become dependent and comfortable with toxic chemicals a long time ago. Several of us wanted to know how this could have happened. A few farmers I knew felt that they had learned a great deal about the history of farming by reading historical pieces in farm magazines and advised that I read their back issues. So I began to read the early journals and almanacs. It was fascinating. My friends and I had all read Jack Pickett's editorials in *California Farmer* for years and knew full well what kind of advice and analysis he was giving, especially about pesticides and chemical fertilizers. He was a chemical industry cheerleader.

Soon I started reading all the old farm magazines I could get from relatives and neighbors. As I read older magazines, I found that Pickett's father, who was the *California Farmer* editor before Jack, also promoted the use of chemicals and defended the farmers' need to use them. I scoured the old journals and found original copies in the University of California, University of Vermont, and Dartmouth College libraries. These magazines told most of the story about how pesticides and fertilizers were first sold, and how they continue to be sold today.

I found that, over the last 160 years, many editors had used their editorial pulpit in the journals to play a major role in promoting and justifying chemicals. The rural magazines told much of the story of how and why farming had changed so dramatically, and how prosperous farmers got comfortable with using highly toxic medicines and other poisons. When there was no competition from radio, TV, or other electronic media, the farm magazines significantly influenced farmers' opinions and decision making. In fact, *California Farmer* was at one time so important a voice in California that it was often used as text material in rural schools.

Many academic and popular studies have concentrated their search for the beginnings of chemical farming on the period after the Second World War, apparently under the assumption that most of agriculture was chemical-free before that time. A few authors have extended the picture to the time of the First World War, and even fewer (especially James Wharton, Margaret Rossiter, and Richard Wines) have understood that the chemical agriculture story has much deeper roots.

I dug further into the magazines and several books, including those mentioned above, and found that the American portion of this story began more than two hundred years ago, with the colonial farmers and the farm crises of the 1700s and early 1800s. Shortly after the beginning of the 1800s, the large-scale farmers in America began to be propagandized by scientists and the mining and manufacturing companies, who proclaimed how newly discovered chemicals would solve both their financial and their farming problems.

The first farmers targeted with propaganda about chemicals farmed large tracts of land, with some cultivating thousands of acres of tobacco, corn, hemp, cotton, and other crops for export. The farming practices on their huge estates and plantations had literally destroyed the fertility on most of the choice farmland on the eastern seaboard even before 1800. The New York State Agricultural Society magazine, *The Cultivator*, conducted the earliest propaganda campaign promoting chemicals, beginning in the 1840s. To conduct the campaign they trumpeted the discoveries of chemical scientists and used testimonials from aristocratic farmers. In spite of their aggressive promotion, most small- and medium-scale farmers opposed the use of harsh chemicals and poisonous metals. Then, in the 1850s, industry developed or discovered several chemical products that they peddled to farmers. Yet farmer resistance to chemicals persisted for most of the nineteenth century. This farmer opposition prompted the chemical corporations to advertise and propagandize more often, more creatively, more fearfully, and more authoritatively.

As a result of seven generations of such campaigns, most American farmers have come to be dependent on chemical pesticides, fertilizers, antibiotics, and genetically manipulated products. Recently, many farmers have begun to compare their chemical and corporate dependency to drug or alcohol addiction.

This story illustrates that, well before the start of the twentieth century, advertising space in rural magazines became an essential platform for chemical corporations. By 1900, the ads were producing more revenue for these farm periodicals than their subscriptions ever could. By that point, the concerns of the reader had become secondary to the concerns of the advertisers. Because of this, the views of the chemical advertisers, not the needs of the farmers, have dominated farm magazines for more than a century, and continue to do so today.

By the 1890s the magazine *Agricultural Advertising* was entirely devoted to the search for farm-journal advertisers. Even by this early date the farm journals knew who paid the bills, and it wasn't their readers. But the publishers still had to keep their subscriptions

WALLACES' FARMER

"GOOD FARMING-CLEAR THINKING-RIGHT LIVING"
A WEEKLY JOURNAL FOR WESTERN FARMERS

Edited by HENRY WALLACE

Wallaces' Farmer does not belong to that too numerous class of agricultural papers that regard circulation as the only or the chief claim to the consideration of advertisers.

This paper is not published primarily as an advertising proposition, but is a business farm paper, and as such reaches more thinking, moneyed farmers than any other western farm paper.

Isn't this the kind of paper that is likely to serve to the best advantage the interests of the advertiser?

WALLACES' FARMER
DES MOINES, IOWA

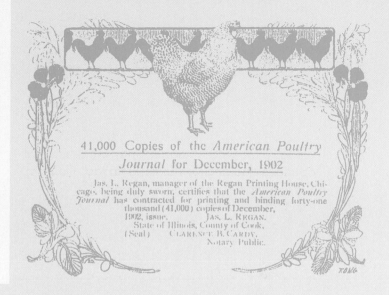

41,000 Copies of the *American Poultry Journal* for December, 1902

Jas. L. Regan, manager of the Regan Printing House, Chicago, being duly sworn, certifies that the *American Poultry Journal* has contracted for printing and binding forty-one thousand (41,000) copies of December, 1902, issue.   JAS. L. REGAN.
State of Illinois, County of Cook.
(Seal)   CLARENCE B. CARDY,
Notary Public.

The *Wallaces' Farmer* ad at top left alleges that the quality of the magazine is more important than the circulation and claims that quality will ensure high circulation. *The Country Gentleman* and other journals are more concerned that the advertisers know that the size of the circulation is large so that they will reach the most customers. From *Agricultural Advertising*, 1903.

The Country Gentleman

A Consolidation of
THE GENESEE FARMER 1831-1839
AND
THE CULTIVATOR 1834-1865

The ONLY Agricultural NEWSpaper

Circulated in 1902 sixteen per cent more papers than in 1901, 22 per cent more than in 1900, 80 per cent more than in 1898, 84 per cent more than in 1896.

The subscription price is $1.50. Some agricultural papers cost ten cents per annum. A good many cost nothing, if that is all that the persons to whom they are sent will pay for them.

Mr. Rowell remarks in a late issue of *Printers' Ink:* "The higher the price at which a publication is sold, the more that publication is worth, per thousand, to the advertiser. The higher the price, the more closely will the actual sale approximate the number issued. The higher the price, the longer the publication will be preserved. The lower the price at which a publication is sold, the less that publication is worth, per thousand, to the advertisers. The lower the price, the more widely will the actual sale fall below the number issued. The lower the price, the shorter the time the publication will be preserved."

The advertising rates of *The Country Gentleman* are not higher than some other agricultural weeklies charge. Send for a sample copy.

FARM JOURNAL.

WIDE-AWAKE MEN WHO ADVERTISE.

[We purpose under this head to print each month a portrait, with a brief sketch, of some prominent man whose name and business are familiar to our readers through their advertisements in the FARM JOURNAL. We think this will prove an interesting feature.]

Eleven years ago one of our good subscribers at Trenton, N. J., told Benjamin Hammond, of Fishkill-on-Hudson, N. Y., that he ought to come and see FARM JOURNAL and arrange for advertising Slug Shot, an article he was putting on the market. Mr. Hammond did call, and being favorably im-

pressed with him and his Slug Shot his card was inserted and has been appearing pretty regularly ever since. Slug Shot is so well known to FARM JOURNAL readers, that although Mr. Hammond advertises in numerous other farm and horticultural papers, he tells us he gets more inquiries from FARM JOURNAL readers than from all the others combined.

Since the time Mr. Hammond commenced advertising, his business has grown wonderfully. Slug Shot is to be found and is known everywhere. He has added various other insecticides and fungicides, —Grape Dust for mildew, Copper Compound for rot, Scrofularia for carpet beetles, etc.

Mr. Hammond is a hard-working, energetic man, who also conducts the largest paint and glass works in his section of New York. He is a young man still and president of his village of Bushkill Landing, secretary of the board of education, and otherwise identified with public life.

HAMMOND'S SLUG-SHOT
INSECTICIDE & FUNGICIDE
I SLAY BUGS
1880 1897
TRADE MARK.

Sold by the Seed Dealers of America. For pamphlets address, B. HAMMOND, Fishkill-on-Hudson, N. Y.

This testimonial and the ad from Hammond are indicative of how pressure was put on farm merchandisers before the turn of the century to advertise. From *Farm Journal*, February 1894.

RENEW YOUR FREE SUBSCRIPTION

Don't wait until it's too late. Act now.

The practice of free subscriptions remains common for many journals today. This ad from *Farm Journal*, 2005 was used throughout the year and sent to farmers as their subscription neared its end.

up because the more subscribers a journal had, the more the advertisers would be willing to pay to place their ads. Many journals began to offer free subscriptions around this time, and most have continued to do so to this day. So, for more than one hundred years, even if farmers didn't pay for the magazines, they usually received them free of charge. The publishers wanted to be sure that farmers got the messages of their advertisers and to make sure that the subscription base was high enough to attract the most expensive ads.

While other factors certainly influenced farmers' decisions to use chemicals besides advertising, a major thesis of this book is that advertising and propaganda campaigns have historically played, and continue to play, a very important role in guiding a farmer's choice of which products to buy.

In the late 1800s Cyrus H. K. Curtiss, the nineteenth-century advertising genius, emphasized the importance of advertising. Curtiss, the owner of the *Saturday Evening Post*, the *Ladies Home Journal*, and the *Country Gentleman* (one of the most influential national farm magazines), once told a gathering of potential advertisers: "The editor of the *Ladies Home Journal* thinks that we publish it for the benefit of American women. That is an illusion, but a very proper one for him to have. The real reason, the publisher's reason, is to give you who manufacture things that American women want and buy a chance to tell about your products."

When asked for the secret of his success, Curtiss always had a one-word explanation: "Advertising!" Curtiss didn't publish the *Country Gentlemen* for farmers any more than he published the *Ladies Home Journal* for women; they were both sales platforms for the product manufacturers.[2]

The story I present in this book is concentrated on farming, but there are occasional glimpses of similar advertising campaigns, including those conducted on military customers, urban landscapers, homeowners, and national and state governments. Uncovering how chemical companies regularly communicated to farmers, homeowners, and businesses reveals patterns and formulas that they have employed to convince people to use deadly and dangerous poisons. It is hoped that this little book, which started in questionings and remembrances with farmer friends, will provide a graphic and interesting outline of the rise of chemical agriculture in the United States.

My goal is to provoke readers with some often-overlooked historical perspectives about food and farming, and to suggest what they can do to ensure that food is produced safely on land that is properly cared for, so that our children and grandchildren will be able to enjoy its bounty and continue to make it productive.

BICKFORD'S PORTABLE PUMP.

Is double-acting, throws a continuous stream 40 ft. Useful for sprinkling lawns and roads, washing windows, extinguishing fires, throwing liquid poison to destroy worms on plants, fruit trees, shrubbery, etc. Very simple, durable, and easy to work. Price, $5.00. Manufactured only by The American Machine Co., Nos. 1916–1924 N. 4th St., Phila., and 128 Chambers St., New York; L. M. Rumsey & Co., St. Louis, Western Agent.

In the 1867 Bickford's ad at left, the manufacturer stresses the ease and safety of spraying operations. From Robert F. Karolevitz, *Old Time Agriculture in the Ads* (Aberdeen, S.D.: North Plains Press, 1970). In the bottom ad, the Rite Company stresses one hundred years later the need for protection from spraying and other industrial jobs. From *Agri Chemical West*, July 1967.

INSIDE...FRESH AIR

A RITE WHITECAP filtered-air helmet lets a man work all day long at high, untiring efficiency in safe, comfortable, clean air. It totally protects head, eyes, lungs and face — without the clamped-in feeling of mask and goggles. Vision is unrestricted. Filters for special conditions, a heater, evaporative or refrigerant coolers are among optional accessories. More information on WHITECAP Helmets? Write:

RITE WHITECAP
RITE
HARDWARE MANUFACTURING CO.
540 West Chevy Chase Dr.
Glendale, Calif., 91204

Introduction

In travels along rural American highways, much of the countryside looks picturesque, tranquil, and bucolic—beautiful farm and forest country. However, once one gets off the highway, anyone can see forestland brutalized by clear-cutting or overgrazing and large areas of farmland crusted over with salt from excessive irrigation, fertilization, and pesticide use. Crop dusters regularly fly over houses, highways, schools, rivers, forests, and canals, leaking and spraying millions of pounds of pesticides, while farmers on the ground spread millions of tons of toxic fertilizer.[1]

On agricultural drainage ponds that formerly were part of majestic lakes on the Pacific flyway the water is so contaminated that air cannons and shotguns are constantly discharged to keep the birds from landing and mating. All matter of salts, heavy metals, chemical fertilizers, and pesticides get dumped into these agricultural sewers. Tulare Lake, in California's San Joaquin Valley, was formerly the largest lake west of the Mississippi. The lake is now an agricultural drainage puddle.

Farming practices such as these contaminate rivers and drinking water with farm chemicals and have increased significantly the number of cancer and birth-defect clusters in farm communities. We haven't reached Rachel Carson's Silent Spring yet, but as a farmer I can tell you that, wherever factory farming dominates the landscape, every spring is more and more polluted. Every spring brings us closer to Carson's nightmare.

For more than twenty years my colleagues and I have been alerting people to this rural disaster by writing, speaking out, and conducting farm tours for more than a thousand representatives from government, the universities, consumer and environmental groups, industry executives, and the press. These tours showcase California, Texas, and mid-South farms where industrial agriculture has polluted, depleted, and destabilized environments and rural communities. The tours also showcase farmers who have figured out how to farm with no toxic chemicals. Most of the tour participants, like most of the general public, have no idea how truly toxic our food production is in the United States.

The most commonly asked questions on these tours were:

"Why do people put up with so many toxic chemicals being used on farms next door or down the street?"

"How do the chemical companies and the large-scale corporate farms get away with it?"

"Don't farmers know these chemicals are toxic?"

"Aren't they worried about the effects on their own family and their community?"

"If they know the chemicals are dangerous toxins and they are worried about their family's health, why do they continue to use such poisonous pesticides and fertilizers?"

This book tries to provide some answers to these questions. Fortunately, the record of corporate, government, scientific, and journalistic advocacy for chemical use in agriculture is plainly and graphically preserved in the pages of the farm magazines, industry pamphlets, and university publications. I use their advertisements and their editorial and "scientific" campaigns to outline, illustrate, and punctuate the story. Much of the narrative, therefore, is dependent on the words and graphics of the advertisers, the publishers, and the scientists.

This story outlines how these powerful entities actively cooperated to promote farm chemicals for more than 160 years. I argue that the chemical corporations, along with the accomplices listed above, are the most responsible parties in the destruction of rural America. Farm journal editors as early as the 1840s, and almost all the editors of farm journals since the 1880s, have served the chemical corporations and other advertisers far more than they served the interests of farmers or the consuming public.

Today, the chemical corporations continue their propaganda efforts to convince farmers that they cannot make a profit without using chemicals, antibiotics, hormones, and genetically manipulated crops and animals. The chemical corporations persistently argue that growers must use their products or suffer crop loss and financial failure. Chemical corporation promoters have convinced U.S. farmers that they are the "bread basket for the world." Farmers have been led to believe that if they don't use chemicals to get the highest yields, then millions around the world will starve—and they will lose their farms.

In the pages that follow, we expose many of the corporate tricks, promises, and promotions and illustrate how farmers who were stridently opposed to chemicals came to believe in and be comfortable with farm poisons and their salespeople. We also draw a chronological outline of the devastating effects on the environment and rural communities wrought by these propaganda campaigns and their toxic chemical

products. Sadly, many of the farmers who believed these promotions and adopted farm chemicals to save their farms lost their farms, their health, and a way of life. At the beginning of the twentieth century, when chemical farming was still viewed with suspicion and fear, more than six million full-time farmers cultivated eight hundred million acres in America. At that time only 8 percent of the land was leased from absentee owners. Today, less than a million full-time farmers and another million part-time farmers graze and cultivate more than a billion acres, and about half of the nation's farmland is leased from absentee landlords, banks, insurance corporations, chemical producers, the U.S. government, and holding companies. Even though the land area farmed has increased over the past century, the number of farmers on the land has decreased, and the loss in farmers continues to this day.

Since the Nixon administration in the 1970s about 650 farms per week have gone bankrupt in the United States. America's farmers, who were supposed to be rescued by one chemical product and one government program after another, have suffered staggering rates of bankruptcy, and the nation's farmland and groundwater basins are more polluted with chemicals now than ever before. As farmers went bankrupt, larger and larger corporate-farming businesses gobbled up the land. Farm size increased exponentially. For example, in 1946 the average size of a U.S. cotton farm was 17 acres. Today the average cotton farm is about 1,000 acres.

With more and more acres to manage, farmers have increased their use of chemicals and bought into the chemical corporation argument that poisons and chemical fertilizers are the most efficient management tools. Advertisers still argue that their chemicals kill more pests than biological control strategies and are cheaper and less bothersome. For years farmers believed that chemicals would be the salvation of their farms; in fact, most U.S. farmers still believe that.

In hindsight, we know that the chemicals were not the promised salvation but instead turned out to be environmentally and economically destructive, not only affecting the environment, but also helping to destroy much of rural American culture. In the process, the banks, the commodity brokers, the chemical corporations, and the large-scale farmers seized control of millions of acres from bankrupt farmers. After their successful land grab in the United States and Europe, big agricultural corporations have focused their efforts on controlling biodiversity and genetic resources or seizing agricultural land throughout the developing world.

This book traces how merchants sold the first commercial fertilizer, Peruvian bird guano. It describes how nineteenth-century miners, merchants, and industrialists tried to convince farmers that mining and industrial-waste products were highly valuable fertilizers. In the following quote from his book *Potash*, J. W. Turrentine demonstrates how early industrial wastes came to be sold as "fertilizer discoveries." The example of German corporations that discovered, in their search for salt, vast natural deposits of potash in 1858 illustrates how scientific data was used to puff up their value, enabling industrialists to recommend their use more authoritatively.

> It should not be understood from this that the agricultural world was anxiously awaiting the discovery of such a deposit to fulfill a long-felt want of agriculture. On the contrary, unenlightened agriculture was quite content with the potash supplies on hand . . . but the Germans, desiring to find some use for a material which at that time was largely without use or value, turned to agriculture as the only industry large enough to absorb a heavy tonnage. Scientific data were easily gotten to support the thesis that potash was a valuable plant food. . . . [T]he present widespread use of potash resulted not from the demand of

agriculture for potash fertilizers but from the German industrialists' demand for a market.[2]

As the Industrial Revolution progressed, even more enormous quantities of waste chemicals accumulated. Finally, after creating mountains and filling valleys with toxic waste, the industrialists were forced by state and national regulators to dispose of it or face serious fines and disposal fees. When the state and national regulators began imposing fines, most of the mining and manufacturing corporations followed the lead of the German potash syndicate and turned to agriculture as the major dumping ground.

The sodium nitrate fertilizer, found in Chile and Peru, was also a waste product of salt mining. Many arsenic and lead pesticides used on food were by-products recovered from flue gas in the smelting of iron and copper or wastes from fabric dyeing and paint manufacturing. Especially after the commercial success of guano, potash, arsenic, and sodium nitrate, other industrial and mining operators saw farming as a potential market. Thereafter, not only did the mining and industrial corporations avoid dumping fines, they made outrageous profits, selling their toxic waste as a fertilizer or a pesticide.

Cyanide gas, a by-product of ammonium-cyanide production, was used both as a pesticide and an antipersonnel poison shortly after the turn of the twentieth century.[3] Waste natural gas and hydrogen, which came along as by-products of gasoline or coke manufacturing, enabled the industrialists to create sufficient heat to liquefy atmospheric gases and produce nitrogen. This form of nitrogen became one of the most widely used chemical fertilizers and is still the dominant form of synthetic nitrogen used today. Later, in the 1930s, fluorine, a by-product of uranium mining, became an important pesticide.

As a consequence, consumers ate increasingly more food grown with industrial waste. What's more, we still do! The use of arsenic continues, industrial and urban sludge is used as fertilizer, spent nuclear fuel rods are used to irradiate food, and all manner of corporate waste is regularly dumped on our food and our farmland.

The War on Bugs outlines how the industrialists changed U.S. agriculture, first with imported guano fertilizer, then with arsenic, lead, synthetic fertilizer, cyanide, DDT, methyl bromide, nerve poisons, antibiotics, growth hormones, and, currently, genetically manipulated products. The facts presented here serve to indict corporate chemical advertisers, farm magazines, university scientists, and the government as enablers of the poisonous changeover in agriculture and the destruction of rural communities in the United States.

This book also tracks the parallel development of organic agriculture and farmer resistance to chemical farming. Because of this duality, our tale exposes more than its share of ironic twists. The first occurred before the 1840s, when farmers first started spreading guano and mining and industrial wastes on American fields and orchards. Well before that time, farming had undergone significant change and was in the midst of a major revolution that resulted in significant mechanical, fertilizer, and land-management improvements.

The mechanical changes included plows, harvesters, combines, steam power, and other products from both the farmer's blacksmith shop and the industrial sweatshops. The structural changes in farming were part of America's first populist movement during and after the Revolutionary War. In these systems, practical land management was the centerpiece of the program: diversity of crops was encouraged and monoculture discouraged. As a consequence, crop rotations, composting, and sophisticated natural fertilization strategies became widely used in the East and Northeast by the 1830s. Farmers who had adopted these innovative strategies voiced strong opposition to

the first use of poisonous chemicals. As chemical advertisers became increasingly aggressive, farmers more vigorously challenged the ads and the editorials from the chemical firms and effectively marginalized the chemical peddlers until the late 1870s.

To counteract farmer skepticism toward toxic chemicals, the ad makers developed increasingly innovative advertising campaigns targeted directly at the farmers' economic fears. These early propaganda efforts led to a flood of advertising and editorials promoting the chemicals and showcasing the aristocratic and most prosperous farmers who used them. This ongoing, 160-year-long campaign attempted to marginalize biological farming strategies while promoting the use of chemicals. The effectiveness of this campaign caused a profound change in how we view both nature and our food production system, and how we treat our bodies and ourselves. Fortunately, today a worldwide movement toward clean foods and sustainable farming practices is growing stronger every year and is once again changing how people perceive themselves and their food.

In telling this story it seemed helpful to provide readers with an overview of the connections between medicine, rat and disease control, chemistry, fertilizer, labor, spray devices, oil, dynamite, and, of course, pesticides. In order to describe these complicated connections, we occasionally are forced to leap back and forth in time. This was necessary to give the individual stories coherence. The interweaving of related stories and the chronological charts throughout the text should help illustrate interrelationships and hopefully lubricate the tracks of time travel.

This outline and synthesis may help explain how chemical advertisers became one of America's most powerful rural voices, attaining an almost de facto authority among most farmers. A big part of their success was due to saturation advertising.

In the process of outlining the effectiveness of the chemical sales strategies, we were struck by the degree and amount of resistance that every generation of activists mustered against federal control, unjust taxes, trusts, cartels, the railroad barons, corporations, poisonous pesticides, poisonous foods, the war on bugs, and the corporate control of everything. Even before the Revolution, the nation was in ferment about the discriminatory practices of the aristocrats toward small farmers as well as toward workers, slaves, and servants. From the North Carolina Regulator revolts in the 1760s to the present there have been significant and repeated populist struggles with the government and the dominating class. In most of these struggles, including the one that is presently under way, food and agriculture have played prominent parts. We hope that the story told in *The War on Bugs* will inspire readers to join in the movement to support biological farming and safe food, to play a part in determining how the world's food is grown and farmland is managed.

The War on Bugs is really the story of two wars, one intended, the other a by-product of chemical use. The intended war using pesticides has been directed against insects, bugs, spiders, disease, and fungus and is designed to KILL. The unintended war comes from the use of chemical fertilizers and pesticides that inadvertently kill highly important soil life, such as microorganisms and earthworms, and their drift that contaminates areas next door and thousands of miles away. Both of these wars have had devastating effects on America's water, farmland soil, wildlife, and rural population.

Union Carbide's Five Decades of Death

Since the Depression, Workers Have Been Dying To Shore Up Union Carbide's Profit Margin.

This cartoon, from the December 28, 1984, issue of *L.A. Weekly*, and the accompanying article attack Union Carbide, one of the largest U.S. chemical firms, which suffered a string of large-scale industrial disasters beginning in the late 1920s. Its worst disaster occurred on December 3, 1984, at a pesticide plant in Bhopal, India, where thousands died and hundreds of thousands were blinded.

Historical Markers

3rd century BC ● **The Greek philosopher Hippocrates advocates a holistic medicine and a balancing of the humors:** blood, phlegm, choler (yellow bile), and melancholy (black bile), with simple herbal potions. He recommends exercise, work, and a good diet as the best medicine. He is considered to be the father and patron saint of Western medicine.

2nd century AD ● The Greek philosopher **Galen serves as a Roman war surgeon**. Like Hippocrates, he also believes in a balancing of the humors (body fluids), but he adds complex potions to purge bad humors and he blisters and bleeds patients to effect his cures. Galen's ideas become the dominant force in Western medicine until 1900. Many linger today. He is considered the father of heroic medicine.

1096–1291 ● **The Crusades—Heroic medicine is introduced to all Europe by the end of the 13th century.**

1300–1400s ● **Arsenic and heavy metals are used to poison rats during European plagues.**

1400–1500s ● **The closing of the European commons begins in Italy and England. This land-grab by powerful landowners enables the development of large-scale commercial agriculture and the accumulation of capital.**

1455 ● Johann Gutenberg perfects his printing press and the use of movable type.

1525 ● **The Swiss chemist-physician Paracelsus advocates the medical use of arsenic, mercury, lead, vitriol (sulfuric acid), and other heavy metals.** These become standard purges and potions in Paracelsian-Galenic medicine. The use of the Paracelsian-Galenic poisons as medicines remained common until the 1940s.
● **First newspaper/magazine advertisement appears in Germany.** Other newspapers are printed about this time in Austria, Holland, and Italy.

1535 ● First printing press appears in the Americas, in Mexico City.

1638 ● First printing press appears in the British American colonies, in Cambridge, Massachusetts (Harvard).

1693 ● **First American newspaper, *Publick Occurrences*, is published in Boston.** Newspapers thereafter are banned until 1704.

1700s ● **Arsenic is used to control pests in France. The French government applies strict regulations, eliminating most of the usage.**

1704 ● **First periodically produced British colonial newspaper in the United States, the *Boston News-Letter*, is published.**

1728 ● Benjamin Franklin purchases and begins publishing the Pennsylvania Gazette in Philadelphia.

1750s ● **Esere** or **Calabar** beans are discovered by Europeans in Africa. This is the source for carbamate pesticides.

1760s ● Benjamin Franklin's *Pennsylvania Gazette* becomes an American advertising innovator, devoting 50 percent of its space to advertisements.

1793 ● **Eli Whitney invents the cotton gin. This invention replaces thousands of jobs and causes cotton to replace hemp as an important American fiber. This sets the stage for monocultural and industrial agriculture in America. It also sets the stage for US industrialization.**

1800 ● Alessandro Volta invents the electric cell.

1803 ● The Louisiana Purchase nearly doubles the size of the United States, adding some 828,000 square miles of territory between the Mississippi River and the Rocky Mountains.

1804 ● First railway engine is produced by Richard Trevithick. The Lewis and Clark Expedition leaves St. Louis to explore and open a route to the West Coast.
● **Western industrialists discover the commercial and agricultural value of Peruvian guano deposits.**

1808 ● **The Panic of 1808 (economic depression) is the first of the economic depressions to hit the United States in the 1800s.** Great loss of personal and public capital results.

1810 ● **First American farm journal, *An Agricultural Museum*, is published.**

1811 ● Luddites wreck industrial machinery in northern England.

1812 ● First canning factory is opened in London.
● **The War of 1812 begins with Britain.**

1813 ● *The Arator*, the most important American agricultural book in the early 1800s, is written by John Taylor.
● Mexican independence is established.
● **Tecumseh, the Shawnee leader, is killed** by William Henry Harrison's troops in the Battle of the Thames, north of Lake Erie.

1819 ● **Second and third American farm journals are started**, first the *American Farmer*, and, two months later the *Plough Boy. Plough Boy* lasts only lasted a short time, but *American Farmer* becomes the most important American farm

publication in the early 1800s and lasts until the 1890s.
- United States purchases Florida.
- Simon Bolivar liberates South America.
- **The Panic of 1819 (economic depression).**

1822 • **First chemical synthesis of an organophosphate (OP).**

1823 • The Monroe Doctrine restricts foreign intervention in the Americas.

1824 • *American Farmer* introduces Peruvian guano for the first time in a farm publication.

1825 • Trade unions are first legalized in Britain. First passenger railway service, in England, begins.

1826 • French chemist Antoine Balard discovers bromine. German chemist Justus von Liebig had earlier isolated this element from the saltworks at Kreuznach but had assumed that it was iodine chloride.

1830 • **Sodium nitrate deposits are discovered by Western industrialists in Chile.**

1831 • Rural journal **American Agriculturalist** is first published.

1832 • Justus von Liebig, R. Brandes, and F. L. Geiger publish and edit the first edition of **Annalen der Pharmacie,** which, after Liebig's death in 1873, is changed to *Justus Liebig's Annalen der Chemie*. This most famous of all chemical journals is the launching pad for industrial chemistry and chemical agriculture.

1833 • Slavery is abolished in the British Empire.

1834 • **Cyrus McCormick invents the mechanical reaper, and the trend toward monoculture and industrial farming is thereby accelerated. Thousands of threshing jobs disappear.**

1837 • **The Panic of 1837 (economic depression) causes great loss of land by small farmers in the eastern states.** This is a much more widespread and severe depression than the Panics of 1808 and 1819.

1838 • First electric telegraph is created Britain.
- Sir Marc Brunel's steamship, *Great Western*, is launched.

1840 • **Justus von Liebig publishes *Organic Chemistry and Its Applications to Agriculture and Physiology*. Scientific farm advice following the publication of this book is highly chemical, and soil tests and farming by the book are encouraged.**

1841 • **John Lawes, in England, develops the first synthetic fertilizer, derived from pouring sulfuric acid on bones, low-grade guanos, and phosphate rock to release the phosphate and make it more available to plants.**

1844 • **First shipment of guano to the United States is made by Pacific Guano Co. (later Union Oil, PureGro Co.). These are products of Peruvian and Chilean slavery. Millions of tons of guano had already been used in Europe and England for more than a decade.**

1845 • United States annexes Texas.
- **First shipment of sodium nitrate is made to the United States and Europe—with no buyers. The search for an agriculture market for this salt by-product begins.**

1846 • **Irish potato famine occurs.**
- United States gains Oregon territory from Great Britain.

1847 • **Sulfuric acid is promoted by *Scientific American* as a fertilizer.**

1848 • Slavery is abolished in French colonies.
- United States gains New Mexico and California from Mexico.
- California Gold Rush begins.

1849 • **First synthetic fertilizer in the United States is manufactured by P.S. Chappel, Kettlewell & Davison in Baltimore, Maryland.**

1851 • **First German state-supported agricultural experiment station is established at Mockern, near Leipzig, by Liebig's students at his urging.**
- Singer sewing machine is invented.

1854 • **175,849 tons of guano are used in the United States.**

1858 • *American Farmers Encyclopedia* **is published,**

1859 • Slavery is abolished in Peru.

1860–62 • Potash is discovered and mined in Strasfurt, Germany. The search for a potash market begins.

1860–70s • Wars between indigenous tribes (American Indians) and white settlers occur.

1861 • Russian serfs are emancipated.

1861–65 • **US Civil War.**
- Slavery is abolished in the United States.

1862 • **The Morrill Act, which provides funding for the land-grant college system, is signed into law.** To settlers this promises to be another source of cheap good land.
- **The Homestead Act is signed into law, creating the Bureau of Agriculture.** It also provides the promise of cheap or free land. The Bureau of Agriculture's first scientist is a Liebig student.

1863-70 • Scientific discovery and analysis reveal the toxic properties of the **Calabar** bean, the source of carbamate poisons. **Carbamate** poison symptoms are known by this time.

*Chronology continues on page 60.

This picture is of common methods of plowing and sowing utilized from Roman times through the early 1800s in both Europe and North America.

Chapter 1

THE ROOTS OF FARMING
IN THE AMERICAS

No realistic understanding of farming in the United States is possible without an understanding of the economic, cultural, and farming systems that gave rise to its development. Examples from several different historical analyses illustrate that "[European-American] . . . colonies were nearly all settled by chartered companies whose purposes were purely commercial and whose success depended on their securing immigrants."[1] Furthermore, "the accepted theory which prevailed in England was that the title to the land in America was vested in the Crown. Indigenous titles to the land occupied by the several tribes were never considered. The title passed directly from the King or Queen to proprietors who disposed of it as they saw fit, subject only to charter provisions. The process of disposing of land in the Crown colonies was through agents to the actual settlers under laws or decrees of the English government."[2]

The London Company, which was chartered by the Crown to settle Virginia, was empowered with both authority and privileges that made it an absolute monopoly. The settlement of New York was carried on by the Dutch West India Company, whose members were the largest landowners in the Netherlands. They became the largest landowners in eastern America.

> Like so many petty monarchs, each had his distinct flag and insignia; each fortified his domain with fortresses, armed with cannon and manned by his paid soldiery. The colonists were but humble dependents; they were his immediate subjects and were forced to take an oath of fealty and allegiance to him.[3]

Even George Washington was a leader of the Ohio Company, a land-speculation firm that obtained more than 6,700,000 acres from 1749 to 1792.[4]

The ownership of property in America guaranteed males the right to vote and gave them power over those who did not have land. In Massachusetts only those who owned a tract of land or other property worth 40 shillings had a right to vote. In Maryland only those could vote who owned 50 acres of land or other property valued at 40 pounds. In New Jersey it was necessary to own 200 acres of land or 50 pounds. In New York only the property owners were allowed to vote. The same qualifications were required in Connecticut and Virginia. In South Carolina only the owner of 500 acres of land or at least 10 slaves could be elected to the "parliament." In Georgia only the owners of 500 acres of land could be elected and only those who owned not less than 40 acres could vote.[5]

When the first votes were taken in America, less than 5 percent of the population was eligible to vote.

Only white males who owned land had any vote or rights of participation. Blacks, indigenous Americans, and women had no votes or participatory rights. Instead of freedom, most immigrants of any color faced years of slavery or indentured servitude on aristocratically owned estates growing mostly export crops.

To understand how American rural communities developed over the last four hundred years, it helps to look at how European farming changed after the Renaissance and the Reformation. From Roman times until the 1500s many European countries had common and church lands, to which a large majority of the population (the peasantry) held hereditary use rights. These use rights allowed peasants to live on, farm, and maintain extensive tracts of both common and parish land.

The great plagues of the fourteenth century ravaged much of Europe and fractured many long-standing

Enclosures in Leicestershire by 1550. From W. G Hoskins, *Essays in Leicestershire History* (Liverpool, U.K.: Liverpool University Press, 1950).

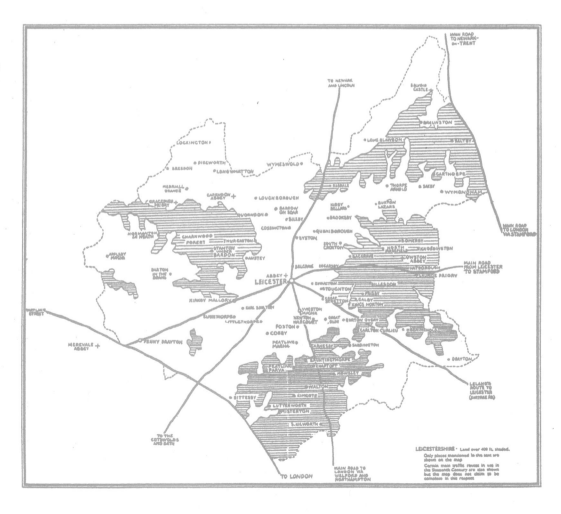

social contracts, leaving many unsure of their rights or obligations. In the early 1500s, in the climate of insecurity that had prevailed since the plagues, the royalty, the nobles, and an emerging capitalist merchant class began evicting the peasants from the common lands and filing documents claiming the land as their private property. This landgrab in Europe has come to be called the enclosure movement. The first enclosures took place in Italy and rapidly spread throughout much of Europe. The merchants and nobles converted the commons to large-scale sheep and grain farms, mostly to produce wool for the rapidly expanding garment industry. In the process, both common and church lands were gradually closed to the peasants. The enclosure movement in England and the rest of the British Isles resulted in a loss to the peasantry of almost all of the commons as well as the loss of Catholic Church land after the Reformation.

Under the enclosure laws, the new owners could evict the peasants from the land, and if the peasants refused work or had no job, the sheriff could arrest them as vagabonds. The peasantry and small-scale farmers were forced to move to the cities and work for textile and other manufacturers, or, if they remained in the countryside, they were forced to work in gangs as farm laborers on land they had previously farmed and occupied.[6]

From the capitalists' perspective, the enclosure movement was their first major success. From an agricultural perspective, productivity declined on the commons as a result of several decades of profit taking without maintaining soil fertility. From a cultural perspective, many rural cultures were eradicated in western Europe.

In 1605, John Stow wrote, "The Commons of Cornwall and Devonshire rose against the nobles and gentlemen, and required not only that the enclosures might be disparked, but also to have their old religion, and act of six articles restored."[7] The rebellion that Stow describes was called the Western Rebellion. Popular unrest began in the west of England when the Cornish opposed the government's religious reforms that took away their common land. Resistance and discontent spread into the neighboring county of Devon and in 1549 grew into one of the largest rebellions in the century, lasting for decades.

As the enclosure movement accelerated, the new entrepreneurial class accumulated huge amounts of capital. European merchants began the colonization of North America, largely with the capital accumulated by the enclosure movement. The Spanish and French entered what is now the United States from the Gulf of Mexico and moved up the Mississippi River. The English and the Dutch began to colonize eastern and northern North America in the early 1600s.[8]

By the 1600s, northern Europe had an abundance of slave and cheap labor, since the populations of the debtor prisons had swelled considerably as a consequence of the enclosure movement and the Reformation. When colonization began, the prisons rapidly emptied as convicts and peasants were shipped as bound servants or slaves to cash-producing colonies in the Americas. European peasants who were not in prison could emigrate and were encouraged to do so, but even the great majority of these "freemen" and women went as indentured slaves, servants, or mercenaries to colonial danger zones in America. By 1800, 40 percent of the English population lived in cities, and immigration was always held out as an escape route from the low wages and deplorable living conditions that prevailed in most of urban England.[9]

In Europe, newly enriched capitalists and the nobles seized large tracts of land by means of laws, armed force, and religious conversion of the Reformation (beginning in 1519) in the years following the great plagues of the 1400s. In North America, the European monarchies and the land speculators duplicated the landgrab of the enclosure movement in Europe. They were able to do this because many of the Native American tribes were decimated by European diseases, just as untold numbers of European peasants had been wiped out by the plague. After seizing the land, they introduced the same agricultural practices that they had perfected in Europe. Thereafter they held these lands by force and imprisoned, enslaved, or killed any survivors who resisted.

As Jared Diamond has shown in his book, *Guns, Germs, and Steel*, the European conquerors mostly subdued the indigenous American tribes with diseases that came from domestic animals brought over from Europe. Often, whole tribes and nations died off after being exposed to such diseases as measles, chicken pox, smallpox, and whooping cough. Slaves, bound servants, forged metal tools, and draft animals helped the new "owners" of America harvest or burn the wooded landscape and turn it into sheep pasturage and cash crop farms, just as they had done in Europe.[10]

The eastern seaboard of America was thickly forested at the time of conquest, and the indigenous tribes had developed a complex form of farming within the forests before the European colonies were established. Archaeologists have proven conclusively that farming was practiced in the Americas for several thousand years before Europeans arrived. Corn had been cultivated in the Tehuacan Valley of Mexico for at least eight thousand years and was grown north of the Canadian border long before white people arrived. Tribal farmers domesticated potatoes and tomatoes in the Peruvian Andes and beans in Mexico about five thousand years ago. The use of both of these crops was also widespread in the Americas before Columbus arrived.

By the time Columbus arrived in 1492, indigenous tribes farmed as far north as southeastern Quebec (around the mouth of the St. Lawrence River).

Throughout the Americas, there were several major agricultural centers, including the eastern woodlands of the United States; the Mississippi Valley; Arkansas; Arizona–New Mexico; California as far north as the Russian River; the Tehuacan Valley; the Valley of Mexico; the Yucatan Peninsula; Oaxaca; Central America; the Caribbean; Peru-Ecuador-Chile; and the tributaries of the Amazon Basin.[11] So, knowledge about farming was old and widespread in this so-called New World.

Only a few of the early cultures survived beyond the beginning of the colonial period. Unfortunately, most of the early recorders viewed the indigenous cultures with disdain. Europeans perceived themselves as civilized, technologically superior, and more culturally advanced than the native tribes. They also believed that their god was superior since they worshiped a single god, whereas the animistic tribes often gave tribute to several gods and felt that everything was part of god. The Hopi had more than ninety kachinas, spiritual, godlike beings who were represented as dolls. Their human advocates dressed and danced as kachinas. Many tribes (especially the Sioux, Mandan, and Arikara) believed in Wakon Tonka as a great spirit but envisioned trees, rocks, animals, and especially places as godlike.

Because of their discriminatory views, missionaries, adventurers, naturalists, and anthropologists documented very few aboriginal systems of farming. Daniel Quinn and many students of native cultures describe how European farmers dismissed tribal farming in America as primitive and backward and rapidly eradicated it wherever it survived the European diseases.[12] Many early researchers also dismissed native farmers as backward and ignorant; however, more recently, ethnographers and archaeologists and surviving indigenous people have demonstrated that the agricultural methods of tribal American farmers were in fact very innovative and sophisticated.

4

Research conducted among several tribal farming cultures in the Amazon Basin of South America provides a glimpse of what much of preconquest agriculture in America's woodlands was probably like. These farmer-fisher-hunters utilized a wide range of vegetable, fruit, and tree crops. Tribes grew annual crops on the riverbanks and planted small slashed-and-burned areas of the forest with perennial crops that fit the natural succession for the ecological niche.

These farming practices preserved habitat, discouraged many pests, and allowed the domesticated crops and the forest to run through their natural ecological progression before they were slashed and burned again (in about fifty years). Amazonian tribes usually had five or six patches that they farmed, and they lived serially in small villages in different locations to fit the seasonal maturity of the plots and the changes in weather. Indications are that similar well-planned and managed slash-and-burn systems were common throughout eastern North America when Columbus arrived.[13]

The earliest European pilgrims to North America (1607 in Jamestown) were forced to borrow certain techniques from these native farmers just to survive. However, almost no European immigrants adopted the Indians' complex tribal farming systems. This is because the survival of the Dutch and English colonies depended on producing commodities for export and for local sale—in other words, on making a capital profit for their investors.

By 1700, both the conquest and European diseases had wiped out a majority of the indigenous tribal farmers (American Indians) along the East Coast of North America.[14] According to Jared Diamond, many anthropologists and geographers now feel that much larger tribal populations existed in North America than we previously assumed and are arguing about the actual numbers. But whether the population was ten million or a hundred million, we know that only a relative handful of people survived both the disease epidemics and the Indian wars.

Because of this population crash, the immigrant aristocrats, merchants, and land-speculation companies from Europe were able to seize most of the good land east of the Appalachians, and they converted former Indian villages and woodlands to plantations and estates. These early farms were slave farms, which employed both European prisoners and native tribal captives. They exported increasing amounts of potash, wood, tobacco, cotton, corn, sheep, wool, and other crops to Britain, Holland, France, Germany, Italy, and Spain.

At about the same time, in the west, the California, Arizona, and New Mexico conquests took place. The Spanish mission system began exporting olives, beef, fruit, wine, and other crops to Spain and the rest of Europe. The Spaniards imprisoned the tribes in gender-isolated barracks and both deliberately and accidentally killed tens of thousands of indigenous peoples by the distribution of smallpox-infected cloth, as well as through other European cattle diseases and forced labor.[15]

From the early 1600s until 1840 the Russians employed Aleuts and other seagoing tribes from Alaska to San Diego to hunt fur-producing animals in order to export pelts back to privileged markets in their homeland. Until the early 1800s the Russians controlled all of the territory of the American and Canadian western coast north of San Francisco. While the Russians did not enslave the Aleuts or any other tribes, their predatory hunt for pelts and shellfish seriously damaged the seal, sea otter, sea lion, and shellfish populations on which the coastal tribes depended. This damage extended as far south as Baja, California, and in some areas the Russians wiped out the marine mammal resource base for the coastal tribes and nations and contributed to the decline of these tribes.[16]

The Revolutionary War and the War of 1812 caused even more cultural destruction and constriction of Native American tribal lands in the Ohio, Wabash, Illinois, Miami, Kaskaskia, Kankakee, Mississippi, and Missouri river valleys. George Rogers Clark and William Henry Harrison conducted scorched-earth campaigns against native farming tribes in all these areas from 1775 until well after Harrison's troops killed the Shawnee chief Tecumseh in Canada in 1813.

This westward expansion (called Manifest Destiny by Europeans) completely destroyed the complex tribal farming systems east of the Mississippi. Unfortunately, the eradication of the sedentary tribal farmers cut the colonists off from thousands of years of complicated agricultural know-how. The farming, gathering, and hunting practices developed by hundreds of tribes were rejected or ignored by profit-focused European merchants and lost in the rush to clear the forests to plant cash crops and graze animals.

Because many tribes were exterminated so soon after conquest, their systems of farming, fishing, and hunting perished with them. Consequently, Europeans adopted many more tribal crops than tribal farm practices. The list of native crops includes not only corn, cotton, hemp, and tobacco but also beans, squash, peppers, tomatoes, potatoes, sweet potatoes, avocados, plums, cranberries, blueberries, blackberries, chocolate, maple syrup, rubber, and more than two hundred other edible or useful crops.[17]

From the early 1600s until the 1850s the eastern colonial estates and plantations and the Spanish missions in the west principally depended on an enslaved and indentured workforce. As in Europe, a very small minority seized control of almost all the good land in both eastern and western North America, and so the enclosure movement that began in Europe continued in America.

The huge European and colonial American farms enjoyed fabulous profitability. With slaves, axes, and oxen, the European land speculators clear-cut the forests along most of the eastern seaboard before the 1800s. The European merchants mined America, using and selling off the timber, burning the lower-quality trees to make potash, and converting the former forests to sheep ranches, tobacco estates, corn fields, hemp farms, and cotton plantations.

As a result, before the Revolutionary War a number of land speculators had grown prosperous. The state governors and almost all of the Founding Fathers were leaders of these real-estate speculation schemes. The Ohio Company, of which George Washington was a leader, was organized in 1749 when King George II granted it a royal charter and 500,000 acres of land along the upper Ohio River, in addition to the company's initial 200,000-acre grant made earlier that year. After the Revolution, agents of this company were buying up land for about 8 or 9 cents an acre. In 1792, the Ohio Company obtained 1,000,000 additional acres from the government, almost free.[18] That gave the Ohio Company a total of 6,700,000 acres—which made Washington one of the richest men in America and one of the most persuasive real estate agents.

Historian James O'Neal described the climate of freedom in eastern North America immediately before and after the Revolutionary War: "From 1682 to 1804 the proportion of white slaves to the whole number of immigrants to Pennsylvania steadily increased, till they constituted two-thirds during the last 19 years."[19] Pennsylvania was politically the most liberal state and did not even require property ownership for voting privilege. But even in this most liberal state, only one-third of the immigrants in the period immediately after the Revolution obtained their freedom when they arrived in America.

To the Right Honourable the Lords Spiritual and Temporal of the House of Lords in Parliament assembled

The humble Petition of Thomas Hunter and Robert Hunter both of Meddomsley in the County of Durham Gentlemen Charles Collinson Clerk and Cuthbert Johnson gentleman both of Meddomsley aforesaid

Sheweth

That your Petitioners Thomas Hunter and Robert Hunter are Seised to them and their heires of and in the Antient Mannor of Meddomsley in the said County and also of and in severall Messuages Lands Tenements hereditaments and large Wastes Comons and Moors lying and being within and part and parcell of the said Mannor within the Parish of Lanchester in the said County of Durham And that they and all those whose Estates they have of and in the said Mannor and premises have severally held and enjoyed and ought to hold and enjoy Comon of pasture and other Rights and priviledges within and upon a large Comon or Fell called Lanchester Fell part of the Mannor of Lanchester aforesaid And your said Petitioners Charles Collinson and Cuthbert Johnson are also severally Seised to them and their severall heires of and in

A successful anti-enclosure petition, for Lanchester, Co. Durham, 1721 (enclosed in fact 1773–81)

Successful anti-enclosure petition for Lanchester, England. But despite the peasants' victory all the Lanchester land was enclosed sixty years later in 1781.

It was the slaves, serfs, and servants that the land speculators imported and exploited to extract the riches of America, even after the Revolution. And it was the slaves, serfs, and servants who had become freedmen who consistently demanded rights that they had been led to believe would be forthcoming when they became free in this "free" society. Sadly, these freedmen were mistaken, and their belief in the system was misplaced. But their optimism led them to demand their freedom and revolt over and over again against the power and decisions of the new American government. As a result of the conflicts between the ruling sector and the workers and freeholders, many historians refer to the sixty-five years immediately following the American Revolution as the Second Revolutionary Period. Between 1785 and 1820, small farmers on the American frontier revolted numerous times. More than twenty large-scale local rebellions, hundreds of minor revolts, and three economic depressions (called "panics") preceded the Civil War in 1861.

Historian Catherine McNichols Stock has illustrated that, for frontier and rural Americans, the object of both the pre- and post-Revolutionary struggles, as well as the Revolution itself, was not simply home rule but who should rule at home.[20] Shortly after the war with the British, the aristocratic Federalists decided that they should rule at home and quickly asserted their power. In this Second Revolutionary Period, conflict flared between the Federalists (the aristocratic large-scale farmers, merchants, land speculators, and manufacturers) and the small-scale subsistence farmers who had endured slavery or servitude and finally had earned enough to buy a small plot of land. Though the aristocratic Federalists took credit for leading the tax revolt in the Revolution against England, the small farmers, laborers, servants, slaves, and shopkeepers were, for the most part, the ones who fought and died.

The Federalists, led by Washington and Alexander Hamilton, wanted the national government to have a significant amount of central control over the various states. The Federalists argued that to maintain order they needed to raise a standing army that would put down local rebellions and revolts. In fact, however, the landed class used the army against tax revolts and money and barter revolts. These were exactly the same issues that had been central to the broad-based revolt against England, but which the Federalists now perceived as threats to their national hegemony.

After the first money and barter revolts, Hamilton argued successfully that the implied powers of the Constitution gave Congress the right to enfranchise national banks with monopolistic power to print money without the backing of gold or silver. But the rural farmers on the frontier were strapped for money, even though, for the most part, they were producing sufficient yields to barter for cash or to pay their taxes and mortgages. The national government was in debt from fighting the Revolution, and the former colony's most important prewar trading partners, England and France, would not or could not trade with the United States. England spitefully refused to trade with its former colony for years after Americans revolted. Meanwhile, France was in the middle of its own revolution and unable to buy anything but war materiel.

As a consequence, the national government wouldn't accept bartered goods because commodity brokers didn't have access to as large an international market to trade with as American merchants did prior to the war. The refusal to barter left the farmers without a profitable means to dispose of their commodities and desperately short of spending money, or money for taxes. From the perspective of the small-scale farmers, the Federalist assaults especially targeted frontier farmers and rural resistance as a way of making a statement that protest and revolt (however legitimate) would not be tolerated.

After the Revolutionary War, the Federalists used the resources of both the banks and the army to fight

Shays's Rebellion, the Whiskey Rebellion, and numerous other significant rural uprisings in New Jersey, New York, Vermont, Pennsylvania, Virginia, North Carolina, South Carolina, and Maine. There should be no doubt that the Federalists were adamant about crushing any "rogue" states or rebellions. The armies that they amassed to put down Shays's Rebellion and the Whiskey Rebellion are examples of their determination to control the newly established state and shape it to their needs.

Shays's Rebellion began because the Massachusetts governor and legislature promoted and passed tax laws to pay wealthy Bostonians the full amount for speculations on soldiers' pay guarantees from the Revolutionary War, which they had bought for 10 percent of face value. The government demanded payment of taxes in specie (gold, silver, and copper currency) in order to pay for soldiers' pay-guarantee taxes. The rebels resisted the tax as a get-richer-quick scheme (which it was). When several rebels agreed to pay the tax, they wanted the state to accept bartered products in replacement for specie as they had in the past. The state refused. The rebels led by Daniel Shays were trying to regulate aristocratic government by resisting tax payments and closing down the courts, as the South Carolina Regulators had done successfully before the Revolutionary War. The Boston aristocracy, however, remained ruthless and uncompromising.

Washington and Hamilton supported the aristocracy and ordered five thousand soldiers to return to duty to subdue Shays's Rebellion in 1787. However, most of the soldiers who were called up had fought in the Revolution and supported the cause of the revolt. Only one hundred troops volunteered to fight against the rebels.

Not to be outdone, the Boston speculators donated money so that General Benjamin Lincoln could hire three thousand mercenary troops to put down regulator revolts in Worcester, Springfield, Berkshire County, and other western Massachusetts towns. A combination of luck and superior firepower enabled these mercenary troops to crush Shays's revolt. Immediately after Shays's Rebellion, a major campaign was spearheaded by the Federalists in an effort to get the national Constitution ratified in a time of alarm about insurgency threats from regulator communities all along the frontier. The frontier in the case of Shays's Rebellion was only forty miles from Boston. Once the Constitution was ratified in 1788 it superseded the Bill of Rights and the Declaration of Independence, which had demanded much more radical civil and economic rights for the revolutionary fighters and the general white male population. With the Constitution in place, control by the aristocracy was firmly established.

The Whiskey Rebellion followed the ratification of the Constitution and was the first real test after its passage of the power of the aristocratic Federalists. This revolt resulted from an onerous and discriminatory federal tax on grain used for whiskey. The grain was turned into whiskey in rural areas because the farmers couldn't afford the overland freight or barge charges for transporting raw grain to Philadelphia, Washington, Baltimore, or any other wealthy port or city. This rebellion was also a protest against the federal courts that were usurping the legitimate and effective power of the state and local courts.

By 1794 the Whiskey Rebellion was gaining support from settlers in many rural areas who also were passionate about regulating the excesses of taxation and other financial schemes. Washington and Hamilton decided that they needed to make an example of the most strident of the rural rebellions and crush it thoroughly in order to discourage other revolts. Washington sent an army of at least fifteen thousand men against the Whiskey Rebels. This force was as large as the army that had fought the Revolution against England. Of course, the army thoroughly crushed the Whiskey Rebellion. The power that the aristocratic Federalists would use against citizens trying to exercise their right of dissent was violently and indelibly imprinted on rural American farmers.

Most of these post-Revolutionary uprisings focused on land title, land loss, corrupt banks, local rule, unjust commodity taxation (such as the taxation that led to the Whiskey Rebellion), and the lack of circulating money (gold and silver) in rural areas. Without gold and silver, paper money, or the acceptance of bartered goods for taxes and other debts, the rural residents couldn't pay their taxes or their land debts. Clearly, Washington and Hamilton did not need to use such force against these mostly veteran revolutionaries, unless they were trying to impress on the freeholders the power of the landed aristocracy. The heavy-handed punitive actions by the federal government against these legitimately aggrieved small land-holders served to solidify for the landed class their domination of rural America. It also served notice that the aristocracy would be continuing a business-as-usual approach to land use (and land and labor abuse) that began with the enclosures in Europe and continued throughout the colonial period.[21]

Almost all of the social uproar of the 1700s was directed at the landed gentry on both sides of the Atlantic. A major portion of this movement focused criticism on privileged tax schemes, large-scale estates, and the labor abuse[22] and destructive farming practices that the landed gentry employed. Especially in Europe, scientists and philosophers began to disparage the enormous sheep and grain farms that followed the enclosure movement. In the two decades before the Revolutionary War, a growing sector of small-scale American farmers also began to condemn the European and American land speculators who were profiting from the large-scale colonial farms, while destroying the land and exploiting the settlers in the process.

Most of the farmland in Virginia and Maryland had been badly abused and suffered long-term damage from nearly two hundred years of continuous tobacco, corn, cotton, jute, and hemp production by aristocratic Europeans and their American managers. Even the most fertile land couldn't tolerate continuous cropping of tobacco, corn, and cotton without suffering serious soil degradation.[23]

Before the war, some of the more brazen colonial farmers argued that soil protection and recovery strategies were essential because the English and Dutch land companies had devastated enormous tracts of land from southern Georgia to as far north as Massachusetts. As the general rebellion against England spread, critics became outspoken and publicly exclaimed, still at considerable personal risk, that the agricultural decline was both a scandal and a crisis that would surely limit the amount of food that the nation's farmers could produce.

Not long after the Revolutionary War, aristocratic American farmers became interested in the new scientific farming programs that were becoming all the rage in Europe. At the same time a precious few American farm philosophers, a few large-scale farmers, and thousands of small-scale farmers joined the search for more sustainable land-management strategies. From the very first days of the new republic, two classes of farmers emerged, each with different attitudes toward farming.

The ad at the top came from the *Pennsylvania Journal or Weekly Advertiser*, August 1754. The ad gives notice that a new tax collector for Philadelphia City and County has been appointed. Such tax collectors worked for the crown and were much despised by regulators and other rebels. The middle and bottom ads came from the *Pennsylvania Packet and Daily Advertiser*. The middle ad was printed on July 3, 1790. The ad offers plaster of Paris to be used as a manure. The bottom ad was printed on July 5, 1790 and is an announcement of a new shipment of drugs, medicines, paints, glass, and dyes from London and Amsterdam.

LEAVITT'S

GENUINE, IMPROVED
NEW-ENGLAND
FARMER's ALMANACK,
AND
Agricultural Register.
FOR THE YEAR
1815.

Being the third after Bissextile, or Leap Year, and the Thirty-
ninth of the Independence of the United States of America.
Calculated for the Meridian of Boston, Mass. lat. 42° 25′ north.

Devoted to Science, Agriculture and the Arts.

BY DUDLEY LEAVITT.
TEACHER OF MATHEMATICS AND NATURAL PHILOSOPHY.

"The spacious firmament on high,
"With all the blue ethereal sky,
"And spangled Heavens, a shining frame,
"Their great Original proclaim".

EXETER....Published and sold, wholesale and retail, by
C. NORRIS & Co.
Sold also by all the Booksellers and Traders in New-England.
Price $44 per groce, 42 cts. per dozen, and 6 cts. single.

THE FARMER'S CALENDAR.

Jan. Settle your accounts for the last year, if you have not done it before. Let the smoke curl out of your chimney before sunrise. Feed your cattle in good season. Be industrious, and see that nothing is wasted.

Feb. See that your cattle are well tended. Haul timber for fencing; sted up more fire wood if you have not a sufficient supply. To be employed about farming affairs will be as profitable as disputing about politicks.

Mar. Dress out your flax; prepare for making maple sugar; repair your farming tools for use; put up your fences and bars in good season, and allow not your cattle to graze and tread up your fields.

April. Plough and dig your gardens as soon as the season will admit. Onions, parsnips, and peas should be sown as early as possible;—so should summer or spring rye — wheat will do a little later. Prepare for planting.

May. Plant corn, potatoes, and beans, as soon as possible. If you put ashes or plaster in the hills when you plant, the crop will probably be the better. See that the birds and squirrels do not pull up your corn.

June. Hoe corn as soon as it is large enough; that which is hoed first generally keeps the start all the year till the corn is ripe. Before you hoe, however, put plaster or ashes round the corn. Keep good fences.

July. Prepare for mowing; see that your barns, carts, rakes, scythes, and pitchforks are all in order. It used to be thought that meadow grass would do to cut late, as well as any time, but it is found to be much better to cut it early. While busy about haying, do not neglect your grain.

Aug. Finish haying as soon as possible. Reap and get in your grain in good fair weather. As soon as you have done reaping, plough in the stubble if you intend to sow or plant the same ground next year.

Sept. Break up land to plant another season. Haul rocks, and build stone wall. Clear up low land and make drains and ditches. Cut corn stalks and have them well dried. Stubble ground that has been ploughed and cross-ploughed, may be sown with winter rye.

Oct. Harvest your corn and potatoes; make cider, and see that nothing is wasted for want of care. When making your cider, pick out and throw away the rotten apples; let your cider mill be clean and neat, and use no water about making your cider.

Nov. Bank up your houses that your cellars may be secured against the frost. Cut and pile wood for your winter's use now, while little else can be done on the farm. See that your summer farming tools are safely laid up till another season.

Dec. See that your school houses are in good repair, and provide a good quantity of fuel for your winter schools. If your school house is inconvenient, with broken windows and a smoking chimney, the scholars may rub their eyes till they ache, but they cannot learn much.

A countryman once going into a city, observed that the most fashionable people, instead of butter, ate oil with fish, from which he concluded it was somehow or rather very excellent, and though he did not know what kind of oil it was, he determined to try it at his own table; so after returning home he sent his boy to the currier, for a quart of oil, and told his family that the city folks used oil, instead of butter, with their fish, and meaning to be as genteel as any of them, he would have some himself. So he had his fish cooked, and having poured the oil upon it, he began to eat. His wife asked him how he liked the oil? Why, said he, if it was not for the *name* of oil, I should much rather have good butter.

Chapter 2

AMERICA'S FIRST FARM REVITALIZATION MOVEMENT

In his 1955 book *Our Vanishing Landscape*, Eric Sloane wrote, "A hundred or more years ago, whether you were a blacksmith, a butcher, a carpenter, a politician, or a banker, you were also a farmer. Before setting out for the day, there were chores to be done that often took as much time as a complete day's work for the average man today."[1] While Sloane's book presents a somewhat romantic view of farmers, he accurately describes an attachment to the land and an ethos of self-reliant subsistence that is woven throughout the tapestry of American farming history. Farming is still the most fundamental human occupation on this earth, and it has been for more than ten thousand years. About 2,850,000,000 people in the world still farm today, and they represent about 45 percent of the world's population. In sharp contrast, only about 1 percent of the U.S. populace still farms, and less than 1 percent own any farmland.

After the turn of the nineteenth century, criticism of America's aristocracy and its farming practices became much more widespread, and even some of the aristocratic farmers became early critics of large-scale agriculture. In 1813, the aristocrat John Taylor published *Arator*. Taylor's strategies, though still dependent on slave labor, accurately addressed the problems caused by destructive agricultural practices in America and pointed out the crisis caused by the loss of organic matter:

> Let us boldly face the fact. Our country is nearly ruined. We certainly have drawn out of the earth three-fourths of the vegetable matter [organic matter] it contained, within reach of the plow. . . . Forbear, oh forbear matricide, not for futurity, not for God's sake, but for your own sake.[2]

Here we see Taylor experiencing the same kind of realization farmers did in this century as they discovered the damages done by industrial-chemical agriculture. Like Jethro Tull, the famous agricultural innovator of the 1700s, he believed that the atmosphere had an effect on supplying nitrogen to the soil. Taylor realized that farmers needed to continually feed the soil with organic matter, mineral nutrients, and animal and plant fertilizers, and to regularly rotate their crops to maintain soil health and productivity—just to replace what the soil had lost in producing cash crops for two hundred years.

Taylor's epiphany helped spark a revolution in farm practices. In order to rejuvenate soil devastated by large-scale farming in the colonies, Taylor advocated the use of deep plowing; the development of a four-field system of rotation; the bedding, composting, and handling of manure; and the restoration of worn-out lands by planting fallow crops and not grazing the fallows.[3] These are very similar to modern organic strategies, yet Taylor wrote about them in the early nineteenth century.

News about strategies like Taylor's appeared in the farmers' almanacs, which were America's oldest rural publications. The oldest almanac was the *New England Almanac*, published in 1639. *Poor Richard's Almanac* was printed and sold by Benjamin Franklin in 1739. The oldest continuously published almanac, *The Old Farmer's Almanac*, first appeared in 1791 and is still around today. Farmers read the almanacs avidly, and many used the farming ideas that they promoted. Also popular after the turn of the century were the agricultural society fairs, and a high percentage of the rural population attended them.[4] The fairs called for agricultural reform similar to those strategies advocated by Taylor.

The Berkshire Agriculture Society was founded in 1811 in Massachusetts, New York, and Vermont and promoted some of the earliest and most successful fairs. Several other agricultural societies copied these early fairs, and soon events such as these were being held in rural areas around the country. Both the almanacs and the farm fairs significantly changed farming practices in the early nineteenth century. But the fairs probably had the most positive influence on farm practices in the first two decades of the 1800s.

While the first American agricultural reformers drew largely on the work of Taylor, agricultural reform was in fact an older phenomenon. Many well-developed soil-recovery strategies actually came from practices employed by ancient European farmers, including those from Roman occupations of late antiquity. For hundreds of years in many European countries, peasant farmers traditionally spread animal manure and several other fertilizers on the soil to revitalize the land and produce better crops. These fertilizers included marl, sand, greensand, kelp, peat, nutrient-rich soil (humus), leaf mold, thatch, stable litter, dried blood, fish scraps, and ground bones, the last often recovered from the battlefields of Europe.[5]

In this earliest of the agricultural-reform movements in America, progressive farmers composted manure and other wastes but also knew that a variety of plants acted like fertilizers and were good for the following crop. Without knowing all of the biological or enzymatic reactions that were taking place, growers included fertilizing cover crops and beneficial plants in field rotations, because their effectiveness had been known for centuries. For example, many growers knew that bean plants, mustards, and certain grasses provided valuable fertility for the next year, and that planting garlic and onions stimulated the crops that followed them. Others knew that tomatoes made the soil more acid and that acid-loving plants, like beans, produced higher yields after tomatoes, or that corn following beans also produced higher yields.

From generations of practice, farmers learned that essential nutrients and the soil conditions for optimum plant growth could be replenished by planting these and many other fertilizer crops, such as buckwheat, Italian rye, barley, and clovers. Well before the mid-1800s many American growers planted a variety of those crops that accumulated nitrogen, phosphorus, potassium, and other soil nutrients.[6]

While the Berkshire Society and the assorted farmers' almanacs promoted promising strategies for soil recovery and health, they also reported on the wide variety of pest controls being used. These techniques came mostly from European farmers but included elements from tribal American farmers and even traveling medicine peddlers. Several of the more popular early-nineteenth-century insecticides were hellebore, quasia, lime, tobacco, sulfur, copper, potash soaps, and soda.

Despite the local successes of the farm societies and the fairs and all the philosophical and practical strategizing, the crisis over soil loss and degenerating farmland continued to worsen in the early 1800s. This occurred largely because the large-scale farms owned by aristocrats were exploitative of resources, and soil degradation on a grand scale was inevitable. Meanwhile, for the small-scale farmer, economic depressions became all too common, and, as a result, avoiding bankruptcy became the farmers' highest priority. For many, it still is.

Though soil revitalization often became a secondary concern during tough times, over the decades it remained a high priority with both small and medium-sized farmers. As early as the 1790s, the northern-based farm societies pleaded with President Washington to create a national Department of Agriculture, especially after England formed its own Department of Agriculture in 1793. They were successful: Washington proposed a Department of Agriculture that same year. Though popular in the North, Washington's proposals were repeatedly killed in Congress by large-scale southern planters who generally saw no need to improve farming techniques or economic practices on the plantations.

In the first decade of the 1800s, farmers and the agricultural societies also advocated for colleges of agriculture within the universities. Several progressive states rapidly developed colleges, departments, and programs to meet the farmers' demands. The societies also promoted the idea of rural journals that could help popularize farm recovery programs. They argued that farm journals could spread the messages of the farm societies to a wider audience than the yearly almanacs or the seasonal fairs.

In the early 1800s, political and military action continued from Canada to the Carolinas as the nation rapidly headed for another conflict with England in 1812. The fighting in the Civil War began in 1861. The conflicts that created the war began much earlier. The unsettled nature of the country resulting from rebellion, war, economic panics, and friction between the North and South greatly inhibited the development of agriculture in the first seventy-five years of the Republic.

The First Berkshire Society Fair ad (1910) and the report of its success (1911). This was a posted bill.

" BERKSHIRE CATTLE-SHOW.

" The multiplication of useful animals is a common blessing to mankind."   WASHINGTON.

" TO FARMERS.

"The subscribers take the liberty to address you " on a momentous subject, which, in all probability, " will materially affect the Agricultural interest of " this county."

After several other remarks, it concludes :

" In a hope of being instrumental in commencing " a plan so useful in its consequences, we propose " to exhibit in the square in the village of Pittsfield, " on the 1st October next, from 9 to 3 o'clock, Bulls," &c. &c. " It is hoped this essay will not be confined " to the present year, but will lead to permanent " annual cattle shows, and that an incorporated " Agricultural Society will emanate from these " meetings, which will hereafter be possessed of " funds sufficient to award premiums," &c. &c.

Signed, 1st August, 1810, Samuel H. Wheeler, and twenty-six farmers, among whom I include myself.

In consequence of this first step, on the 1st October, 1810, I find the following notice in the Pittsfield Sun.

" The first Berkshire Cattle Show was exhibited " with considerable eclat on Monday last. This land-" able measure cannot fail to be highly beneficial, con-" sidering its novelty in this part of the world, and " that many had their doubts, and even a dread of " being held up for the finger of scorn to point at. " The display of fine animals, and the number exhi-" bited, exceeded the most sanguine hopes of its pro-" moters,* and a large collection of people partici-" pated in the display : the weather was delight-" ful—the ice is now broke—all squeamish feelings " buried—and a general satisfaction evinced. It will " now be impossible to arrest its course ; we have " every thing to hope, and to expect, the year ensu-" ing."

A committee of fourteen respectable farmers, from different parts of the county, was appointed to take preparatory measures for a real exhibition in October, 1811. As I had thus far taken the lead in every thing, by common consent, I was placed at the head of a senseless procession of farmers, marching round the square, without motive, or object ; having

* The farmers held back their animals in the vicinity, for fear of being laughed at, which compelled me to lead the way with several prime animals.

have Ten Shillings Reward, paid by Isaac Warner.

RUN away laſt Night, from on Board the Dianna, of Dublin, Richard M'Carty, Maſter, a Servant Man, named Valentine Handlin, aged about 30 Years, a luſty rawbon'd Fellow ſmall round Viſaged, is of a dark Complexion with ſhort Black Hair; Had on when he went away, a brown bob Wig, old Felt-Hat, an old lightiſh colour'd cloth great Coat, a blue grey Waiſtcoat, old leather Breeches, yarn Stockings, broad ſquare toe'd Shoes; and perhaps may have taken ſome other Cloaths with him. He is remarkably hollow Footed, and ſeems crumpfooted when his Shoes are off. Whoever ſecures the ſaid Servant ſo he may be had again, ſhall have Twenty Shillings Reward, paid by

Philad. December 3. 1740.   William Blair.

N. B. A ſmall Boat belonging to the ſaid Sloop was loſt on Monday laſt, Whoever will bring her to the ſaid Sloop ſhall have Ten Shillings Reward, paid by   William Blair.

TO BE SOLD

A Likely young Negro Woman, can Waſh or Iron, or do any kind of houſhold Work, and is fit for either Town or Country; with two Children. Enquire of George Harding, Skinner, or the Printer hereof.

A TRACT of Land containing about a Thouſand Acres, lying on *South River*, about Four Miles from *Brunſwick*, and about Eight Miles from *Amboy*; it is well timber'd, and has about 60 Acres of good Meadow, a good Dwelling Houſe and a Store-houſe, and a good Landing very convenient for Trade and bringing of Goods from *York* to the *Jerſeys*, and excellent for taking of Shad at the Time of the Year when they are in Seaſon, as hath been proved. Any one who hath a mind to purchaſe the ſame, may apply themſelves to *Gabriel Stealea*, Eſq; at *Perth Amboy*, or Dr. *Brown* near *Burlington*.

CHoice Maid Servants, fit for Town or

Chapter 3

THE BIRTH OF THE RURAL JOURNALS

Johannes Gutenberg began experimenting with movable type and printing presses as early as 1438 in Germany. In 1450, Johannes Fust financed Peter Schoffer's designs of movable type and the construction of Gutenberg's innovative printing presses. By 1452, in addition to Bibles, they began to print sibylline books, letters of indulgence (sort of a "Get Out of Hell Free" card from the Pope for the wealthy), and reprints of religious works that had previously been available only in handwritten script.

Shortly after Gutenberg and his partners published the Bibles in the 1450s, merchants, nobles, and manufacturers began to sell products and to enforce their newly enacted land and labor laws through printed advertisements, propaganda, and announcements on billboards and periodicals. By 1477, the first advertisements were printed with movable type.

The oldest European printed-news pamphlet appeared in Germany in 1525 and contained the first newspaper advertisement. Similar pamphlets appeared in Italy, Holland, Belgium, Switzerland, and England at about the same time. Then, Luther's German Bible was printed, and it became the most important new Bible since Constantine. Several thousand copies were sold all over Europe. This caused a century of religious reform, religious revolt, religious wars, the slaughter of millions, and the conversion of half of Europe's population to Protestantism. In1532 Machiavelli published *The Prince*, the first mechanically printed book to appear

A page from the forty-two-line Gutenberg Bible, printed about 1452–55. From the Laubach copy at the Gutenberg Museum, Mainz, Germany. This first of the movable-type-printed Bibles is said to be one of the most beautifully printed books in the history of printing. Reportedly, 180 copies were printed. Parts of 48 copies are known to exist. Only the New Testament remains from this Laubach copy.

without biblical quotations or references to writers of antiquity.

Mexico imported America's first printing press in 1535, and Lima, Peru, imported one in 1585. None existed in the New England colonies until Harvard University imported a press in 1639. By this time, printed billboards were common in Germany, Italy, and Britain. Often whole walls in these cities devoted their space to printed advertising bills.

In 1690, the first printed newspaper appeared in Boston, with the publication of *Publick Occurrences*. But immediately after it was published, censorship was imposed on all New England printers, and no other news bill appeared for fourteen years.[1] The first American newspaper published after the ban, *The Boston News-Letter*, began on April 17, 1704, and the first advertisements followed less than a month later in the same paper. In 1708, the *News-Letter* printed the first advertisement for a patent medicine.

Among the most successful and influential of the early American publishers was Ben Franklin. In 1728, Franklin took over his brother's job of publishing the *Pennsylvania Gazette*. Franklin helped change the course of printed journalism in America. He adopted the European newspapers' excessive emphasis on advertising, and by 1760 he was devoting half of the *Pennsylvania Gazette* to advertising. Several of his early ads were for farm implements, farm products, and farmland, but Franklin also devoted a large percentage of his advertisements to white and black slave sales or rewards for escaped slaves. As time went on, the *Pennsylvania Gazette*, under Franklin's control, came to advertise almost anything that could be sold, no matter how odious its nature or questionable its value. Though Franklin currently enjoys a deserved reputation as an innovative scientist, inventor, educator, and statesman, his approach to publishing was very commercial.[2] Before he became publisher, the *Gazette* had had about 10

percent of its space devoted to advertising and 90 percent to literature, news, "how-to" articles, recipes, and local infrastructure improvement ideas.

In spite of the glaring commercialism in city papers and on city walls, farmers, who composed more than 90 percent of the population, still hoped that quality information could be gotten from periodicals that were devoted exclusively to rural issues and farm problems. Consequently, the demand increased for magazines that directly served the farm community. In response to these demands, the first American farm magazine appeared in 1810, titled *An Agricultural Museum*. This journal was short-lived and poorly financed and had very little impact except to further whet the appetite for rural-affairs journals. *American Farmer* and *The Plough Boy* followed in 1819. *The Plough Boy* also appeared for only a short time, but *American Farmer* lasted until the 1890s under different publishers.

Most historians agree that *American Farmer* was the most influential of the early American farm magazines. For the first fifteen years almost all the rural journals copied its format and style of reporting. Between 1819 and 1834, the editor of *American Farmer*, John Stuart Skinner, promoted a sophisticated form of planned and managed soil fertility, similar to that advocated by John Taylor and the Berkshire Agricultural Society. These strategies were presented in the first issues of the American farm magazines and continued to be promoted vigorously for decades to come.

In these earliest farm journals the editors promoted many of the same pest controls as had the Berkshire Society and the farmers' almanacs, but they also advocated for other solutions, including pyrethrum powder, a biological insecticide derived from chrysanthemum flowers. Editors also promoted dormant oils (derived from whale blubber, animal fats, cottonseed oil, and hempseed oil) to control a broad spectrum of pests. Before the mid-1800s, the

Oldest surviving print advertisement, by William Caxton, in 1477, to promote his publication *The Pyes of Salisbury*. In T. R. Nevett, *Advertising in Britain* (London: Heineman Publishing, 1982).

The first Outdoor Advertising was on Covered bridges

C.A. Rennacker CLOTHIER
33-41 ASYLUM ST. HARTFORD

BURNS & TAYLOR CLOTHIERS

at Windsor, Conn
over the
Farmington River

at Bridgeville, N.Y. 1817

WICOMA — THE PERFECT CURE

the Medicine Bridge
at Lexington, Virginia

JUST SUITS TOBACCO

"Just suits" tobacco.
Over East Creek, Rutland, Vt.

Coca-Cola
Delicious and Refreshing
Coca-Cola
5¢ Coca-Cola

Coca-Cola bridge
at Portland, Pennsylvania

ALL LOTTERIES END FOR EVER 18th OCTr SIX of 30,000 EACH ALL LOTTERIES BLOCK

This drawing of covered bridges is from Eric Sloane, *Our Vanishing Landscape* (New York: Wilfred Funk Inc., 1955). The first covered bridges were built in the early 1800s, along the most well-traveled routes. Sloane's book contends that outdoor advertising began on bridges. This may be accurate for parts of the rural United States but certainly not for the cities, where posted bills appeared in the early 1700s.

first widespread searches for beneficial predators were attempted successfully and reported on in the farm magazines.

John Stuart Skinner was a postmaster, as were many of the editors or agents for America's earliest farm magazines. In the rural culture at the beginning of the 1800s, farm publications finally became possible because efficient mail service made mass publication more practical and because the self-interested postmaster-editors facilitated prompt delivery. Still, the price for mail delivery was very high relative to the value of money, as this rate schedule from the *Farmer's Almanac* of 1813 shows: "'Every letter composed of a single sheet of paper not conveyed above thirty miles, six cents. From thirty miles to eighty miles, ten cents. From eighty to one hundred and fifty miles, twelve and a half cents. From one hundred and fifty to four hundred miles, eighteen and three-fourths cents. Over four hundred miles, twenty five cents.' For each extra sheet of paper used you were charged an additional postage rate."[3]

When first printed, *American Farmer* was an eight-page journal, using four pages on both sides; consequently, delivery was expensive and often exceeded the cost of the subscription. In *Our Vanishing Landscape*, Eric Sloane reminds us of why the postal deliveries were so expensive when he notes that it took seventy-six horses to complete the stage run that connected New York City and Boston. Today that trip can be made in a four-hour car or train ride. But in the 1800s, almost the entire trip was rural, with little towns sprinkled along the way. Farm magazines were dropped off at rural post offices and delivered to farms.

With postage so costly, the earliest farm journals concentrated on text and limited their advertisements to one-quarter of a page. In the beginning, *American Farmer* and *The Plough Boy* accepted an ad for one-time publication only, under the assumption that the farmers were saving each issue, so that if they were

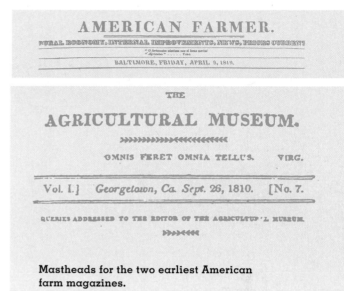

Mastheads for the two earliest American farm magazines.

interested in the advertised product, they would find it in a back issue.

Gradually, this advertising rule of "once only" broke down under the persuasive pressure of dollars from advertisers, who knew that repetitive ads meant higher sales. As a consequence, ads increased, in spite of objections from subscribers who declared that the advertising material was infringing on the reading matter.[4]

Though revolutionary farm advice dominated the first farm journals, many of the large-scale farmers ignored the farm advice from *American Farmer* and the farm societies. Most kept farming as they always had, without any thought of the future, since they believed that virgin land in America was endless.

Because of their economic clout, the dominant farmers and land companies felt that their access to this seemingly endless land was guaranteed. And of course it was, since they controlled almost everything, including the courts, the banks, the army, Congress, most of the commerce, and most of the already developed land. A common practice was to use up the fertility of a farm and then move on and open up a new tract of land when that piece wore out. As this process accelerated, the land companies overran many outlying small farm settlements.

When the federal and state governments planned new roads, canals, and railroads, the aristocratic farmers and merchants, sensing the potential for profit, expanded their operations by seizing farmland near the transport lines and then getting the federal government to back their claims. The first canal was completed around the falls of the Connecticut River in South Hadley, Massachusetts, in 1793. The canal systems were very successful for about fifty years, but the railroad systems developed rapidly, and they eventually put the canals out of business. Many small-scale farmers fought long-term battles before finally selling out or losing their land to large-scale farmers, federalist bankers, cattle kings, land schemes, and speculators who fought desperately to gain access to transportation of any kind.

The large-scale farmers won most of the battles, and the size of their farms increased. The land became even more exhausted by having the same commercial crops continuously planted for several decades or by destructive grazing practices. So, in spite of the success of the farm societies, fairs, almanacs, farm literature, and farm journals in introducing innovative agricultural strategies, soil fertility continued to decline and pasture land became overgrazed. At the same time, the wealthy land speculators concerned themselves more with profit and expansion than with revolutionary farm practices or ideas.

While Ben Franklin was changing advertising and American journalism, chemists were working to obtain scientific credibility for their much-maligned area of study and research. Chemists and alchemists

were still the laughingstock of scientific societies throughout the 1700s, and scientists and other intellectuals frequently referred to them as quacks. By the start of the 1800s, however, chemists had made significant discoveries that proved to have economic significance and that catapulted them to a previously unimaginable level of respectability, as we shall see in the next few chapters.

Right: This advisory from *American Farmer* (1819) spells out that advertisements will be inserted only once unless the law requires that a greater number be printed.

Left: Mastheads for farm journals.

☞ ADVERTISEMENTS, which are, in their nature and objects suited to a paper of this sort, such as the sales of land, seed, live stock, implements of husbandry, new inventions, &c. &c., will be inserted *once only*, at the rate of $1 per square, to be paid in advance. The very extensive circulation of this paper among landed men, throughout the United States, makes it an eligible medium for giving such public notices, and one publication is as good as forty, unless in cases where the *law prescribes* a greater number of times.

PRINTED EVERY FRIDAY AT $4 PER ANN.

FOR JOHN S. SKINNER, EDITOR,

At the corner of Market and Belvidere-streets,

Collage of nineteenth-century ads for whale oil, pyrethrum, and potash.

PERUVIAN GUANO.

Fig. 1.

Front view of Genuine No. 1 *Peruvian Guano*, with 10 per cent Ammonia.

There are many grades of Peruvian Guano in the market. Guano sold for fertilizing purposes by the regular appointed agents is branded as follows :

No. 1 Peruvian Guano—Standard.
No. 1 Peruvian Guano—Lobos.
No. 1 Peruvian Guano—Guaranteed.

Each of these are genuine Guanos, excepting the *Rectified*, which is treated with sulphuric acid.

Fig. 2.

No. 1 Peruvian Guano—*Lobos.* The other side of bag branded same as the Standard No. 1 Peruvian.

The brand Lobos, which is put lengthwise on the bag, in large letters, designates the island or location from which it was taken, namely, *Lobos Island* or *Lobos Point*, in *Peru.* This Guano is becoming very popular with those who have used it. It is a splendid fertilizer, and is far richer than those ingredients which vegetation feeds upon than was the Genuine Chincha Guano sold a few years ago. It is guaranteed to analize *six* per cent. of *ammonia*, and is very rich in *phosphoric acid* and *potassa.*

It is put up in bags and weighed by a United States Weigher. The bags vary in weight from 160 to 190 pounds.

A Lobos Peruvian Guano ad from 1851. Lobos was the best Peruvian guano.

Chapter 4

GUANO AND THE BIRTH OF THE RURAL-INDUSTRIAL MOUTHPIECE

The entrance of large-scale commercial interests into the field of improving soil fertility can be traced to the "guano craze" in the middle of the nineteenth century. For thousands of years bird manure, mostly from cormorants, pelicans, and terns, had collected on the Chincha Islands off the coast of Peru. This manure was called *huano* by the indigenous Peruvians and *guano* by the Spaniards. Forty to a hundred feet of guano covered these few small islands. The guano was a well-composted, highly usable fertilizer that supplied most of the nutrients that plants needed.

The trademark of Pacific Guano. From Richard Wines, *Fertilizer in America* (Philadelphia: Temple University Press, 1985).

Guano had been used as a fertilizer by the Peruvians for thousands of years. A succession of empires carefully controlled the mining of guano. The Inca Empire so highly valued guano that even disturbing shore birds was an offense punishable by death. The only really effective guanos during the 1800s were from these few islands off the west coast of Peru. Because these islands received almost no rainfall, the nitrogen, phosphorus, potash, and other nutrients were not leached out of the composted bird droppings and carcasses.

In 1804, the naturalist/scientist/adventurer Alexander von Humboldt learned about the agricultural value of Peruvian guano on his expedition to South America. Humboldt returned to Europe with guano samples and testified to their amazing fertilizer properties and their soil-recovery potential for worn-out land. He also proposed that merchants and cargo carriers could profit from the mining and shipping of guano. This set off a flurry of activity by mining and shipping companies as they lined up to exploit this miracle fertilizer.

In 1824 John Stuart Skinner printed the first report in *American Farmer* about the discovery and effectiveness of Peruvian guano. Thereafter, a barrage of reports in the urban American journals propagandized that guano was a miraculous fertilizer. Within a few years, shipping merchants and their scientists claimed that guano would be the quick-fix salvation for Europe and America's damaged soils. The commercial mining of the Peruvian guano deposits began shortly after this propaganda campaign in the late 1820s.

Peruvian guano was a very effective natural fertilizer and was the first successfully marketed commercial fertilizer. However, guano mining and shipping was a grim business. Since no one would volunteer for such abominable labor, Chinese and Peruvian Indian slaves and prisoners did most of the difficult and dangerous dynamiting, shoveling, and picking. Slavery was not abolished in Peru until 1859, but by then much of the guano was mined out. In his 1944 book, *The Great Guano Rush*, J. M. Skaggs wrote:

> Respiratory problems, asphyxiation from ammonia vapors, coughing or spitting up blood, "embarrassed breathing" and fainting spells were widely reported among miners. Disease transmitted from the guano to the workers was common.
> Those who did not work diligently were set upon by half-starved dogs employed to instill fear.
> After the guano was mined, the ships' crew became victims as they worked to fill the holds. To protect themselves from dust and ammonia vapors, they wore improvised oakum respirators (loose hemp and jute fiber masks plastered together with tar) . . . over their mouths and nostrils. Even then they could remain below deck no more than twenty minutes at a time. [1]

Since guano was a totally natural bird fertilizer it was similar in the farmers' minds to chicken, duck, or goose manure. However, when magazine editorials began to promote this first of the commercially mined fertilizers, angry letters from the farmers and the farm societies argued that many of the guanos were fakes and that the use of such miracle fertilizers couldn't and shouldn't replace the sophisticated farm planning that was being promoted by John Taylor, the farm societies, and the seasonal agricultural fairs. Farmers, since antiquity, have depended on a long-term history of success before accepting different fertilizers or new or innovative farm tools, so most of

them were suspicious of guano, even though the fertilizer merchants advertised it as a miracle cure for worn-out soils.

Another factor that stopped many farmers from using it was the cost of guano versus the cost of manure from their own pastures and barns and a widespread belief in the self-sufficiency ethic that frowned on buying off-farm products—even ones alleged to be miraculous. In post-Revolutionary days, how long something had worked well still influenced farmers far more than the newest innovation on the market. Farmers were openly critical of "book farming" ideas from city dwellers and "quack" products that had no track record. Growers vented their criticism of guano and other new products at the fairs, in the almanacs, and in the dozens of new farm journals that were started between 1819 and 1845.

Farmers remained suspicious about any advertised product because of fraudulent promotions that had appeared in urban papers before and after the Revolution. Consequently, very few ads appeared for any products in the early issues of America's farm journals. Some of the first farm-magazine ads pertaining to agriculture sold equipment, plants, seeds, animal medicines, patent medicines, slaves, and real estate. The articles concentrated on farm problems and solutions, travel, current events, philosophy, infrastructure development, and a celebration of the rural lifestyle.

By the late 1840s and early 1850s, however, many of the most influential farm magazines had begun to sell fertilizer, farm equipment, or seed. Dealing and promoting a publisher's own commercial products became a common practice a short twenty years after the birth of the rural magazines. The following passage from *The American Agricultural Press* is worth quoting at length because it shows both the freedom of the press and the gross commercialization of the farm press at such a very early date.

Since most of the editors and proprietors were forced to seek additional sources of income, they frequently established agencies for the sale of horses, cattle, sheep, hogs, seeds, agricultural machinery, implements, and books. Thus, the seed store or agricultural warehouse was a common adjunct to an agricultural journal office. For example, the proprietors of the *New England Farmer* [started in 1822] not only owned a large seed store, but were the manufacturers and owners of the patent right of the Howard's Plow; the editors of the *American Agriculturalist* [started in 1841] were proprietors of the New York Agricultural Warehouse and Seed Store, sole agents for Premium Plows, and manufacturers and holders of patents on agricultural machinery. The editor of the *Farmer's Cabinet* [started in 1837] sold a patented fertilizer called Poudrette, and for a time J. S. Wright of the *Prairie Farmer* [started in 1840] manufactured the Hussey Reaper and Atkins Automation; while the editors of the *Southern Planter* [started in 1840] and *Ohio Cultivator* [started 1844] were agents for the McCormick Reaper.[2]

Because of well-grounded suspicions about publisher bias, only a small percentage of farmers subscribed to the early farm magazines. Many felt that the magazines had already given themselves over to product promotion instead of agricultural improvement. In the late 1840s, after twenty years of guano promotions in the urban newspapers and journals, several farm magazines finally began to advertise guano. When these ads appeared, a new flood of letters from growers expressed skepticism about both the products and the farm journals that promoted them. History has shown that the farmers' skepticism about such products and their promoters was justified. Very few of the guanos proved to be effective after the Peruvian deposits were gone, and most of the mixed fertilizers that followed guano for the next sixty years were either worthless or marginally effective.

For several years after Peruvian guano was introduced it was an economic bust in America, even though it had great initial success in England and other parts of Europe. In order to engineer the success of guano in the United States, the merchants had to overcome the skepticism of those farmers that Taylor and the agricultural societies had deeply influenced. To accomplish this, the Pacific Guano Company adopted two very clever advertising strategies.

Building a belief in the naturalness of the product became one essential ingredient in the selling of guano, since farmers were committed to using only natural fertilizers. So fertilizer merchants repeatedly promoted guano as a natural product, which it was, and claimed that it was similar to chicken manure, only better, more concentrated, and faster acting. Richard Wines, in *Fertilizer in America*, has shown how Pacific Guano's fertilizer ads always featured a cormorant standing on a bag of fertilizer with a fish in its mouth. Pacific Guano's fertilizer had maybe a handful of guano in every one-hundred-pound bag. The guano fertilizer was fish scraps with sulfuric acid poured over it and mixed with mined phosphates and potash. But the company's ads were clever and well thought out. Wines discusses how Pacific Guano's iconography played to the farmers' long-term acceptance of both chicken manure and fish as natural fertilizers.[3]

The effectiveness of guano provided the second major piece of the Pacific Guano propaganda campaign. Advertisers and editors focused on dramatic increases in yield from using Peruvian Guano. Since guanos had immediately improved yields in Europe and America on some of the most impoverished soils, the advertisers leaped on this as one of their central messages: "if guano can cure these soils it can cure any soils." Substantial harvest increases after applications of guano in the Tidewater area of Maryland and Virginia had a major impact on farmers' acceptance of guano, because the Tidewater soil was badly damaged from growing cotton and tobacco for more than two hundred years.

Advertisers of, and editorial campaigns for, fertilizer concoctions used the dramatic early successes with Peruvian guano for sixty years after it was no longer available. The fertilizer merchants claimed that each subsequent fertilizer product was Peruvian guano, or the equivalent of or superior to the Peruvian guanos long after they were mined out. Today, fertilizer merchants continue to claim miraculous power for each new product and, unless forced, never discuss any negative effects.

Between 1840 and 1880, editorials and experts trumpeted their praise for guano, but they also tried to belittle the value of manure and composts. A typical report from the *Country Gentleman* for May 26, 1853, praised the value of guano when compared with barnyard manure. The article cites Dr. Keene from Rhode Island, who alleged that in all cases the guano outperformed the stable manure. The author concluded: "We have elsewhere stated that a hundred pounds of guano may generally be reckoned as worth one [wagon] load of good stable manure." (One wagon load held about a ton of manure, so guano, it was claimed, would be roughly twenty times more effective than manure.)

In a letter to the *Country Gentleman* from January 27, 1853, A. J. Clinton from North Carolina provides advice on the use of guano on corn and cotton as follows: "The quantity should be from two to four hundred pounds of guano per acre—four hundred pounds for poor soils. If the soil is quite poor, an addition of one hundred pounds applied in the hill before planting would be of use."

Articles and letters such as these, coupled with aggressive advertisements, encouraged farmers to buy guano, even though Dr. Keene goes on to note later in the same article that "guano is often badly applied, and produces no valuable results, and spurious or worthless articles are sometimes sold under this name."[4]

In spite of concerns like Dr. Keene's, the twenty-year propaganda blitz worked. Guano use became more common in the eastern United States by the late 1840s. The demand was so great that the highest-quality Peruvian guanos were completely mined out before the 1870s and sold out by the 1880s. A resource that several Peruvian tribes and kingdoms had protected and utilized for more than a thousand years was exploited, exported, and exhausted in less than sixty. Companies and nations scrambled to find other shorebird deposits, and the U.S. government even seized ninety-four guano islands by congressional action, but the age of high-quality bird guanos was over, and so was the Peruvian government's earliest flirtation with free trade.

Innovative American companies such as Chappel, Kettlewell & Davison (in 1849) knew that the Peruvian deposits would eventually run out. They began copying the Englishman J. B. Lawes (1847), who saturated low-grade guanos, phosphorus, and ground

Above: Mapes was not the only quack in the marketplace. This April 1891 *Farm Journal* ad from Christian & Co. illustrates that even at this late date these charlatans were still selling guanos claimed to be Peruvian. Notice also the promotion of spraying.

Below: Allen's Marine Guano ad from *Farm Journal*, April 1880. Advertisements for Peruvian guano appeared at least ten and twenty years after all the good Peruvian guano had been mined out.

Mapes's son continued producing shoddy products and dishonest advertising well into the 1890s. Peruvian guano was part of the Mapes company name even though all the Peruvian guano had been used up by the time this ad was published in the *Farm Journal*, January 1891.

The ads here were common from the mid-1850s to the late 1870s. The Ichaboe Guano and potash ads were printed in the *Country Gentleman* on January 3 and February 21, 1856, respectively. The three ads below the potash ad are from *Farm Journal*, March 1877. The fish guano ad is from the *Country Gentleman*, January 24, 1856. The ad below it is an envelope for an advertising circular from the 1860s.

bones with sulfuric acid. All of these firms claimed that the sulfuric acid liberated the available nutrients, making them accessible to plant roots. Fertilizer merchants advertised these mixtures as identical to guano in all respects and claimed that, like guano, they were "complete" fertilizers.

The fertilizers produced by Lawes and by Chappel, Kettlewell & Davison, however, were definitely not complete. Eventually, scientific analyses proved them to be significantly less effective than the Peruvian guanos. Of course, this didn't stop most of the fertilizer merchants from advertising that their low-grade guanos were Peruvian guano, or at least equivalent to it.

Gradually, and largely because of the commercial success with Peruvian guano in Europe and on the East Coast of America, farmers became more interested in the products advertised in the journals, and publishers began including more advertising. From the late 1840s until the 1860s, advertisements for guano and editorial puffing of other fertilizer products appeared over and over again in several rural journals.

One of the leaders of the advertising blitz on farmers was Robert Mapes, the editor and publisher of *Working Farmer*. Mapes was mainly a fertilizer salesman who used his magazine to sell products and promote chemical agriculture. Mapes understood that repeated advertising was a key to his success, even though his products and many others that he sold were inconsistent, ranging from worthless to mediocre. What was consistent, however, was the frequency of the advertisements.

Unfortunately for the growers, Mapes promised everything but delivered almost exclusively worthless fertilizer products. The Mapes's family fertilizer firm was one of the merchants that continued to sell "Peruvian guano," and even included it in their business name, long after the real Peruvian guano was mined out. Though Mapes's company produced low-quality fertilizers, he prospered on these inferior products because he not only advertised constantly but also fought every battle against his adversaries, no matter how wrong he was or how many experts and chemists denounced his products. His secrets for success were notoriety, combativeness, and a refusal to admit publicly that he sold quack products. Even the *Country Gentleman* caught Mapes selling quack products and conducted a long-term fight with him in the journal.

As farmers became aware that the good guanos had run out, complaints about bogus guanos and useless mixed fertilizers increased. The high level of farmer protest in the journals would seem to indicate that many of the fertilizer claims were groundless. To address the farmers' criticisms, almost all subsequent fertilizer ads and editorials took great pains to demonstrate the naturalness of the fertilizer and the miraculous capability of their concoctions. The manufacturers claimed that each of their products contained abundant fertilizing elements, and many claimed that their products were complete fertilizers—containing all the necessary nutrients for good plant growth, just like Peruvian guano.

Scientists at the country's first state research station in Connecticut examined all synthetic fertilizers and proved that almost all of the "completeness" claims were false. While the low-grade guanos, bonemeals, and phosphates that fertilizer merchants doused with sulfuric acid were the first widely sold synthetic fertilizers in American history, the question of marketing fraud will forever dog their reputation. Because of the lack of consistency and the ineffectiveness of these manufactured fertilizers, most small farmers remained wary of the products and their "quack" promoters. Partly as a result of such suspicions, as we will see in later chapters, U.S. farmers continued to depend mostly on animal fertilizers and fertilizer crops for their fertilizer supplies until the middle of the twentieth century.

Peruvian guano's early effectiveness as a commercial fertilizer, however, helped break down farmers' pessimistic attitude toward advertisements, farm journals, and purchased fertilizer. It was seductive for farmers to think that they could get better yields by spreading a smaller amount of these highly concentrated fertilizers more quickly and cheaply than they could spread manures, ground bones, greensand, kelp, and lime. Any strategy that saved labor was tempting to the large-scale farm businesses. Peruvian guano's success and quality made farmers hope for other high-quality quick-fix products.

Many desperate farmers suffering from depression, economic panics, and war were duped by the constant flow of clever propaganda and advertisements for largely worthless products. Many farmers bought the altered, fortified, and synthesized guanos, hoping them to be equivalent to the miraculous Peruvian guanos. Sadly, for the farmers, they were not.[5]

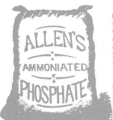

The ads for Poudrette and phosphate fertilizer are from the *Farm Journal*, September 1880. The Brown Chemical homemade fertilizer ad below is from the *Farm Journal*, July, 1886. Notice the use of marine guano in the fertilizer and seed ads—this is from the 1880s, when all the quality guanos were long gone.

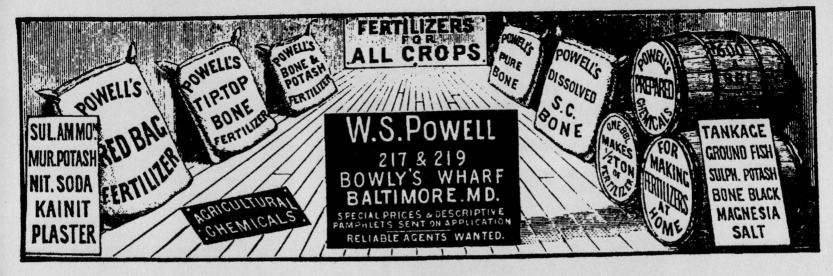

Farmer politics combined with a collage of interesting tidbits and Powell fertilizer ads in the *Farm Journal* from the mid-1880s, with the "1/2 Price" ad. Also included is the 1894 "Hard Times" ad. Notice that the Columbiana Pump ad implies that spraying is so safe and easy a young girl could do it.

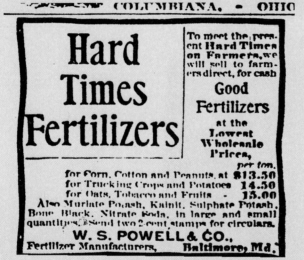

We're a-winnin' lots of glory and we're settin'
things to rights ;
We're a-showin' all creation how the world's
affairs should run.
Future men'll gaze in wonder at the things that
we have done,
And they'll overlook the feller, jist the same as
we do now,
Who's the whole concern's foundation—that's the
man behind the plow.—From Chicago News.

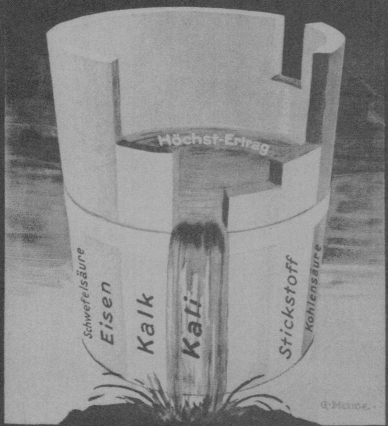

Liebig's Law of the Minimum. This theory alleged that any deficient mineral nutrient decreases soil fertility and reduces yields. This led to the attempt to *narrowly* analyze soil quality. From Forest Ray Moulton, ed., *Liebig and after Liebig* (Washington, D.C.: American Association for the Advancement of Science, 1942).

Chapter 5

LIEBIG AND THE INDUSTRIALISTS

The foundations of modern chemistry had been developed in the late 1700s. The Frenchmen Lavoiser and Gay-Lussac and the Englishman Priestly finally went beyond alchemy and founded chemical science. It was in Germany, however, that discoveries in organic chemistry would be most spectacular. One of the German pioneers in industrial chemistry was Justus von Liebig, who had studied in Bonn, in Erlangen, and finally in Paris with Gay-Lussac. After his studies, Liebig went back to Germany and spent his life teaching students from several countries how to be chemists. Liebig laid the basis for agricultural chemistry and captured the attention of American industrialists, aristocratic farmers, and Harvard and Yale. He also captured the attention of German mining and industrial interests. One of Liebig's students, August Wilhelm von Hoffman, discovered the value of coal-tar-derived organic chemicals. It was this research that provided the breakthroughs that Bayer, BASF, Hoescht, and several other companies used to turn coal tar into drugs, pesticides, and finally fertilizers.[1]

Much of the justification for the exaggerated fertilizer ads and puff-piece editorials that began to appear in the 1840s came from the continuous series of claims made by the German organic chemists, the European aristocratic scientists, and their industrialist adherents in the United States. In the 1830s, Liebig began asserting that the most essential plant nutrients were nitrogen, phosphorus, and potassium. His theories fueled the development of chemical fertilizers and ushered in a new age of agricultural science and soil chemistry in the 1840s and 1850s. Though many of Liebig's theories were wrong, he was the first great propagandist for chemistry and for chemical-industrial agriculture.

Liebig attempted to narrow the extremely complex study of soil fertility. But his analyses frequently excluded important elements, such as soil texture, tilth,

microorganisms, soil type, organic matter, the effect of crop rotations, and, most importantly, the uniqueness of place. Liebig felt that humus (organic matter) was an unimportant component of the soil, and that the minerals supplied by the humus could be more profitably and conveniently supplied with synthetic chemicals and mineral additives. Liebig was first and foremost a propagandist, a very innovative hands-on teacher, and what we would call today a lab rat. Because he focused most of his energies on laboratory research in the early years of his career, he often scoffed at farmer experience and rejected highly respectable farm theories and practices.

Liebig and most of his contemporaries (including his American students) seemed incapable of appreciating traditional peasant or tribal farming practices. Liebig and other scholars were trapped in the white European chauvinistic mind-set of the time. They believed in the evolutionary vision promoted by the anthropologist Lewis Henry Morgan, who divided human cultural development into Savagery, Barbarism, and Civilization. All these men believed that European civilization resided at the top end of the developmental pyramid, and that aristocratic scientists like themselves came from the upper level of civilization's cultural strata.[2]

Many of the European philosopher-scientists viewed Native Americans, Africans, Indians, Southeast Asians, and all the other conquered tribes and cultures as barbarians and savages. Because of such attitudes of superiority, tribal and peasant farming systems were usually ridiculed or ignored because of their sub-sistence focus and their difference from the capitalist market orientation of the Europeans. The approach of the "civilized" scholars was that only scientific systems made intellectual sense. Consequently, Liebig adopted mechanistic methods in preference to the more holistic approach employed by the traditional farmers of his day, whose ideas descended from tribal and peasant farmers of antiquity, with the addition of techniques advocated by Jethro Tull, John Taylor, and the agricultural societies.

In 1940, the renowned agricultural scientist Sir Albert Howard remarked that Liebig was only half a scientist because he ignored farmer knowledge and thus didn't know agriculture except from a laboratory standpoint.[3] Unfortunately, Liebig's (often erroneous) agricultural theories became the basis for the propaganda that the fertilizer merchants used to convince farmers to replace animal, mineral, and plant-based nutrients with highly toxic nitrogen, phosphorus, sulfuric acid, and potash chemicals derived from mining and manufacturing wastes.

In addition to their negative impacts on the science of agriculture and practical farming, Liebig and his

army of students served as pioneers for a controlling partnership between the German chemical education program and the private chemical industry. Both the industrial and educational sectors became dependent on the creation or betterment of German chemical products.[4] By the mid-1800s this relationship between the German educational system and the chemical and mining industries had already begun, as can be seen in Liebig's pamphlets from the 1830s and his famous agricultural treatise in 1840, *Organic Chemistry in Its Applications to Agriculture and Physiology*. By the 1850s, he had begun actively to advocate for the agricultural use of new and untested chemical discoveries and industrial and mining by-products.[5]

Not unrelated to Liebig's theories and pronounce-ments was the fact that European and American industrialists discovered naturally occurring deposits of nitrogen, phosphorus, and potassium (NPK) between 1804 and 1865. Several corporations rapidly formed to mine, ship, and market these finds. Liebig and his legion of students perfectly served these emerging chemical and mining corporations by claiming that their scientific tests proved that soils needed certain amounts of nitrogen, phosphorus, potassium, and sulfur. Subsequent ads by the industrial chemists always used scientific testimonies alleging that several chemical elements were essential for plant growth and that farmers should use the cheapest source of these chemicals to remain profitable.

Using such theories as the law of the minimum, Liebig blamed a decrease in yields on the depletion of one or more essential minerals or nutrients in the soil. In other words, there needed to be a minimum amount of nitrogen, phosphorus, potassium, calcium, sulfur, magnesium, boron, zinc, or any one of several nutrients in the soil to prevent a decline in yields. This idea was one of his great achievements. But it told only part of the story about soil fertility.

Liebig promoted testing soil for the essential mineral nutrients in his early publications. While the idea of soil testing is a good one, Liebig's analytical scheme was misleading and wrong. Still, many unscrupulous businessmen entered the marketplace and made every dollar they could by making scientific claims about soil testing to trusting farmers who were anxious to improve their yields by correcting their soil deficiencies. Unfortunately, these early ideas of soil testing preceded the full flowering of soil analysis, which matured into a valuable technique much later, when labs finally tested for organic matter content, soil mineral balance, and pH levels, along with mineral content.

Eben Norton Horsford was the first American student of Liebig, in the mid-1840s, and he became a slavish devotee. While studying with Liebig in Germany, Horsford wrote several articles for the *Cultivator*. After his studies with Liebig, he returned to design and run the science and chemistry laboratory at Harvard that was funded by and named after the industrialist Abbott Lawerence. Horsford's goal was to recreate Liebig's German lab in America and train a generation of American chemists. Horsford remained at Harvard until 1863 and did train a handful of agricultural chemists. But Horsford was not as wealthy as most of his colleagues at Harvard, and he left the teaching and academic research profession to produce industrial chemicals with hopes of becoming more wealthy.[6]

John Pitkin Norton also went to Europe to study chemistry, but he studied with James Johnston in Edinburgh and with Gerhardt Mulder in Utrecht. Both Mulder and Johnston were vocal critics of Liebig, and the publication of their criticisms and Liebig's rebuttals continued throughout Liebig's life. After Norton returned from Europe, he was instrumental in starting the Yale Analytical Laboratory and in creating a program for budding agricultural chemists.

Though he was critical of many of Liebig's theories, Norton provided some of the most important help in promoting the Liebig soil-sampling craze that swept the eastern United States. When Norton conducted his soil analyses for struggling farmers he charged them only for the cost of the chemicals, while he contributed his own time. Norton passed away at an early age and was replaced at Yale by John Addison Porter and Samuel Johnson, both of whom had studied with Liebig and were dedicated believers in chemical/scientific agriculture.[7]

Porter married the daughter of Joseph Sheffield, the New Haven Railroad magnate. Through Porter's urgings, Sheffield generously endowed the Sheffield Scientific School at Yale, including a very substantial chemistry laboratory. Johnson was elected professor of analytical chemistry at Yale and spent forty years teaching agricultural and analytical chemistry at the Sheffield-funded school. As professor of chemistry Johnson helped start the first government-supported agricultural experiment station in the United States in 1875.[8]

Horsford, Porter, and Johnson promoted American agricultural science in the ways that Liebig had prescribed. Though America generally was loath to depend on European products or ideas for more than fifty years after the Revolution, European science greatly attracted American industrialists, aristocrats, and students of chemistry. Because of Liebig's charisma and the more than forty American students who studied with him, his scientific theories greatly influenced the direction of agricultural science in America, and his approach continues to influence chemical agriculture today. After Horsford, Norton, Porter, Johnson, and dozens of others returned from their studies abroad and asserted the need for agricultural chemistry, and after Harvard and Yale hired chemists, all agricultural schools and experiment stations in the country followed their lead

.

In spite of the aristocratic and university support for Liebig and for chemistry in general, "book farming" fell out of favor with American farmers after Liebig's soil-sampling craze was debunked in 1852. Liebig remained unpopular in most American farm communities until the

1890s. His fall from favor in rural America was prompted by soil analyses conducted by David Wells from the Lawrence Scientific School at Harvard. Wells compared the soils of western Massachusetts with soils from Ohio. Wells, who was a Liebig adherent, expected to find that mineral deficiencies in the Massachusetts soils had caused the dramatic drop in yield there. Instead he found that the mineral compositions of the two soils were similar. Wells concluded that the mineral component of the soils could not explain their differences in yield, though the quantity of organic matter in the Ohio soils might. Wells's report in the July 1852 *American Journal of Science* concluded that organic analyses should be trusted more than mineral analyses of soils, and that mineral fertilizers should not be allowed to displace the traditional barnyard manures, which contained organic matter.[9]

Some chemists argued that the attempt to popularize science for the masses of farmers and to get them to accept scientific discoveries was a mistake. T. Steery Hunt, for example, who conducted soil analyses for the Canadian geological service, responded to a letter from Eben Horsford as follows: "You ask what predilections I have for Agricultural Chemistry & the line pursued by poor Norton. Very little, for I am a great skeptic as to those things. I look upon them as the charlatanism of our science and as tending to degrade it, making it a plaything for tyros [novices] who analyze soils for 2 dollars, etc.—Science for the millions is humbug! True science, like true nobility, is essentially aristocratic. Far be it from me, however, to underrate the many practical benefits which flow from chemistry." In other words, instead of criticizing Liebig and the quack soil scientists who believed that everything could be reduced to a mineral analysis,

Hunt and other chemists criticized Norton for trying to help the small farmers.[10]

Even if Liebig was wrong and his soil tests useless, most chemists pleaded with the farmers not to reject science or synthetic fertilizers just because some mistakes had been made. Instead they urged farmers to be more careful in their fertilizer purchases. And they urged chemists to do additional scientific research, arguing that they could fix Liebig's analysis and fertilizer recommendations with more, not less, science.

In 1853, after the soil-sampling frauds were exposed and amid extensive criticism from farmers for the magazine's support of Liebig and the chemists, the *Cultivator* changed its name to the *Country Gentleman*. But, as before, the magazine with its new name promoted the "scientific" practices of the large-scale landowners.

Liebig kept up his endless self-promotion and squabbling with other scientists throughout his life. His famous controversy with Pasteur over microbial life, fermentation, and putrefaction led to Pasteur's institute developing the science of soil microbiology. By 1883, Pasteur's institute had discovered the cause of typhoid fever, and importantly for our story it had established that bean plants (legumes) could collect and store nitrogen from the atmosphere. The institute established how microorganisms break down organic matter into ammonia and then into nitrates. It also demonstrated that humus and organic matter were very important to soil fertility. Liebig was wrong in this controversy, as he was in so many others, but the positions he took stimulated Pasteur to discover the true workings of nitrogen accumulation and important breakdown processes in the soil.

This view of the interior of Liebig's laboratory in Geissen, Germany, illustrates both the hands-on nature of the lab and its aristocratic character. Nice lab coats. From Margaret Rossiter, *The Emergence of Agricultural Science* (New Haven: Yale University Press, 1975).

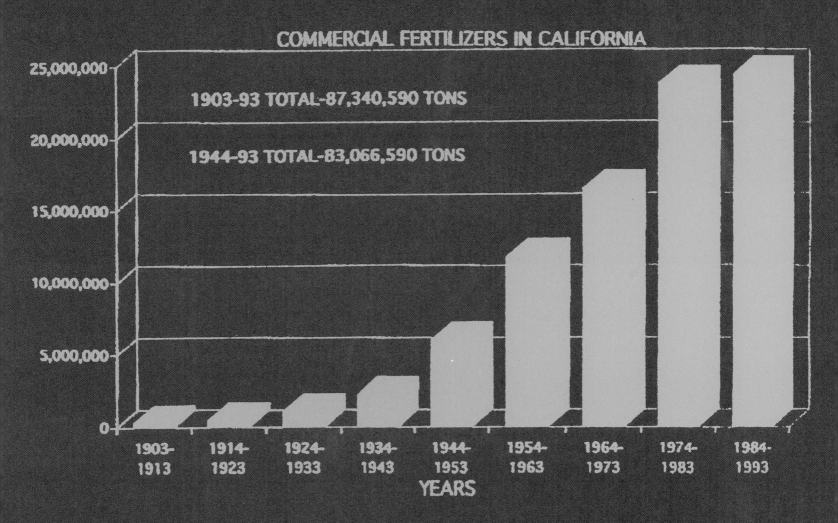

COMMERCIAL FERTILIZERS IN CALIFORNIA

1903-93 TOTAL-87,340,590 TONS

1944-93 TOTAL-83,066,590 TONS

YEARS

Table 2. The growth in the use of commercial fertilizers
(nitrogen, phosphorus, potash) in California 1903–1993. From
the California Dept. of Food and Agriculture, Fertilizer Reports.

Chapter 6

THE FERTILIZER MERCHANTS

Jared Diamond in *Guns, Germs, and Steel* theorizes that most inventors and innovators actually discover something different from what they are trying to discover. Consequently, many by-products become as valuable as the original sought-after discovery. We have already alluded to this in our discussion of the search for salt deposits, which in turn led to the discovery of natural reserves of potash and sodium nitrate. The history of a better-known product, gasoline, is similar. "As chemists and engineers began the process of petroleum distillation in the 1800s they focused their attention on the middle distillate fraction [kerosene] . . . because of its usefulness as fuel oil for lamps. For decades the chemists discarded gasoline, the most volatile fraction, as an unfortunate waste product."[1]

The chemists discarded gasoline until they convinced Henry Ford to use it to power his mass-produced cars because it was cheaper than alcohol and because it represented a profitable use for a material that was otherwise considered useless. Once communities started restricting the dumping of gasoline into lakes, rivers, and oceans the petroleum companies began to search for a market. The emerging automobile and tractor industries became the perfect dumping ground for gas.

The widespread use of gasoline as motor fuel was described at the time as an elegant solution to the problem of how to dispose of gasoline. Ford had built his cars to run on alcohol fuel or fuel oil but was able to convert them to run on gasoline when Rockefeller offered to sell gas at a cheap price.

For the farmers, Rockefeller's elegant solution meant economic disaster. Farmer collectives had planned to build alcohol-distillation plants and expected to supply part of the auto and tractor industry's fuel needs. When the gasoline makers sold gas for a penny or two a gallon, the ethanol producers on the farms saw their hoped-for market for motor fuel evaporate.

As the nineteenth century progressed, scientists and industrialists began to worry out loud about food shortages. They argued that if the world's population grew exponentially, we would soon be short of agricultural nitrogen and other fertilizing materials and thus headed for starvation. These fears led to even more intensive searches for sources of phosphorus and potassium and additional nitrate deposits; they also allowed the fertilizer merchants to create even more innovative and often misleading advertising campaigns.

After guano, the most important fertilizer discovery of the 1800s was Chilean nitrate. At the peak of the great guano rush, the first boatload of Chilean nitrate arrived in California. The date was 1845, four years before gold was discovered at Sutter's Mill in the Sierra foothills of California. Skeptical American and European farmers refused to buy this salty fertilizer, preferring stable manure or guano because they were seen as more natural products.

This new fertilizer, also known as sodium nitrate or nitrate of soda, was a waste by-product of the search for salt in northern Chile and southern Peru. For the first twenty-five years after its introduction, American farmers showed no interest in sodium nitrate. By 1870, however, after the industrialists had mined out and sold off all of the high-quality Peruvian guano, the British fertilizer merchants, and to a lesser degree the Germans, widely propagandized and marketed nitrate of soda.

As a consequence of advertising, propaganda, and economic depression, the farmers' interest in a concentrated quick-fix fertilizer increased after the Civil War, when farmers returned from the battlefields. Even during the panics of 1873 and 1894, the sales of sodium nitrate continued to climb as farmers struggled to make a profit. At that time, these nitrate salts were the only known long-term supply of commercial nitrogen in the world. These deposits ultimately became important not just to farmers, but to explosives manufacturers and users, since sodium nitrate was one of the most important ingredients in bombs and gunpowder.

R. A. Holcombe and Co. nitrate of soda ad from *Pacific Rural Press*, 1890s.

The Irish shipping tycoon W. R. Grace and the William S. Myers Company of New York became major promoters, shippers, and merchandisers of sodium nitrate, as the ads and book liner at right show. The DuPonts were also prominent in the mining of sodium nitrate around the turn of the century. In Chile, DuPont expropriated and developed the Oficina Delaware mine, which produced 50 percent of the nitric acid that was used in explosives during World War I.

While the sodium nitrate deposits were extensive, the industrialists argued that in time of war Chile could be blockaded, and also that large quantities of Chilean nitrate would be diverted to explosives manufacture. This meant that it could not be a dependable source of agricultural nitrogen for the exponential expansion of the world population that English economist Thomas Malthus had predicted back in 1798.

Grace sodium nitrate ad from *Farm Journal*, 1894. This is the year that the last panic of the nineteenth century began. During this depression, W. R. Grace was so successful with sodium nitrate that the company was looking for agents in its ads while most businesses were laying off workers and salesmen.

Though nitrogen, or the lack of it, commanded most of the scientific and popular media attention, it made up only a small portion of the commercial fertilizer industry in the late 1800s and early 1900s, and consequently a smaller portion of fertilizer advertising. As we shall see, despite all the hoopla, synthetic nitrogen remained a minor fertilizing product in America until much later. Synthetic nitrogen use accounted for less than 5 percent of the nitrogen used in the United States at the beginning of the twentieth century. However, use increased somewhat after World War I as promotions and supplies increased.

The bulk of the nineteenth- and early-twentieth-century fertilizer industry developed around the

Gift notification pasted inside the cover leaf of the most important fertilizer book of its time, *Fertilizers*, by E. B. Voorhees, in a 1916 edition (it was first published in 1898). In this edition, which Voorhees's son edited, the value of Chilean nitrate is praised, just as his father praised it in previous editions. Tens of thousands of these books were distributed free by corporations. The *Farm Journal* sold them for one dollar.

The potash ads above are from the German potash monopoly the German Kali Works. It was this firm that first began placing repetitive ads in each magazine issue before and during the planting season for each crop. Its strategy of targeted saturation ads, developed in the 1880s, has continued to be used by chemical advertisers and many other advertisers ever since. Top: From *Farm Journal*, May 1893. Middle: From *Farm Journal*, 1894. Bottom: From *California Cultivator*, January 9, 1908.

At left is text from the 1916 edition of *Fertilizer*, praising sodium nitrate. The gift of such books greatly increased the sales of sodium (Chilean) nitrate and deferred criticism about leaching into waterways.

Nitrate of soda. — There are a few substances found in nature containing nitrogen in the nitrate form. The most important is nitrate of soda, a chemical compound composed of sodium, oxygen and nitrogen. The occurrence of this material is limited to the rainless districts of South America, mainly Chile, where the crude nitrate of soda salts called Caliche are found in vast quantities. These crude salts contain from 5 to 30 per cent of nitrate of soda. In the process of refining for market they are dissolved and recrystallized in order to remove as far as possible the impurities associated with them. There are great quantities of a lower grade containing 3 per cent or less of nitrate of soda which are not at present considered sufficiently rich to refine. The chemically pure salt, nitrate of soda, contains 16.47 per cent of nitrogen, and the commercial article, called "Chili saltpeter," contains from 15.5 to 16 per cent. The impurities which remain in it consist mainly of sodium chlorid, or common salt, which, together with moisture, causes a lower percentage in the commercial product. Because nitrogen in nitrate of soda is in the nitrate form and, therefore, soluble, it is often advanced that there is greater loss from leaching into the drainage waters. Experiments show this is untrue or at any rate the efficiency of this material is greater than that of any other because more is returned in the crop

These ads for acid phosphate fertilizers from an 1880 *Farm Journal* and 1888 and 1890 *Pacific Rural Press* illustrate how advertisers used the text-focused ads, with elaborate claims and the promises of cheap prices, liberal loans, and low freight rates. In the 1860s, the *Pacific Rural Press* had criticized the collusion of banker, broker, large-scale farmer, fertilizer merchant, and railroad baron. Their control of everything was a centerpiece of the complaints printed constantly in the populist press. Under the journal's new ownership in the 1880s that collusion was encouraged. The Newhall Corporation was not just a fertilizer merchant. It was also Newhall Land and Cattle Company, one of the largest farming operations in California and Baja California.

mining and marketing of potash, phosphorus, and sulfur, not nitrogen. In the first few decades of the twentieth century, the most important commercial fertilizers, in terms of tonnage applied, were phosphorus, potash, lime, gypsum, and sulfur.

The dominance of these fertilizers resulted mostly from several early discoveries of extensive mineral deposits. German miners located potassium overlying deposits of salt near Strassfurt in 1858. This enabled the German firms to dominate the world potash market for seventy years. These mining interests formed the first potash syndicate by 1870 and began promoting potassium as a valuable source of agricultural potash. The syndicate discovered additional potassium reserves on the French-German border and added these supplies to their already huge holdings before 1900.

Rock-phosphate deposits were discovered in South Carolina in 1859, and the mining and marketing of these deposits started in 1867. Huge deposits of phosphorus were then discovered in Florida in 1888. Additional resources found in Tennessee in 1894 enabled American industrialists to increase their supplies and become a major force in the national and worldwide fertilizer industry. These discoveries would dramatically change the mixed-fertilizer industry in the Western world.

As previously noted, the first synthetic fertilizers, made in England in the early 1840s and in America by 1849, were made by pouring sulfuric acid on ground-up bones and low-grade guanos. This practice continued into the 1900s. But after the discovery of extensive deposits of rock phosphate, the phosphates began to serve as the basic ingredient for most of the mixed fertilizers, and subsequently acids were poured on phosphorus. Before the start of World War I, the fertilizer chemists replaced the bone and guano fertilizers with several sulfuric- and phosphoric-acid mixtures that they usually advertised as being the most complete fertilizer ever known to man.

Enormous supplies of by-product sulfur from coal and coke processing also became available in both Europe and America as industrialization expanded and as coal and coke replaced wood as the major source of industrial heat. From the 1840s through the 1860s, magazines like *Country Gentleman* and *Practical Farmer* promoted the virtues of these mined and processed fertilizers. So did *Scientific American*, as seen in the following puffed-up editorial for superphosphate:

> Sulphuric acid, invaluable for many purposes, is coming into common use among the English farmers. . . . Crops remove the phosphate of lime from the soil—bones dissolved in sulphuric acid produce the phosphate, and the phosphoric acid so produced has been brought to bear upon the land with the utmost beneficial effects. [2]

Several years later Standard Oil added sulfur deposits to its already enormous natural resource inventory by pulling out huge deposits of sulfur from Calcasieu Parish, Louisiana, as a by-product of oil exploitation. *Scientific American* described the company's hydraulic mining discovery:

> The Standard Oil Company has finally solved the problem . . . of getting at the immense mass of sulphur. . . . Super heated water is forced through a 10-inch pipe on the sulphur, melting it, and the liquid sulphur is then pumped up. A little exposure to the air, so as to evaporate the water, leaves almost pure sulphur. [3]

The production of waste minerals like sodium nitrate, potash, and sulfur and the discovery of enormous phosphorus deposits, combined with Justus von Liebig's promotion of mineral plant nutrients, enabled corporations and cartels to create an enormous worldwide commercial fertilizer industry. The fertilizer barons fueled the industrialization of agriculture with any commodity they could transport and sell, no matter how toxic, useless, or damaging.

Consequently, uncounted tonnages of toxic waste fertilized both barren and fertile farms all over Europe and America.

Slave labor, guano, and sodium nitrate fueled the initial agricultural expansion in the first half of the nineteenth century. After the Civil War, the development and rapid growth of the commercial fertilizer industry served as an important stimulus to the increase in large-scale commercial farms. The completion of transportation systems (especially canals, roads, and railroads) and the seemingly endless influx of immigrants also played major parts in increasing the size of industrial farms and the number of acres under cultivation. The farmers needed a way to get goods to market, but they also needed fertilizers to compensate for the continuous decline in soil fertility caused by several generations of cash cropping or the farming of marginal lands.

As farm size grew ever larger, so did the mechanical challenge of moving and spreading increasing quantities of manure. Large-scale farmers searched for materials that were more concentrated and thus more convenient to apply, requiring less labor and time than animal manures. Guanos, sodium nitrate, and mixed fertilizers containing potash and phos-phorus were more concentrated than manures and thus faster and easier to apply, so they became very attractive, especially to those farmers who were managing hundreds or thousands of acres.

Fertilizer corporations conspired with each other to fix prices before 1900 and dominate the market for fertilizer, with the Germans controlling potash and the British ultimately controlling sodium nitrate. J. W. Turrentine's analysis of the potash industry illustrates

how effective the industrialists were at both marketing waste by-products and financing the scientific justification for using such materials on our food.[4]

All of these mineral and waste discoveries were peddled to the farmers as essential for success—for profitability. The exhaustion of Peruvian guano in the late 1860s led to the use of sodium nitrate. But sodium nitrate provided only nitrogen, whereas the Peruvian guanos had contained nitrogen, phosphorus, potassium, and several minor elements in forms that were available to plants. After the discovery of potash, rock phosphorus, and sulfur deposits, the fertilizer merchants convinced farmers to use endless mountains of these products mixed with Chilean nitrate in an attempt to replace guano. As it turned out, however, no mined or synthetic fertilizers could replace the rich materials produced in the natural Peruvian laboratory of composted seabird deposits or composted manures from the farmer's livestock.

While rock phosphate fertilizers dominated the market by the 1880s, these ads for bone-based fertilizers still had appeal at this late date. J. T. Roberts and Bro. ad from *Farm Journal*, September 1881. Walter Stratman and Co. ad from *Farm Journal*, August 1889.

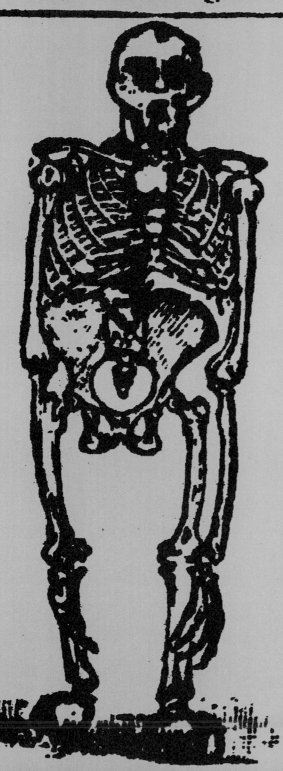

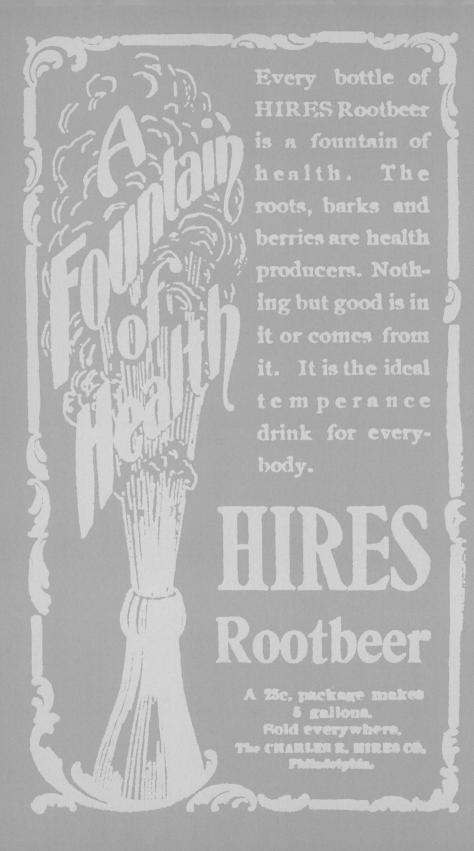

A Fountain of Health

Every bottle of HIRES Rootbeer is a fountain of health. The roots, barks and berries are health producers. Nothing but good is in it or comes from it. It is the ideal temperance drink for everybody.

HIRES
Rootbeer

A 25c. package makes
5 gallons.
Sold everywhere.
The CHARLES E. HIRES CO.
Philadelphia.

This 1893 ad for Coca Cola from *For God, Country, and Coca Cola* by Mark Pendergrast (New York: Orion Books, 1993). The Hires Rootbeer ad is from *Farm Journal*, May 1896. Advertisers claimed that these soft drinks were health food tonics, even though critics had shown they were dangerous and unhealthful. Health claims such as these were made for a wide variety of patent and over-the-counter medicines. These claims confused ailing customers and made billions for the quack drug and tonic makers. The Coca Cola of 1893 still contained cocaine. Yet the ads claimed Coca Cola was okay for and desired by kids. No doubt!

Chapter 7

THE MEDICAL ANCESTRY OF PESTICIDES

The discovery, promotion, and use of synthetic fertilizers preceded the advertising and use of chemical pesticides. No pesticide ads appeared in the earliest farm magazines, but many carried ads for both patent medicines and rodent-control strategies. The city magazines also carried many patent-medicine ads long before the rural journals existed. A review of both provides us with old and interesting medical connections to modern-day chemical corporations and later-day pesticides. James Wharton's book, *Before Silent Spring*, discusses how toxic medicines are linked to early agricultural pesticides, and how both are linked to chemical-pharmaceutical-dye manufacturers. Wharton's account also provides us with important clues for what to look for in the ads and promotions in the farm journals.[1]

Western medicine proudly traces its roots to the fifth-century-B.C. Greek philosopher Hippocrates. Besides being responsible for the Hippocratic oath (which obligates doctors to honor their patients' confidences and needs), Hippocrates advocated a holistic and preventive medicine, including the use of exercise and good diet, to maintain health. Hippocrates believed that there were four humors that resided in the liquids of the body: blood, phlegm, yellow bile (or "choler"), and black bile (or "melancholy"). He believed it was the physician's job to help balance the humors with simple herbal cures (called simples), which were derived from plants both wild and domesticated.

Even though Hippocrates is Western medicine's patron saint, Western medical practices from the second century A.D. until the 1930s descended more directly from the so-called heroic practices developed by Galen than from the holistic medicine of Hippocrates. Galen served as physician and surgeon to the Roman emperor Marcus Aurelius in the second century A.D. and developed many of his medical innovations in

battlefield situations. Equally concerned with balancing the four humors, Galen introduced dozens of new and complex potions and practices that had been developed to deal with the most horrible battlefield injuries. These included bleeding, blistering, and purging with highly toxic substances, all designed to equalize the humors—and, of course, extract hard-earned cash from rich patients. These heroic practices, potions, and prescriptions came to be known as the materia medica. The most complex medicine required dozens of ingredients, including snake flesh and feces. The seemingly disparaging term "snake oil remedy" refers to patent-medicine copies of Galen's potions that contained both snake flesh and manure (especially the potion known as Galena).

Roman imperial medicine withered in Europe after the collapse of the Empire in the fifth century A.D., but several Roman cures survived in the Near East. There, many practiced Galen's techniques, along with those of Hippocrates, until the ninth century A.D. However, in the four centuries of almost continuous warfare beginning with the ninth-century Crusades and ending with the Mongol invasions in the mid-thirteenth century, Arabic physicians came to rely on the more interventionist practices of Galen. This battlefield medicine especially impressed the European Crusaders, and they carried it back to Europe when they returned home. By the end of the twelfth century, Galenic medicine dominated the Western world and was being taught in the best universities in Europe.

Ad claiming tobacco is a panacea alongside a King James I broadside against tobacco. From Frank Presbrey, *The History and Development of Advertising* (Garden City, N.Y.: Doubleday, Doran and Co., 1929).

In the early 1500s, Paracelsus (a Swiss-German doctor and alchemist) successfully promoted the use of highly poisonous metals as healing drugs and purges. Paracelsus attacked Galenic doctors for their failure to use such innovative medicines. Paracelsian medicine—which still used bleedings, blistering, purges, and many of Galen's complex potions—now added the highly toxic heavy metals to the doctor's proverbial little black bag.

Though his cures were highly toxic, there is no doubt that Paracelsus's research was well intentioned. There is also no doubt that Paracelsus shifted the alchemists' focus from turning lead into gold toward the exploration of pharmaceutical concoctions. Thereafter, chemists and alchemists concentrated on turning lead, chemicals, and other metals into medicine and, ultimately, for our story, into pesticides.

Common toxic metals in Paracelsian medicine were mercury, arsenic, antimony, lead, and vitriol (sulfuric or hydrochloric acid). Paracelsus significantly

changed both chemistry and scientific medicine. Partly because of its supposed scientific stature, Galen and Paracelsus's poisonous blend of alchemy endured as the officially accepted medical practice in post-Revolutionary America. Some of the worst elements from their macabre medical kit continued to be prescribed in Western medicine until the 1940s.

Over the last several hundred years, alchemists, quacks, chemists, and corporations have concocted literally thousands of cures and remedies for human and animal ailments. All advertised the superior effectiveness of their particular potion in urban and rural journals. By the mid-1700s corporations and quacks developed advertising campaigns to sell both effective and worthless medicines to customers with serious health or pest problems. Even tobacco was alleged to be medically beneficial when it was first advertised in London in the 1600s.

Cottage-produced medicines had been one medical answer for people's ailments at least since the ninth-century development of pharmacy in Baghdad. In 1793, the U.S. Congress passed the law that established the Patent Office and allowed for the granting of patents to inventors, including aggressive cottage-medicine manufacturers. Thereafter, promoters of all sorts were able to secure patents on medicinal concoctions designed to solve every physical or psychic ailment. Some patent medicines were over-the-counter imitations of Galen and Paracelsus's potions; others were simple herbal remedies. While the simple herbal medicines were effective, lamentably, most of the other concoctions were useless fakes, like Coca Cola and Hires Rootbeer.

European pharmaceutical firms aggressively sold medical concoctions in America as well as Europe. Drug-peddling firms claimed that nearly every ailment, from impotency to pregnancy to obesity, could be cured by a few teaspoonfuls of their medicine. Manufacturers of patent medicines knew from long experience that these potions provided the only medical hope for untold millions who could not afford a physician, so there was literally an endless market.[2]

Among the earliest chemical manufacturers of patent and prescription medicines were J. R. Geigy, Hoechst, Bayer, Merck, DuPont, Ciba, BASF, Sandoz, Nobel, Monsanto, General Chemical, Graselli, and Dow. Each originated as a drug, explosives, or dye manufacturer. Many of these companies still rank today among the top chemical corporations or conglomerates in the world.[3]

In *Before Silent Spring* Wharton argues that by the late 1800s most Americans were disgusted by the claims and counterclaims of all of the medical "experts" and the patent-medicine advertisements. He shows how criticism of heroic medicine and disbelief in patent medicines were common, and that skepticism had increased significantly after doctors

Right: Thinness and obesity remedies from the same company, Loring and Co.

Left: Collage of patent medicine advertisements. From Frank Rowsome, *They Laughed When I Sat Down* (New York: Bonanza, 1959).

REDUCED 33 POUNDS BY Dr. EDISON'S OBESITY REMEDIES.

RELIEF FOR THE FAT.
Dr. Edison's Pills and Salt Take Off 20 to 30 Pounds a Month.

LORING & CO., DEPT. 119. No. 42 WEST 22d STREET, NEW YORK CITY.
No. 5 HAMILTON PLACE, BOSTON, MASS.
No. 115 STATE ST., CHICAGO, ILL.

Get Plump!

LORING'S Fat-Ten-U and Corpula Foods Make the Thin Plump and Comely and impart Vim to the Debilitated — They Cool the Blood and Prevent Unpleasant Perspiration

These foods cure nerve and brain exhaustion, which you know as general debility or nervous prostration. They make pale folks pink and thin folks plump and weak folks well and despairing folks happy. Women have learned that they more than take the place of all female remedies and regulators. They will make you young all your life. You know it is better to be a young old man or woman than a prematurely old young man or woman

The portrait above is that of Mrs. Sara Montgomery Wade, Vincennes Avenue, Chicago, who writes: "In six weeks Loring's Fat-Ten-U and Corpula Foods increased my weight 32 pounds, gave me new womanly vigor and developed me finely. They should be used in hot weather, as they keep you cool and make you well."

No "tonics," "nervines," "sarsaparillas" or other medicines are necessary when Fat-Ten-U and Corpula are used. Corpula, $1.00 a bottle. Fat-Ten-U, $1 and $2 a bottle. Send for a free copy of "How to Get Plump and Rosy" and instructions for improving the bust and form.

A Month's Treatment, $2.00

Write to our Chicago Medical Department about your thinness or about any other medical question. Our physicians will advise you free of charge.

Send letters and mail, express or C. O. D. orders to Loring & Co., Proprietors. To insure prompt reply, mention Department as below. Use only the nearest address:

LORING & CO. DEPT. 119.
No. 42 West 22d Street, New York City.
No. 5 Hamilton Place, Boston, Mass.
No. 115 State Street, Chicago, Ill.

SICK HEADACHE POSITIVELY CURED BY CARTER'S LITTLE LIVER PILLS.

ONLY TWENTY-FIVE CENTS.

KIDDER'S PASTILLES.

AYER'S CHERRY PECTORAL

NO MORE RHEUMATISM
TAPE-WORM
DEAFNESS

ELY'S CREAM BALM CURES CATARRH
PRICE 50 CENTS, ALL DRUGGISTS

Prepare for Cholera.

DEAFNESS & HEAD NOISES CURED
FREE

greatly expanded their medical and anatomical knowledge in the Revolutionary War, the War of 1812, and the Civil War. After the Civil War many patients, given the choice between dying or heroic medical cures, like arsenic or mercury, chose to die, because the side effects from these poisons were considered to be worse than death.

Although both heroic and patent medicines became readily available in the cities by 1850, 90 percent of the American population still lived in rural areas. For many of these people, it often required several days' journey to reach a physician, an apothecary, or other sources of patent medicines. Some medical historians claim that many of the medical solutions for nineteenth-century farmers did not come from physicians, quacks, apothecaries, or the general store. Instead, farmers' gardens and forest gatherings provided the ingredients for a complex pharmacopoeia that included natural cures from both European and Native American cultures.

Barbara Griggs contends that natural cures—those herbal cures called simples—were common in colonial America for all ailments. Griggs outlines these cures in her book, *Green Pharmacy*, including historical documentation for their age-old use. Medical historians like Griggs have shown that country folk maintained well-stocked herbal gardens and medical cabinets throughout the 1800s.[4] But patent medicines continued to be promoted and large volumes were sold in the towns and cities where people had less access to herbal gardens.

Between the Civil War and the 1890s, the struggle to eliminate heroic medicine intensified as the luster of these ancient remedies rapidly began to fade. But many heroic advocates remained powerful, within both the medical profession and the world of chemistry. Those advocates believed that only these highly toxic metals worked on certain illnesses, such as syphilis, dyspepsia, and depression.

The partial success of, and grudging dependence on, these known killers muddled the debate. Because of their short-term effectiveness (usually affording only symptomatic relief), long-term health and safety issues were ignored, even by some supposed experts. By the end of the nineteenth century, however, many people were fed up with the dangers and refused to use them as medicines.[5]

Despite all the condemnations and resistance, the use of highly toxic heroic medicines survived well into the twentieth century in both Europe and America as one of the only hopes for the desperately ill and the desperately poor. Up until World War II arsenic was still prescribed as a cure for syphilis and many other ailments.[6]

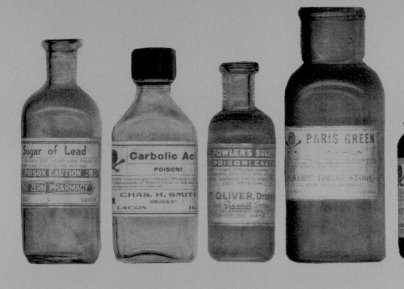

Patent medicine bottles from the eighteenth and early nineteenth centuries. From the collection of the author.

Beecham's Pills ad from *Advertising in Victorian England, 1837–1901*, p. 5. Notice that it is "sold by all medicine vendors."

Carbolic ads from the 1880s: The ad at left, from T. R. Nevett, *Advertising in Britain* (London: Heineman, 1982), claims that the carbolic smoke ball will cure coughs, colds, asthma, sore throat, whooping cough, hay fever, bronchitis, snoring, loss of voice, influenza, and more. The smoke ball ad at right is from Diana and Geoffrey Hundly, *Advertising in Victorian England 1837–1901* (London: Wayland, 1972).

DOGS—RATS.

To the Editor of "Life in London:"

SIR—Observing in your paper of last week a paragraph stating that 50*l.* was deposited at Cribb's, to be laid against my dog China drawing a badger in ten minutes, I called with a friend to cover it, but found it was "no go," and, probably, the offer was a mere bounce of *little Charley's.* Now, to stop this little gemman's *chaff*, I am ready to stake, when and where he shall appoint, and further, to bet him 50*l.* that his dog Billy does not kill a hundred rats, finding rats myself, in eight minutes.

Yours, &c.   FRANK REDMOND.

A challenge to Billy the champion rat
killer, published in *American Farmer*, 1828.

Chapter 8

RATCATCHERS, QUACKS, AND EARLY PEST CONTROL IN AMERICA

In 1800, 97 percent of the United States population still lived in rural areas. By contrast, only 60 percent of the British population lived in rural areas. The settling of people in European and Middle Eastern cities had begun centuries earlier. Although many Old World cities were centers of disease and death resulting from filthy living conditions, several of them had devised multiple strategies to deal with pests such as rats, mice, flies, and roaches. Rat-catching dogs was one solution used by all the early pest exterminators.

Those cultures in Europe and the Middle East fortunate enough to survive the fourteenth-century plagues did so because of improved sanitation techniques and the control of fleas and vermin. Jewish, Arabic, and Germanic peoples were especially successful at combating disease and controlling pests during plagues and epidemics that ravaged Old World cities, even though both Jews and Arabs were accused of spreading the plague through Europe. Each of these cultures was more fastidious about sanitation than most Europeans. Each had guilds of pest-control experts who enjoyed praise and financial rewards but also endured scorn and second-class citizenship.[1]

The ratcatchers of Pied Piper fame descended from these guilds. Such pest-control specialists often doubled as barbers and surgeons or peddled patent medicines, purges, and poisons. Many also cut hair, pulled teeth, and bled and blistered their patients. They loudly shouted and hawked their wares and services as they roamed through the markets and neighborhoods. A commonly told story holds that ratcatchers, barbers, and surgeons became known as "quacks" in medieval England because their sales pitch resembled a duck's constant quacking: "Kill your rats. Kill your rats. Kill

your rats. Kill your rats"—such was heard over and over again throughout the neighborhoods of London.

Those ratcatchers who were not barbers and surgeons purchased their poisons from them. Technical pest-control knowledge and lore generally was not recorded or shared but passed down within families from generation to generation as oral history and trade secrets. Many ratcatchers used ferrets as well as dogs to find and kill the mice and rats.

In spite of some localized advancements in domestic pest control, much of the European citizenry still relied on God, the saints, and clergy when pest-control problems spiraled out of control. In 1497, the bishop of Lausanne, Switzerland, excommunicated June beetles as a means of pest control. As late as 1822, peasants in the Lower Rhine Valley left offerings for the seventh-century saint Gertrude of Nivelles to enlist her help against field mice.[2]

The exact inner workings of pest control have always been something of a mystery, because most people didn't (and still don't) understand the interactions between pests and predators or even the effects of

Ratcatcher for Her Majesty the Queen, circa 1850. From Robert Snetsinger, *The Ratcatchers Child* (Cleveland: Franzak and Foster, 1983). Notice that the ratcatcher is accompanied by a dog and carries a ferret in the cage.

poor sanitation. So any kind of control was considered valuable, even if one didn't understand why or how it worked. Even asking for divine intervention wasn't, and still isn't, outside the realm of normal behavior.[3]

Since so much of America was still rural after the Revolution, many of the earliest pest problems on the frontier went unreported or were too far from the urban areas to be seen as a market for the pest controllers. What got the attention of the pesticide and chemical manufacturers were the growing pest problems in the few urban areas that existed in America at the end of the eighteenth century: New York, Washington, Philadelphia, Boston, and Baltimore.

When the United States began to urbanize and industrialize in the nineteenth century, disease and pest problems common to Europe increased dramatically and came to haunt New World cities and towns. Built in unplanned ways, American cities became a nightmare of muddy streets, garbage, and filth. Some cities released pigs in the streets to clean up the garbage, but that wasn't enough. European rats and mice, imported with the colonial immigrants, rapidly increased their populations in the cities. As was the case in medieval Europe, a lack of general sanitation in America's newly crowded areas caused serious pest outbreaks and epidemics.

Old World ratcatchers migrated to America in the 1800s to exploit the ever-growing market. Because of the severity of this problem and the relative sophistication of the ratcatchers, the use of rat-control dogs, rat traps, and rodent poisons became the first pest-control solutions to appear in the almanacs and in the early rural magazines. These mostly household and barn pest-control ads continued to dominate the farm journals until the late 1880s. A great deal of ad space continued to be devoted to these pests after the turn of the twentieth century, and, of course, there is no shortage of bug- and rat-control ads today in every medium from telephone books to TV.

Some European and Asian cities devised recycling schemes to control garbage, sewage, and household waste in order to deal with disease and rats. Many early American cities also operated urban recycling and sanitation systems. In New York City, Philadelphia, Baltimore, Boston, and other cities, food scraps and food waste were recycled back to farms, as was an enormous quantity of horse manure and oxen manure.

This manure from city stables and pens was composted with sawdust, spent tanner's bark, charcoal, and ashes. Also collected was street dirt, which was manure, and any food scraps. But the cities grew so fast in the late 1800s that these programs either couldn't or didn't expand fast enough to deal with the growth. Thereafter, instead of preventing rodent populations and other pest problems from becoming a health and safety issue in the first place, urban and rural America went overboard with traps, poisons, and disinfectants in an attempt to clean up the mess and fight the pests that resulted from the lack of sanitation and recycling.

America's love affair with disinfectants and the Orkin Man stems from its inefficient urban sanitation facilities and legitimate paranoia about rats, mice, lice, roaches, filth, and plague. To this day, the United States still has numerous unsophisticated sanitation schemes, and rat populations are large in many cities, including Washington, D.C., New York City, and even beautiful Santa Barbara, California.[4]

The bishop of Lausanne, Switzerland, excommunicating June beetles in 1497. From Robert Snetsinger, *The Ratcatcher's Child* (Cleveland: Franzak and Foster, 1983).

RATS.—A writer in the New Gen. Farmer recommends the following bait for traps:—

"Oatmeal one quart, one grain of musk, and six drops of the oil of rhodium. Put the musk and oil into sufficient sweet milk to moisten the meal; then mix all together in a stiff paste. The oil rhodium can generally be procured at a druggist's store; and seldom fails, together with the musk, to draw rats into any place. Caution is requisite to guard against the common cause of traps failing, which is the smell of the hand. This can be avoided by using an old knife or spoon:"

"Farmers may sometimes drive away rats from their premises in the summer season, by blocking up their holes with broken glass, or blacksmith's cinders, and plastering them with mortar, repeating the process wherever new holes appear."

The ad below is from the *Genesee Farmer*, May 26, 1832. The ad at left is from the *Union Agriculturist*, April 1841. The second ad promotes the same control strategy as the first but provides more how-to advice. Not much change in that decade.

DESTRUCTION OF RATS.

The following are among the means recommended for the destruction of these vermin:

Take one quart of oatmeal, four drops of oil rhodium, one grain of musk, two nuts of nuxvomica powdered; mix the whole together, and place it where the rats frequent; continue to do so while they eat it, and it will soon destroy them.

Another mode of destroying rats.—Take equal quantities of unslacked lime and powdered oat meal; mix them by stirring, without adding any liquid, and place a small quantity in any place infested by rats. They will swallow the preparation, become thirsty, and the water taken will swell the lime and destroy them.

Another mode of destroying rats.—A friend in Salem, Mass. informs us, that rats are easily destroyed by sprinkling a little of the powder of Spanish flies on some buttered bread, or other food of which rats are fond —*N. E. Farmer.*

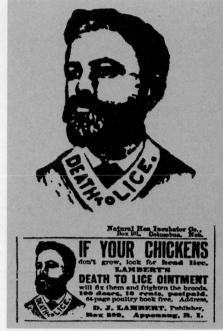

The example above advertises the availability of an effective pest control applicator. Its model wears a "Death to Lice" collar. From *Farm Journal*, September 1897.

The Rough on Rats ads are examples of the pitches that farmers saw in the 1880s and '90s for rat control. The ad on the left is from *Farm Journal*, September 1887. The ad on the right is an excerpt from the 1890s as seen in several journals. This ad shows the racist overtones that accompanied the profession of ratcatching early in the United States.

Historical Markers

1865   ● **The Colorado potato beetle, transported east by the transcontinental railway system, destroys much of the eastern US potato crop.**

1867   ● **Paris green, a common arsenic paint pigment, is first used successfully by Brian Markham on the Colorado potato beetle. There are no regulations on its use at this time, and food often carries high arsenic residues.**
● **South Carolina phosphate deposits are first mined.**
● United States buys Alaska from Russia for $7.2 million, or 2 cents an acre.
● **First use of the worldwide search for beneficial controls by the US Bureau of Agriculture.**

1869   ● Completion of the Transcontinental Railroad.

1870–80 ● **This was a farm populist decade.** The Grange movement in the West and Northeast was followed by the populist movement, stretching from Texas, through the South and up to New York and the Midwest. Both movements were anti-monopolist and pro-labor.

1871   ● Paris Commune rises and is suppressed.
● **The rural journal *Country Gentleman* promotes the arsenic pesticide known as Paris green.**

1873   ● **DDT is first synthesized by a German scientist. No commercial value is immediately recognized. The Panic (economic depression) of 1873 causes widespread farm bankruptcies.**

1875 ● **Connecticut establishes the first state-supported agricultural experiment station in the United States. By this time there are more than a hundred such stations in Germany (74), Austria, and Italy.**

1876 ● Alexander Graham Bell invents the telephone.
● Custer is killed at the Battle of Little Big Horn.

1878 ● Swan invents the light bulb.

1879 ● First electric train is launched in Berlin, Germany.

1880–90 ● **This too was a farm populist decade.** The Texas National Alliance movement grows into the populist movement and the People's Party and begins to elect candidates.

1883 ● **The National Fertilizer Association is formed by a conglomeration of dominant fertilizer corporations, in part to fund research at land-grant colleges and promote the use of agricultural chemicals.** It later becomes the Fertilizer Institute.

1884 ● Steam turbine is developed by Charles Parsons.

1885 ● Karl Benz makes the first practical petrol-burning car. Gottlieb Daimler produces the first motorcycle.

1887 ● **The Hatch Act is passed, providing money for entomo-logical work at land-grant colleges. The act is also the basis for the establishment of at least one agricultural experiment station in each state.**

1889 ● **Cottony cushion scale on California citrus comes under control after worldwide search for and use of a predator, the vedalia beetle.** This is a major success against a destructive pest.

1890s ● **This decade sees the culmination of the rural populist movement.** Four previous economic depressions and a civil war in this century had accelerated the concentration of land into fewer hands.

1892 ● Ellis Island immigration processing center opens.
● **Lead arsenate is introduced as a pesticide to control the gypsy moth. It becomes the most popular pesticide in the United States until the early 1950s.**

1893 ● **Sales of South Carolina phosphate exceed 600,000 tons. Fertilizer ad campaigns are finally proven effective.**

1894 ● **The Panic (economic depression) of 1894 causes additional farm losses.**
● **900,000 tons of sodium nitrate are used in Europe alone.**
● **More than 165,000 copies of the US Bureau of Agriculture Farmer's Bulletin 127, *Important Insecticides: Directions for Their Preparation and Use*, are distributed by 1900. An important recommended poison is arsenic. Arsenic poisoning of apples is reported in the *Boston Globe*.**

1895 ● **Standard Oil discovers a process for mining sulfur deposits in Louisiana.**
● *Scientific American* helps sell farmers on the importance of using sulfur.

1899 ● *Fertilizers* is published by E. B. Voorhees. The book sells for one dollar. Fertilizer merchants buy and distribute thousands of copies to their customers.

1900 ● **There are 6,400,000 farms in the United States. These farms occupy about 400,000,000 acres.**
● Labour Party founded in Britain.

1903 ● Wright Brothers make the first controlled powered flight in Kitty Hawk, NC.
● **World standard for arsenic and lead residues is set by the Royal Society of London and agreed to by France at 0.01 grains of arsenic and 0.02 grains of lead per pound or gallon of product.** Big US agriculture will fight these tolerances for the next fifty years.

1904 ● **First German chemical cartel is formed** — often referred to as Baby Farben.

1906 ● National concern over food adulteration results in the watered-down **Pure Food and Drug Act of 1906.** Upton Sinclair's book *The Jungle* (a description of the filthy meatpacking industry) and pesticide poisonings prompt much of the public outcry. Administration of the 1906 act resides within the Department of Agriculture, so unfortunately public and environmental safety issues become secondary to big agriculture's demands.

1907 ● **American Cyanamid synthesizes ammonium cyanide in Niagara Falls, Canada.**

1908 ● **First reported pest resistance to lead arsenate, sixteen years after its first use.**
● Model T car is produced by Henry Ford.

1909 ● **American Cyanamid begins selling cyanide gas and ammonium cyanide products.**
● **German chemist Fritz Haber perfects a process for extracting nitrogen from the atmosphere and synthesizing ammonia.**

1910 ● **First National Pesticide Act. The law regulates the composition of arsenic and other pesticides, not their safety. It is only a truth-in-advertising law.**
● Mexican Revolution (1910–1911).
● **The Industrial Workers of the World (IWW) begins organizing farmworkers.**

1911 ● F. H. King publishes *Farmers of Forty Centuries.* **This book describes organic farming practices in Asia.** Officially, the book is ignored by US agricultural institutions, though many US farmers own it and use many of the practices it describes.

1912 ● **Federal Trade Commission is established; pesticides and fertilizers are supposed to be regulated in the marketplace for the first time.**

1914 ● First World War begins in Europe.

1916 ● Easter Rising in Dublin.

● **The BASF scientist Fritz Haber, working with Robert Bosch, further develops his process for extracting nitrogen by using large feedstocks of natural gas, a cheap by-product of petroleum and coal/coke. Haber and Bosch win the Nobel Prize for the discovery.**

● First birth control clinic is established in Brooklyn.

1917 ● Bolsheviks storm the Winter Palace.

● United States declares war on Germany.

1918 ● First World War ends.

● **I. G. Farben, the German chemical cartel, forms and includes Bayer, BASF, and Hoechst. I. G. Farben will build chemical plants next to concentration camps during WWII, in support of the Third Reich.**

● **Basle A. G., the Swiss chemical cartel, forms to counter I. G. Farben. It includes Ciba, Geigy, and Sandoz. They will dissolve the cartel after World War II but later merge to create Novartus.**

● **William A. Albrecht scientifically demonstrates the validity of soil inoculants. Albrecht's findings greatly increase farmers' knowledge of organic soil management and the value of legume fertilizer crops.**

1919 ● Alcock and Brown cross the Atlantic by air, only sixteen years after the Wright brothers' flight.

● **Boston Health Department destroys arsenic-contaminated apples. Though a few articles appear in the newspapers, the poisonings are mostly hidden from the general public.**

1920 ● Prohibition begins.

● **Boston Health Department destroys more arsenic-contaminated apples.**

● Ghandi's resistance begins in India.

1921 ● **First use of an airplane as a crop duster.**

● **Boston Health Department destroys arsenic-contaminated apples for the third year in a row. Little news of this is leaked to the public.**

● Communist Party organizes in China.

1922 ● Mussolini forms Fascist government in Italy.

● USSR is established.

● Irish free state is established.

1923 ● Unsuccessful Nazi uprising in Germany.

● **Rudolph Steiner (founder of the Waldorf schools) gives his first lecture on agriculture.**

1924 ● **Rudolph Steiner publishes *Agriculture*, which describes the basics of biodynamic agriculture.**

1925 ● Vietnamese Nationalist Party is founded; Ho Chi Minh is the spiritual and political leader.

● Adolph Hitler publishes *Mein Kampf*.

1926 ● **Standard Oil signs a twenty-five-year working agreement with I. G. Farben, the German chemical cartel that includes BASF, Bayer, and Hoechst.**

1927 ● **The US Food and Drug Administration is established.** The USDA is the administrator of the FDA.

1928 ● **Standard Oil hires Theodor Seuss Geisel (aka, "Dr. Seuss") to sell the insecticide Flit.** This move proves to be a stroke of marketing genius, leading to the most successful advertising campaign up to that time.

1929 ● **Twenty-nine million pounds of lead arsenate and twenty-nine million pounds of calcium arsenate** are applied on food and cotton. *The American Journal of Health* suggests arsenic spraying should be reduced.

1929 ● **Panic of 1929. The Great Depression begins.**

1931 ● Strong winds begin the "Dust Bowl" process of soil erosion in the western Great Plains of the United States.
● DuPont scientist discovers the pesticidal properties of **carbamates**.

1932 ● First use of an **organophosphate** as a pesticide.

1933 ● President Franklin D. Roosevelt introduces the New Deal.
● Hitler is elected German chancellor.
● Falange, the Spanish fascist party, is created.
● **Arthur Kallett and F.J. Schlink publish their book *100,000,000 Guinea Pigs*. The chapter "A Steady Diet of Arsenic and Lead" chronicles the hazards of spray residues. The book becomes a best seller.**
● **A fifteen-year-old Montana girl dies from poisoning by fruit sprayed with arsenic.**

1934 ● The Long March by the Communist Chinese begins.
● Hitler is declared German Fuhrer; he exercises dictatorial control until 1945.

1935 ● Land redistribution in Mexico.

1936 ● **Dust Bowl ends.**
● **Methyl bromide is introduced to the United States by the California Department of Food and Agriculture.**

1938 ● **The USDA *Yearbook of Agriculture*, titled *Soils and Men*,**
provides an excellent account of organic farming.
● **Passage of the Federal Food, Drug and Cosmetic Act which frees the FDA from the USDA's direct control by 1940.**

1939 ● Germany invades Poland; World War II begins.
● **Swiss entomologist Paul Muller, working for Geigy Chemicals, discovers DDT's insecticidal properties. For this discovery he wins the Nobel Prize in Medicine in 1948.**

1940 ● **Sir Albert Howard's book, *An Agricultural Testament*, is published. It describes many of the principles of organic agriculture. He should have won a Nobel Prize.**
● **The FDA finally moves out from under the supervision of the Department of Agriculture; it is now placed in the Federal Services Administration.**

1941 ● United States enters World War II against Germany and Japan.
● **The Nutrition Foundation is formed by food and chemical companies to fund research into nutrition.**
● **Agriculture in the United States becomes even more industrialized, using chemicals, mechanization, and monoculture to increase production and feed war-torn Europe (1941–50).**

1942 ● J. I. Rodale launches the magazine *Organic Farming and Gardening*.

1945 ● Germany surrenders.
● United States drops atomic bombs on Hiroshima and Nagasaki.
● **First use of DDT and Telone (DD or 666) on American farms.**

*Chronology concludes on page 134.

Pages such as this one from the *Indiana Farmer,* June 1858, blanketed the farm journals with ads for reapers, mills, and threshers. After the 1830s, American farming underwent a mechanical revolution. But that changeover was not enough to help small and medium-sized farmers deal with credit, shipping, and processing costs that swallowed up most of their profit.

Chapter 9

THE POPULIST FARMERS' MOVEMENT

By the late 1860s any prosperity caused by mechanical advancements on the farm had turned into economic panic. Though terrible economic conditions had prevailed in rural areas since the revolution against the British, they worsened after the Civil War, and by 1873 the country was experiencing its fourth depression since the turn of the century. This one was called the Panic of 1873.

This was a full-blown economic depression that saw periods of partial recovery but worsened in the late 1880s, before the country plunged into another deep depression in 1894. The late 1860s to the late 1890s were thirty years of very hard times for farmers and for any American who was not enfranchised with significant wealth or significant property. The panic of the 1870s caused profound suffering in rural communities, resulting from local bank insolvencies, a further tightening of the money supply, and even more depressed prices for farm products. All of this translated into farm foreclosures and land loss for small and medium-sized farmers trying to get their lives back together after the war.

Many farmers moved west in search of a new life and cheap land from a government program like the Homestead Act or the Swamp Land Act, or land granted to the state agricultural colleges. Some of the first arrivals were lucky and realized their dreams, but by the time most of them went west after the war, a majority of the good land had been gobbled up by land schemes while they were busy fighting. Often, people had already figured out a loophole that allowed them to acquire more than their share of the best land.

The widespread corruption of the government programs during and immediately after the Civil War was a bitter blow to the national psyche. These land schemes dashed the hopes of bankrupt landholders who were praying for a new start and an escape from the economic and political calamities in the eastern part of America. The congressional land grants had been enacted after much populist agitation before the Civil War. Small-scale farmers had hoped that they would get a piece of land that wasn't worn out through the Homestead Act or that they could drain a slightly swampy piece of land and get it for $1.25 an acre through the Swamp Land Act. The population had demanded these acts in an effort to equalize the economic pie in America. Sadly, that equalization was not realized through railroad grants, homestead grants, swamp recovery grants, mining claims, Spanish land grants, or any other government giveaway program.

In one blatant example of government rip-offs, Henry Miller and Charles Lux were able to acquire one million, two hundred thousand acres in California, Oregon, and Nevada from the 1850s to the 1870s. A large portion of that came from schemes to obtain land from government land programs and Spanish land grants. At one point Henry Miller allegedly bolted a row boat on a wagon and drove it across relatively dry ground. He then falsely filed the papers to prove that he had drained swamp land that was so wet he had to cross it in a boat. As a consequence, he and his partner became eligible to buy, for $1.25 an acre, through the Swamp Land Act, tens of thousands of acres that were already dry ground. As they grew richer they used one scheme after another, using the courts and the legislature to obtain more land and more cattle.

Charles Lux ran the Miller and Lux meat business in San Francisco and was a long-established figure in the meat business in the city before his partnership with Miller. He knew from his wealthy contacts and government connections when programs were going to emerge or change, so he and Miller could take advantage of the government giveaways as soon as possible. Henry Miller, on the other hand, worked in the country and knew where the land was that could be obtained from the government, leased from the railroads, or bought from farmers who could not compete with them. Bankrupt farmers traveling west to get a fresh start in life from one of the government programs were no match for the team of Miller and Lux or other land schemers who had insider information.[1]

In these difficult economic times, farmers mounted a populist effort to reform agriculture and society once again. Some of their frequent targets were Miller and Lux, the Newhall Land and Cattle Company, and other large-scale farmers who were also fertilizer and shipping companies. Many of their criticisms focused on the destructive farming practices of the large-scale land schemes. For example, Miller and Lux completely destroyed the natural flow of the San Joaquin River system and contributed to the collapse of the San Joaquin, Tulare Lake, and many of the river systems feeding the San Joaquin Valley.

In the 1860s and 1870s, most farm journal writers and editors promoted the value of a cooperative rural lifestyle, as they had earlier in the century. But this time many journals also reported on and promoted the farmers' collective strategies to fight the trusts,

brokerage houses, shipping schemes, and federal and state land rip-offs. Several rural journals showcased the programs of the new populist farm society movements, the Patrons of Husbandry (the Grange) and the Farmers' Alliance (sometimes called the Farmers League).

The Patrons of Husbandry, a national fraternal order of farmers, originated in 1866 with inspiration and organizing help from within the United States Bureau of Agriculture. The first Grange was established in 1867 from Patrons of Husbandry farmers' clubs and grew into the National Grange movement. The Grange movement was passionately interested in farmland and rural revitalization efforts, similar agriculturally to the Berkshire Society fifty years earlier, but focused also on mechanization, since reapers, plows, and other tools had wrought a mechanical revolution in day-to-day farming. As the economic depression worsened, the Grangers, as they were called, began to speak with a more organized political-economic voice than the populist farmers from the 1800s to the 1850s.[2]

A second, much larger and more successful populist movement, the Farmers' Alliance, began in the 1870s. This movement started out as a revolt of Kansas farmers against land rip-offs by the railroad trusts. The railroads were granted checkerboard sections (a section is one square mile/640 acres) of land all along the course of the new railroads. Many farmers had already claimed and developed that land and found themselves losing it to the railroads. The Kansas revolt almost immediately spread to New York, where farmers had also been dispossessed of land they had cleared and developed. Subsequently the movement spread to Texas, where energetic and activist-minded farmers helped form the National Alliance, which organized northern and southern farmers and even urban workers.

As the rural crisis worsened, several rural magazines became proactive participants in these farm revitalization movements. In fact, both the Grange and Farmers' Alliance organizing drives were very successful, largely because of the efforts of local magazines like the *Pacific Rural Press* in California, the *Progressive Farmer* in North Carolina, the *Kansas Farmer*, the *Alliance Sentinel* in Michigan, the *National Economist*, and a new national magazine, the *Farm Journal*. Probably the most important journal was the *Southern Mercury* in Texas, which served as the national voice since Texas played such a pivotal role in the National Alliance.[3]

During Reconstruction, the populist movements formed buying and marketing cooperatives to counteract the railroad trusts and the tight federal control of the money supply and to gain access to modern farm equipment and commodity markets. In addition to high freight rates, low commodity prices, a fixed currency, and shortage of money, farmers also suffered from heavy mortgage debts, high interest rates, local grain monopolies, corrupt local governments, and cloudy titles for land and water rights. Again, as in the earlier rebellions, a central issue with these populists was money (specie) for use versus money (capital) for profit, power, and control. Since before the Revolution, small farmers had been commodity rich and capital poor. Any surplus yield that the farmers produced was bought cheap by brokers in frontier settlements and sold dear in the cities. Consequently, most of the profit went to middlemen who produced none of the commodities.

As a result of all this instability, Grange and Farmers' Alliance halls dotted the rural American countryside, with the organizational leaders often acting as local farm advisors, social organizers, and redressers of grievances. The populist organizations were avowedly pro-labor, but the Farmers' Alliance was the most inclusive, opening its membership to people of color as well as white workers and farmers. This was a very radical civil rights movement for such an early date.[4]

Promotional page for the Grange with the Grange Cardinal Confession of Faith. From *Pacific Rural Press,* August 1, 1874.

WHAT THE FARMER GIVES AND WHAT HE GETS

"Look on this picture, then on that."

These two pictures fairly represent what the farmer gives to the world and what he is getting now in return. The products of his toil nourish and sustain his fellow citizens of the United States, and are sent abroad to feed the inhabitants of Europe; and for all this bounty he is being shabbily paid in dollars so few that he is cramped and hampered in his life, and advised to economize, to work harder, to keep quiet, and to be contented.

WHAT HE GIVES

Within the last few months several hundred million dollars have been borrowed from abroad, which must be added to the $6,000,000,000 already owing, and the farmer is expected to foot the bill, at least seventy-five per cent. of it, for he furnishes this proportion of foreign exports.

When the civil war closed our National debt was about $3,000,000,000 and probably only one-third of it due abroad; now, while this debt has been reduced to $1,000,000,000, our municipal, railroad, mortgage and other obligations have been piled up until it is believed that they now amount to over $4,000,000,000, all due to foreigners, and on which interest has to be paid in gold, and of which the farmer must pay three-fourths of the whole, both principal and interest.

WHAT HE GETS

Introduction to "What the Farmer Gives and What the Farmer Gets," from *Farm Journal*, February 1896. In this article *Farm Journal* outlines the farmers' discontent with national debt, gold, and paper money. This lament is reminiscent of the complaints that led to Shays's Rebellion and the Whiskey Rebellion as well as that of today's demonstrators against corporate factory farms. The editorial delivers a broadside against the aristocratic and corporate power that controlled farming and most of the country.

Throughout this populist period, several farm journals and the Farmers' Alliance movement promoted biological agriculture, complete with natural fertilizer advice and biological pest-control strategies. Many populist farmers used natural forms of pest control and land management strategies advocated by the Patrons of Husbandry and the Farmers' Alliance. It is surprising that at a time of so much economic depression these farmers and a few scientists were able to increase significantly their knowledge of biologically based farming techniques and strategies.

Before and especially after the Civil War, populist farmers in California struggled to establish agricultural programs for the 1862 land-grant agricultural college that became the University of California. The farmers and the scientists who supported them advocated the development of a long-term fertility plan and the use of a well-developed pharmacopeia of naturally occurring pesticides as well as some promising new pest-control strategies. The university initiated major biological farming programs, which resulted from the requests and demands of populist farmers. Articles describing the bureaucratic struggles and the scientific breakthroughs at the land-grant universities were highlighted in the Farmers' Alliance and Grange farm magazines.

For more than two decades after the University of California was established as a land-grant college a struggle raged between the populist farmers on one side and the university administration and the academic senate on the other. This struggle was waged over how the agriculture program fit the university's image, how it would be structured, who would be hired, and how many would be hired. To a large degree the regents and the academic staff attempted to marginalize the agriculture department, even though agricultural research and education provided the major political prop for the creation of land-grant colleges, while the selling of the granted land provided a significant amount of the money that the states used to create the colleges.

Though harassed by the administrators and the politicians, individuals within the University of California developed many techniques and practices in biological agriculture and created and operated the Division of Biological Control for more than one hundred years before the university administrators finally terminated it in 2000. While many in the University of California Department of Agriculture created valuable programs, the department also promoted the ideas of Liebig and hired Liebig-influenced chemists to lead the agricultural program—because they were perceived nationwide to be scientific and academic. These Liebig adherents received extensive criticism from the farmers, however, because California farmers felt that all they knew was book and laboratory farming and little if anything about real farm problems and real-life solutions.

By the 1860s new pesticides began to appear on the market, some safe, others incredibly dangerous. We will discuss the safer pesticides and pest-control strategies here and defer the discussion of the dangerous pesticides from the nineteenth century,

ORGANIZATION.

Farmers should stand by each other and pull together—Other people combine for mutual help and protection, why not we?—Notes and news of Organization among farmers—The Grange—The Alliance—The League.

We believe that farmers should have seats at the First Table, and partake of the good things that abound, without crowding other worthy people away.

The farmers of this country are not going to smash things as some people fear. They are trying to keep things from going to smash : this is all.

This call for organizing was published in the *Farm Journal* in August 1891.

such as arsenic, lead, mercury, and sulfuric acid, until later chapters. The Grangers and the Farmers' Alliance movement endorsed the safer biological pest controls including those described in the following paragraphs.

Pyrethrum powder, a biological pesticide used in the United States since the middle of the eighteenth century, comes from ground chrysanthemum flowers. Pyrethrum was discovered before the 1700s in the Caucasus. Farmers in the area applied this pesticide for more than one hundred years but kept its pesticide properties a secret from the outside world until the early 1800s.[5] After the secret source of pyrethrum leaked out and seeds were obtained, the U.S. Bureau of Agriculture promoted its production in America and so did the populist press.

The Buhach Company of Stockton, California, conducted one of the most prosperous pyrethrum operations in the United States, but it was short-lived. United States–produced pyrethrum never lived up to its promise as an insecticidal panacea, even though it is still widely used today and is deadly to thousands of insects and mites. The lack of spectacular success in the 1800s resulted from the high cost of the product and irregular availability, in contrast to the cheap and deadly waste chemicals that became widely available.[6]

Another extremely promising pest-control technique, developed in the 1860s, concentrated on searching for the natural enemies of certain crop pests, the so-called beneficial insects, predators, and parasites. The natural enemies were usually located in the crop's country of origin, so researchers traveled to a crop's presumed hearth-land to hunt for natural pest controls. The first worldwide search, conducted in 1867, proved the value of the technique. Ironically, this is the same year that experimenting chemists used arsenic to kill Colorado potato beetles—but that is getting ahead of ourselves, since there will be more

about that important turn of events in the next chapter. Twenty years later researchers and chemists were struggling to control the cottony cushion scale (*Icerya purchasi*) in the orange and lemon groves of Southern California. In 1887, an invasion of cottony cushion scale caused serious damage to the orchards. The growers, desperate to avoid economic disaster, invited Charles Riley, chief of the U.S. Division of Entomology in the Bureau of Agriculture (now the USDA), to be the principal speaker at their annual Convention of Fruit Growers in 1887.

Riley proposed a plan of action for the anxious growers. He believed that cottony cushion scale came from Australia or possibly New Zealand. Albert Koebele, an employee of the Bureau of Agriculture, was studying the history and habits of insects in California. Riley sent Koebele to Australia in early 1888, and within months after his arrival he located the cottony cushion scale and found parasites and predators feeding on its eggs. Specimens of both were sent back to California: 12,000 parasites and a few handfuls of vedalia predator beetles (Chrysopa), 129 to be exact.

The 129 predator beetles became the story's heroines. These 129 beetles multiplied exponentially in the lab and in the field. The vedalia beetles completely controlled the cottony cushion scale.

The successful handling of the cottony cushion scale by the use of natural enemies gave the practice of biological control such prominence that this method of controlling pests soon spread throughout the world, and other successes almost rivaling that of the cottony cushion scale have since been accomplished in several different countries.

The California citrus industry was saved for under $2,000. By the next season, Los Angeles County orange shipments jumped from 700 to 2,000 cars. The control of cottony cushion scale with the vedalia beetle was

recognized as an entomological miracle.[7] Because of such successes, Grange and Farmers' Alliance editorials about biological farming appeared often in farm magazines, such as the *Farm Journal*, the *Pacific Rural Press*, the *Southern Mercury*, and the *California Cultivator*, until well into the 1880s.

Until the mid-1880s, the populist press thought Liebig as a scientist and Mapes as a manufacturer of fertilizers were both phonies, given more to book farming and the selling of products than to a sound agricultural program. The progressive farmers attacked Mapes as a quack in the journals and actually helped catch him knowingly selling bogus fertilizers. As a result, from the mid-1850s through the early 1880s, many farmers and most of the rural magazines constantly questioned the scientific "hoohaw" behind both Liebig's theories and Mapes's products.

Many of the farm journals remained wary of advertisements and bogus products because of the puffed propaganda from Liebig, Mapes, and the low-grade fertilizer quacks. To reassure their readers that they wouldn't sell such worthless products, the publishers of the populist magazines stood behind all claims made by their advertisers and printed very few ads in the early years of their existence. Instead of selling the publisher's products, the articles in the populist press were designed to enhance rural communities, improve farming practices, and educate and delight the readers. To counteract the book-farming, chemical-advocacy journals like the *Country Gentleman* and *Working Farmer*, advice from other farmers and the farm societies appeared regularly in the populist farm magazines from this period, warning about their experiences with dangerous and useless products.

The rural journals continued to publish negative critiques of book farming, large-scale agriculture,

and Liebig's theories. At the same time the agricultural colleges and administrators in the Bureau of Agriculture were hiring European-trained chemists who believed in book farming and Liebig. The agricultural chemistry professors encouraged the use of industrial fertilizers, large-scale farming, and experimentation with toxic pesticides. Before the 1890s there was a distinct difference between the academic and government promotion of chemicals and the farmers' rejection of toxic chemicals and synthetic fertilizers.

The panic that hit the country in 1894 brought with it even more disastrous times for small and medium-sized farms. The Grange influence diminished somewhat, probably because the Farmers' Alliance had grown into a national political organization, known as the People's Party, and had taken over center stage from the Grangers. The People's Party held national conventions and actually elected congressmen, legislators, mayors, and city councilmen. By the end of the century, however, this version of populist revolt in the United States had been beaten back or co-opted by other issues. Significant rural political agitation subsided once again. Subsequently, many changes occurred in farming in America, some caused by decades of economic depression, others caused by rampant commercialism.

After the mid-1880s, the chemical corporations began to accumulate significant wealth and political power. Their advertising dollars and growing political clout changed the attitudes of many of the publishers, and many other publishers sold out. Progressive messages about biological approaches appeared less frequently. They did not disappear entirely, however, since there continued to be a determined and faithful market. But the ads and editorials for biological products appeared less frequently in the 1890s and in the first decade of the twentieth century.

The chemical corporations had huge advertising budgets, and the publishers of formerly populist magazines were anxious to have them spend it for ads in their journals. Advertisements and editorials for chemical solutions to pest problems finally came to dominate farm publications by the close of the century. They still do. Once that happened, the publishers were loath to criticize pesticides and fertilizers that provided so many advertising dollars. But they were anxious to preserve the public image that they were still the populist press that had the interests of the farmers as their paramount objective.

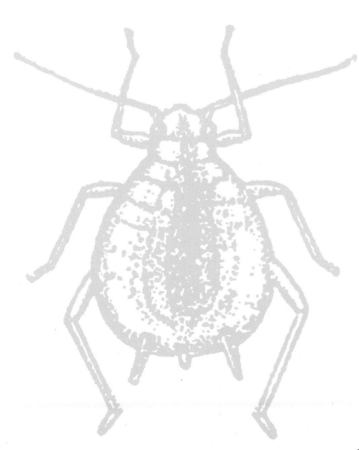

The farmer took little Weevil

And put him in Paris Green

The Weevil said to the farmer

"It's the best I've ever seen.

I'm gonna' have a happy home, a happy home."

Late-nineteenth-century traditional American folk song about the arsenic pesticide Paris green.

Chapter 10

PAINT PIGMENTS AND OTHER LETHAL POISONS

After the Civil War, American farmland expanded rapidly. Immigration from Europe caused both the population living in cities and the ranks of new farmers to swell. Many of these immigrants who took up farming knew little about it and apparently even less about the natural pest and fertility solutions that many farmers used. As the nation's cultivated acreage expanded, pests began to plague potatoes, cotton, corn, wheat, oats, barley, and tobacco on the commercial farms that now stretched from Georgia north to New England and as far west as the trains ran.

In the mid-1860s, before the transcontinental railroad was completed, a particularly hungry pest, the Colorado potato beetle, was munching its way through the nation's potato fields and spreading rapidly to the east and north, along the new canals and railroads. Commercial potato farmers were desperate to find something—anything—to stop the beetle. For a time, "improved" beetle-control products appeared in magazine ads more frequently than patent-medicine ads or "improved" mousetrap solutions. One comical cure was a mail-order kit that contained two short pieces of board, along with instructions that explained how to squash the bugs between them.

In 1867, after years of such frustrating hoaxes, a chemist poured several arsenic mixtures on the beetles. One of them, named Paris green, worked very well. By 1868, several farm journals began to promote the amazing insect-killing capability of Paris green, a widely known and widely available green-paint pigment.

Some of the arsenic used for colored paints came from the waste products of mining and manufacturing. If the early industrial firms could not find a market for their waste material they usually dumped it in creeks, rivers, lakes, and oceans or created mountainous piles of toxic waste on land. Ultimately, after decades of abuse, many towns, states, and nations regulated the dumping of toxic waste and forced corporations to dispose of their waste legally. When that happened, the corporations were forced to find acceptable dump sites or a market willing to buy their waste.

The dawn of toxic pesticide use on a large scale in America began in the late 1860s with the use of arsenic to control bugs on potatoes. As with guano, which opened the door to the commercial fertilizer market, arsenic paved the way for all the poisonous pesticides that followed. The chemical merchants learned several merchandising and advertising lessons from promoting arsenic, which they used to greatly increase sales for all of the other poisons that followed it. With arsenic, they also learned how to defend chemicals from victims, consumer advocates, and regulators. A larger problem for the applicator, however, was how to defend against the pesticides.

In the decades following the Civil War, other poisons followed arsenic into the marketplace. Some had colorful names and deadly capabilities; others had

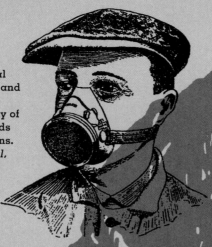

Typical dust and chemical protectors from the 1880s and 1890s. At right: from the Wilson catalog, University of California at Davis Shields Library Special Collections. Below: From *Farm Journal*, August 1897.

A Paris green antidote. From *Pacific Rural Press*, August 22, 1874.

Don't breathe injurious chemicals!
Guard your nose and throat!

Fruit sprays are injurious. They irritate your nose and throat. They can seriously damage your lungs and impair your health.

Why risk your health with them? Willson Dustite Respirator for Dust and Spray offers you complete protection. It allows free breathing. It keeps out the harmful chemicals that are used in spraying compounds. Wear it when you're threshing and treating seed! Whenever you have to work with poisons or in dust!

The Willson Dustite Respirator is easy to adjust. Comfortable to wear. Gives years of service. Recommended by manufacturers of argicultural chemicals. Priced at $2.25. If your dealer can't supply you, write to us direct and we will send it C. O. D. Address Willson Products, Inc., 203 Washington St., Reading, Pa., U. S. A.

The Willson Dustite Respirator
for DUST *and* SPRAY

Dust Protector. The "Perfection" has an Improved **Automatic Valve. Keeps all dust out of the lungs and is just the thing for Threshermen.** Nickel-plated Protector, postpaid, for $1. Circular free. **AGENTS WANTED. H. S. Cover, 157 Paris St., So. Bend, Ind.**

At right: a respirator to protect against arsenic pesticides, lead poisoning, and other dangers. From *Pacific Rural Press*, August 4, 1885.

Patent Life-Saving Respirator.

PREVENTS LEAD POISONING AND SALIVATION.

Invaluable to those engaged in dry crushing quartz mills, quicksilver mines, white lead corroding, feeding thrashing machines and all occupations where the surrounding atmosphere is filled with dust, obnoxious smells or poisonous vapors. The Respirators are sold subject to approval after trial, and, if not satisfactory, the price will be refunded. Price, $3 each, or $30 per dozen. Address all communications and orders to

H. H. BROMLEY, Sole Agent, 43 Sacramento Street, San Francisco, Cal.

colorful names and no pest-killing ability. After that time, the chemical corporations spent considerable effort trying to convince both citizens and regulators that their toxic-waste products were efficient at controlling pests on the nation's food supply and were safe to use. In order to sell these poisons to farmers already suspicious of medical arsenic and mercury (knowing they were not safe to use), the chemical and mining corporations concocted several clever sales campaigns. Most importantly, they also conducted a continuous advertising blitz.

Many farmers vigorously opposed arsenic, just as they had earlier opposed guano, sodium nitrate, and mixed fertilizers. They ridiculed the book-farming experts who promoted arsenic for use on food. But the chemical corporations and their experts kept propagandizing with claims that farmers were getting higher yields when they used arsenic—while deliberately avoiding a discussion about the farmers' fear of the poison and fear of polluting their soil.

Some of the rural magazines, including the *Pacific Rural Press* in 1871, warned farmers to avoid the use of toxic chemicals and even printed antidotes for arsenic poisoning. This populist magazine and others like it fought against the use of arsenic and other poisons until the 1880s. Three centuries of medical misuse of arsenic, lead, mercury, and sulfuric acid, which Paracelsus introduced, had left rural people incredibly fearful of these medicines. While these fears persisted and were widely publicized, those farm magazines catering to growers of potatoes, apples, pears, and berries for export jumped on the arsenic bandwagon almost immediately after the discovery of its effect on Colorado potato bugs, coddling moth, and apple worms.

By the mid-1870s, *Country Gentleman* editorials regularly promoted arsenic as the centerpiece of the magazine's pest-control advice. The following letter to the editors from a farmer in Wayne, Michigan, titled

"Poisoning the Colorado Potato Bug, July 11, 1872," illustrates the early arsenic support from university scientists. Also described in the letter are a primitive pesticide duster and the recipe for arsenic preparation:

> Enclosed find three years' experience with the Colorado Potato beetles. I doubt much if they will be any detriment to the farming community. After all, it is stated by those well posted by careful experiments that common land plaster or gypsum increases the crop of potatoes greatly. By mixing one pound of Paris Green with three pecks [3 pecks = 24 dry quarts], by measure, of land plaster, and sifting it on the hills, within 24 or 48 hours every bug will be dead. Three pecks will be sufficient for three-fourths of an acre; and a very active man can spread it in one morning before the dew is off. The cheapest sower is a small box, or, better still, a four-quart wooden measure. Nail to it a wood handle, knock out the bottom and tack on the coarsest brass wire milk strainer cloth. . . . It may be necessary to go over the field three or four times during the season. Farmers need not fear that the green [Paris green] will injure the vines or bulb. I have applied it just five times as strong as above described, and two families of fifteen persons used the potatoes from July fifteen to November without any ill effects; this was in 1870. Prof. Kedzie of the Michigan Agricultural College says that to injure the bulb the Paris Green would first kill the vine, which I have never seen done.

The use of expert testimony, as in this letter, coupled with claims of easy application, helped convince many a farmer to sift arsenic onto his potato plants. But strong criticisms about this dangerously toxic pesticide still persisted. And while reports of praise dominated the pages of *Country Gentleman*, news of arsenic accidents and criticisms of arsenic use also appeared, such as in the following letter to the editor of *Country Gentleman*, dated July 19, 1877:

> Is it safe or prudent to make use of this most poisonous substance, even to its application for the purpose of destroying bugs, insects, &c, preying on growing plants? That there is a need of the utmost caution in its use, where used needs little demonstration to any one who knows the composition of Paris Green, arsenic being the active principle. I notice by my paper that a farmer of Colchester, Conn., lost two head of cattle by their eating potato vines, where Paris Green had been sprinkled upon them for the purpose of destroying the potato bugs; also another farmer in North Stonington lost a yoke of oxen from the same cause.

The author of this letter, M. H. White, also illustrated that the dangers of arsenic were already well known in Europe and cites Dr. Edmund Davy's paper from the *London, Dublin and Edinburgh Philosophical Magazine*, dated August 1859, eighteen years earlier: "[P]lants may . . . take up a considerable quantity of this substance, though its proportion in the soil may be very small. Further, as arsenic is well known to be an accumulative poison, by the continued use of vegetables containing even a minute proportion of arsenic, that substance may collect in the system till its amount may exercise an injurious effect on the health of men and animals."[1]

So, even in the *Country Gentleman*, which was the magazine most supportive of arsenic in the day, arsenic use was strongly opposed in letters to the editor. But although legitimate and widely shared in the farm community, White's fears and those of

thousands of populist farmers didn't prevent the formerly anti-arsenic magazines from finally supporting it.

After the *Pacific Rural Press* reversed its attitude toward toxic pesticides, the publisher printed two advertisements in consecutive issues in 1882 for London purple, a waste dye-fixing reagent from England's garment industry that contained arsenic. Thereafter, the magazine printed editorial, scientific, and farmer testimonies in favor of arsenic. To justify its reversed position on arsenic *Pacific Rural Press* claimed that previous and current reports of arsenic accidents and ineffectiveness were due to farmer errors and not a result of the inherent dangers of arsenic. By 1889, arsenic ads appeared frequently in the formerly populist magazine.

A similar about-face on the issue of arsenic took place in almost all the rural journals, including the *Farm Journal*, as the ads on the following pages show. As previously noted, the control and sometimes ownership of many of the farm magazines shifted from farmer subscribers and populist publishers to an ownership

Country Gentleman **supported arsenic early, as the editorial at right, from May 16, 1872, illustrates. Though the applicator's mouth tasted like a copper mine and his nose and throat ached, the** *Country Gentleman* **continued to promote arsenic as the best destroyer of Colorado potato beetles. Most of the rural journals eventually sold arsenic as the best killer for many pests and ended up trivializing its dangers.**

The best agent for the destruction of the potato beetle has been found to be Paris green, a mineral paint compounded of copper and arsenic. Last year's experience added to our valuable information, the fact that Paris green, mixed with fifteen or twenty times its weight of calcined plaster or even flour, was nearly as effectual as the pure mineral. The compound when dusted on the vines by the means of a common dredging box, kills at a blow the larvæ of the potato bug, but the old ones, the regular mature hard-shelled parents, resist the poison some hours. Great care should be taken in the use of this remedy, as it is a virulent poison. This morning I mixed a pound of the Paris green with ten pounds of calcined plaster, and then dusted a half pint of the mixture over a hundred hills of potatoes, more or less, all the while taking care to keep to the windward and to avoid the dust, but nevertheless my mouth tastes like a copper mine and my nostrils feel a good deal irritated.

that courted advertising dollars, and the chemical corporations had advertising dollars. While the populist papers and magazines held out for more than ten years after *Country Gentleman* began promoting arsenic, the pressure to focus on advertising revenue instead of populist advocacy and land stewardship forced the more idealistic publishers to change their approach and ultimately sell out. But they still wanted to be perceived as the populist press that protected the interests of farmers; unfortunately, their allegiances had shifted to the manufacturers.

Arsenic had long been used as a paint pigment because its metallic strength guaranteed long life and it was poisonous to mold and insects. While highly purified arsenic was usually used in medicines, the arsenic applied on farm fields and as house paint usually came from industrial waste. Most of the waste arsenic came from garment dye water or flue ash from copper, iron, or lead smelters. The arsenic preparations Paris green and Scheele's green tinted most of the green paints in the nineteenth and early twentieth centuries and were derived from flue ash and mining. After they were shown to be effective killers of potato bugs, the Sherwin-Williams paint company, Hoechst Chemical and Dye Works, Graselli Chemical Company, and several other chemical merchants aggressively marketed Paris green and Scheele's green as deadly pesticides.

The garment industry had long disposed of waste dye-fixing mordents, like London purple, by dumping them into the Atlantic Ocean.[2] However, after discovering the pesticidal nature of Paris green and London purple, the large-scale commercial farmers instead began to dump huge amounts of these poisons on the U.S. food and fiber supply. Powdered Paris green, which worked well as a dust, did not mix well as a

"The Favorite Potato Bug Exterminator"—a flour-sifter pesticide duster from J. S. Eddy & Sons, Eagle Mills, N.Y. From *Farm Journal*, 1884.

liquid in the newly developed wet spray pumps. London purple, on the other hand, stayed suspended in a liquid form better than the greens. By the 1890s, London purple had become the arsenic of choice.[3]

London purple left telltale purple stains on plants, which was advertised as a convenience—a sort of early-warning device. When the plants were no longer purple, it was time to spray again. However, because it stayed in suspension, farmers complained that plants frequently absorbed as much London purple as the pests and that many crops were killed. Within ten years, almost no one was using London purple anymore because of this problem, but the advertisements continued for several more years

In 1869 Leopold Trouvelot, a French-born Harvard astronomer with an abiding interest in silk worms, imported the gypsy moth to breed with silk-producing moths in an effort to get higher yields of silk from a larger array of plants than the mulberry tree, which was the only plant silk worms would feed on. The experiment was a failure since the moths were not close enough relatives to mate. Several moths escaped and in the last decade of the 1800s, the gypsy-moth caterpillar defoliated much of New England's fruit, shade, and forest trees. Nothing seemed able to kill it, including Paris green and London purple, until researchers tried applying lead arsenic. As one can imagine, this poisonous concoction was very deadly, and the lead in the formulations helped the arsenic stick to the leaves. After the gypsy moth triumph the chemical firms launched a nonstop propaganda campaign that trumpeted lead arsenic's killing power, its cheap price, and its success against gypsy moths, Colorado potato beetles, and several other pests. Many farmers growing wholesale crops for market began to use large amounts and ultimately applied more lead arsenic products than any other pesticides.[4]

Since most farmers and householders in the 1800s knew about the dangers of arsenic and lead, it is often hard for people today to understand how such well-known killers could have been so widely applied and accepted. But it is helpful to recount that arsenic was not only a pesticide; it could be found everywhere and in almost everything in the 1800s that was manu-factured. More than a hundred products had arsenic as an ingredient, including but not limited to wallpapers, paint, cosmetics, medicines, fruit preserves, candies, bakery goods, cocoa, sweetmeats, tobacco, book coverings, book bindings, lampshades, decorated plates, toy decorations, cardboard boxes, labels, carpets, watercolors, dental fillings, stockings, veils, stuffed animals, and candles. And this widespread use of arsenic persisted until the 1930s. Besides food and manufactured goods, arsenic was common in many medicine chests in the form of a heroic patent medicine known as Fowler's Solution, which was concocted in 1786 by Thomas Fowler, a London physician. Fowler's Solution was alleged to cure everything, and it was widely used until the 1930s, even though people feared its effects and became chronically ill or died from taking it.[5]

Though arsenic was controversial and feared by the general populace, its effectiveness as a medicine and as a pest destroyer enabled the manufacturers to develop several arguments in its favor. A common one was the following: "If it is not too dangerous to be taken as a medicine or used in cosmetics, then how could it be more dangerous to dust on food plants a small amount that the rain will wash off?"

Arsenic advocates frequently quoted university agriculture experts, such as the professor named Kedzie mentioned in the letter to the *Country Gentleman* above. These references to academic "experts" provided scientific credibility for their propaganda. This claim of credibility using Kedzie as the academic prop persisted even though Dr. Kedzie had grown lukewarm to using arsenic well before 1890, and many other scientists had begun to express concern about the safety and effectiveness of arsenic as a pesticide.

Early arsenic ads. Top ad from 1870s *Country Gentleman*. London purple ad from *Farm Journal*, June 1881. Paris purple ad from *Farm Journal*, May 1881.

Below: a contrast in adjacent ads. The London purple ad promotes a chemical solution, while the "injurious insects" ad is for a pest identification manual promoting a natural solution. Both are from *Pacific Rural Press*, 1889.

Arsenic and other pest-control products advertised in *Farm Journal* in the last years of the 1890s indicate how the promises of the chemists were pervasive and persuasive. Bug Death! Death Dust! Quick Destruction. All advertisers promised amazing results for their products.

But the arsenic advocates continued their saturation advertising and promotions using any scrap of credibility they could employ, however suspicious the claims might be.

Another regularly cited claim of safety told of the existence of a population of Austrians in the former duchy of Styria who like many ancient farmers ate dirt to see whether it was sweet (alkaline) or sour (acidic). The Styrian dirt was rich in arsenic but apparently caused no ill effects to the farmers when they tasted it. The chemical-corporation experts argued that the Styrian example proved that arsenic was safe enough to eat. But scientific investigators finally discovered, after analyzing the Styrian farmers' stomach remains, that they in fact ate very small amounts of dirt that contained some arsenic, which passed through their bodies without being assimilated. This claim of arsenic safety continued to be used long after independent scientists had debunked the Styrian myth.[6]

All of the promotional tricks for arsenic certainly increased sales. However, arsenic's effectiveness as a killer of almost all farm and garden pests provided its most convincing advertisement. For the "show me" farmers and householders, the proof was in the results: arsenic killed almost everything. Unfortunately, that included many thousands of farm children.[7] Nevertheless, since this toxic metal also proved very effective against so many hard-to-kill pests, many users overlooked its lethal dangers to their children. Instead of eliminating arsenic from their chemical sheds, farmers threatened their children with harsh punishment if they went anywhere near the shed.

Armed with testimonials and editorial arguments, companies aggressively promoted the use of arsenic against everything that crawled, flew, hopped, or rotted, in spite of the known risks that emerged in the wide-spread food poisonings during the 1890s. As a result of these sales campaigns, arsenic became the most widely

used pesticide from 1880 until the 1950s, greatly outdistancing total DDT usage in the United States.

Though DDT is often seen as the first war-related pesticide, arsenic ranks as an even earlier war hero. Foresters and landscapers used arsenic in the gypsy-moth wars before the turn of the century, and then the Army used it somewhat effectively to combat syphilis and several other pests in World War I. Because it was felt to be indispensable to the war effort and war reconstruction markets, high arsenic use was tolerated on food and fiber crops both during World War I and afterward. Arsenic's warrior status also helped prevent a meaningful public dialogue in the United States about its regulation until quite some time after World War I had ended.

The most important propaganda breakthrough occurred when advertisers realized that they had to trivialize arsenic's public health and environmental damage. Instead of discussing real concerns about health and safety and the poisoning of farm kids, the chemical companies ignored those issues and focused their propaganda and advertisements on the toxic

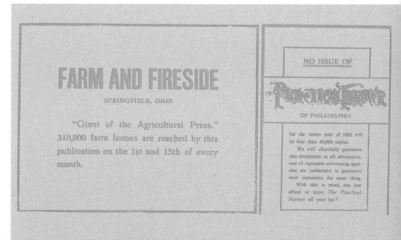

From *Agricultural Advertising*, 1903.

potency and consistency of their products. They developed a strategy that highlighted successful farmers who had won county and state fair prizes for perfect fruits, vegetables, grain, cotton, or tobacco by spraying arsenic on them.

They began portraying arsenic and the other farm poisons as tools or weapons in the farmer's war to survive. The chemical companies and the advertisers successfully magnified farmers' fear of pests and dramatized the problems in such a manner that the farmers felt forced by bankers and the fear of crop loss into using this terrible poison on their farms and families. That the chemical advertisers were able to sell these poisons to farmers already frightened of arsenic and lead was pure advertising genius. It also required deceptive, manipulative, dishonest, and endless promotions.

The availability of such massive amounts of cheap by-product arsenic and lead helped advertisers win the hearts, minds, and pocketbooks of farmers. Arsenic, for example, was much cheaper than the pyrethrum pesticides derived from chrysanthemum flowers, like Buhach powder. Ads and propaganda for arsenic appeared everywhere, whereas the advertising scope and frequency of Buhach powder was much more limited. The conflict between the biological approaches and the chemical solutions became much more dramatically unbalanced in favor of cheap arsenic, especially in the hard economic times that dominated the close of the nineteenth century. The chemical battle with bugs had begun, and arsenic manufacturers rapidly shifted their energies from the medicine cabinet to the farm.

The new market on the farms proved to be a boon to chemical-supply houses, which previously had sold thousands of pounds of arsenic for medical applications per year but whose sales were sagging. Now they would sell millions of pounds of waste arsenic instead. One of the major innovators in arsenic products was the German chemical corporation Hoechst. After arsenic proved to be an effective pesticide, Hoechst alone sold hundreds of thousands of tons of arsenic to control farm pests.

Some of the earliest toxin marketing successes were fashioned by German chemical and mining firms. In the late 1800s, German chemical and fertilizer firms mimicked the successful model of Rockefeller's Trust and began to form huge fertilizer syndicates and chemical cartels. Their interlocking connections enabled them to control the manufacture and marketing of farm chemicals before the end of the century. Their collusion and resultant wealth also enabled them to advertise constantly.

The German and Swiss chemical corporations then began conspiring with each other and ultimately divided among themselves the world market for organic and inorganic chemicals, dyes, pesticides, pharmaceuticals, and chemical-industrial processes.[8] Copying the German Kaliwerks model, the pesticide manufacturers advertised constantly. Copying earlier fertilizer advertisers, they also advertised irresponsibly.

This collection of ads on the same page was common by the early 1890s. Clearly the message was SPRAY. From the *Farm Journal*, May 1892.

Even the *Ladies' Home Journal* in 1892 encouraged its readers to spray. Cyrus H. K. Curtiss, the publisher of the *Ladies' Home Journal*, also owned the *Country Gentleman*, the leading journal in pushing Liebig's theories and ads for chemical fertilizer, arsenic, and lead. Notice that the *Ladies' Home Journal* spray ad is the same ad as the one published in the 1892 *Farm Journal* above.

THE BEETLE
POTATO DUSTER

A mobile spray rig that was used to apply Paris green on the potato crops. It is a forerunner to the tractor-driven spray rigs that are common today. A late 1800s Leggett and Brother implement pamphlet from the University of California at Davis Shields Library Special Collections.

The late 1800s was a time when pesticide, fertilizer, and patent-medicine manufacturers enjoyed an unregulated marketplace. The corporations and their scientists claimed whatever they wanted for their products, since no regulatory agency existed to evaluate the claims of advertisers or prosecute offenders. Customers would have to wait some years before a federal pesticide act was created, and even longer for a trade commission to take any action. Unfortunately, their wait was mostly in vain, as we shall see, since government agencies have consistently failed to protect consumers or farmers.

After the success of arsenic pesticides became well known in the farm community, the ethics and practice of pest-control self-sufficiency began to slowly erode, just as fertilizer self-sufficiency eroded after the promotional campaign for Peruvian guano. The attitude of almost all of the farm journals changed with respect to the scientists and their products. By the mid-1890s Liebig's scientific form of chemical farming was being accepted by a growing number of farmers.

By the turn of the century chemical corporations, including Bayer, Hoechst, DuPont, Grace, Ciba, Geigy, Sandoz, Monsanto, Dow, BASF, Nobel, Graselli, ICI, Rhone-Polenc, American Cyanamid, Wheeler, Reynolds and Stauffer, Standard Oil, Caselli, AGFA,

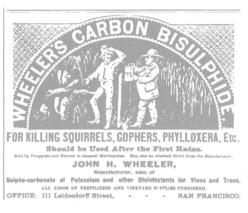

An early ad for carbon bisulphide. From the *Pacific Rural Press*, October 28, 1882.

and literally all the tobacco corporations, began saturation advertising campaigns for their chemical concoctions in U.S. farm journals.

At about the same time that arsenic and lead pesticides became more commonly accepted by the large-scale farmers, the resistance to poisons by many growers lessened, and the corporations began to peddle additional toxic substances to farmers. Numerous products were introduced and promoted, including carbon bisulphide.

Ads appeared often for carbon bisulphide, one of the first successfully advertised fumigants. The Wheeler Company began marketing this chemical in California in 1882. A very effective but dangerous pesticide, it was most commonly employed as a soil fumigant for deep soil diseases and to kill soil worms (nematodes) on grapevines and fruit and nut trees. The advertising campaign for this product was the earliest intensely repetitive sales pitch for a pesticide. Carbon bisulphide actually received advertising space in several populist magazines much earlier than arsenic products, since they opposed arsenic.

Farmers also destroyed plant lice (aphids) with carbon bisulphide, as described in the following 1896 *Farm Journal* editorial from Professor Smith of the New Jersey Experiment Station:

> The melon louse can be destroyed . . . by evaporating a small quantity of bisulphide of carbon over the plants. As soon as a hill is seen to be infested put one or two teaspoonfuls of the liquid on a bit of sponge or rag tied to a stick. Stand this up in the center of the hill and put a cover of paper or other material over all to keep the fumes near the plant. . . . The fumes being heavy sink down and find every insect. . . . Those who have late pickles now should try this remedy, and all melon, cantaloupe and pickle growers should be prepared to give it a trial next June when the lice first make their appearance.

In spite of such glowing advertisements, editorials, and endorsements by the Bureau of Agriculture and the state experiment stations, many farmers refused to use this pesticide because explosions frequently occurred when the applicator hit buried rocks or a chunk of steel or iron. In addition to its explosive and pest-killing abilities, all the tests on carbon bisulphide indicate that it is a neurotoxin and fetotoxin that causes thyroid and adrenal changes and heart, liver, and kidney damage.[9] In spite of its considerable toxic drawbacks, it is today a widely used fumigant for insect control in stored grain. When used as a grain fumigant it is usually combined with carbon tetrachloride to reduce fire and explosive hazards.[10]

Not all of the advertised pesticides were as deadly as the advertisers claimed; some were worthless. In the last decades of the nineteenth century, advertisements for Ongerth's Insecticide Powder and Ongerth's Liquid Tree Protector were common, and these products gained considerable popularity for a time. After years of complaints about their effectiveness, both state and federal government tests found them to be totally ineffective. They were quack products, but even after long-term public criticism, farmers continued to buy them. In 1890, *Insect Life* reported that a farmer from Alameda, California, applied some of Ongerth's compound to his trees, and in a short time more than one hundred of his prize citrus trees died. In many cases the active ingredient in products such as Ongerth's remains unknown, because they were deemed to be company secrets—and held as intellectual property

In the fine print, Taylor claimed that his carbon bisulphide was "indorsed" by the U.S. Department of Agriculture and by the state experiment stations. From the *Southern Cultivator and Industrial Journal*, July 1895.

This ad for Ongerth's appeared in the Pacific Rural Press, on February 5, 1885.

These ads were gleaned from The Pacific Rural Press, Farm Journal and Country Gentleman, from 1885 to 1899. This page greatly resembles a typical advertisement page from this era

rights. Subsequent tests by the state and federal bureaus of agriculture determined that several other "miracle" products were also worthless.

Though pesticides and fertilizers were aggressively advertised and promoted, farmers commonly were both skeptical and fearful of them because of dangerous and useless products like those promoted by Wheeler, Ongerth, Liebig, Mapes, chemical corporations, the farm journals, and the U.S. Bureau of Agriculture. That skepticism about synthetic fertilizers and pesticides continued for decades, especially among farmers operating small and medium-sized farms. Consequently they consumed very few of these products. The large-scale farmers were also concerned about shoddy fertilizers and poorly fabricated pesticides and eventually forced government regulators to set up standards for chemical products.

POISONED GRAPES ON SALE

QUANTITIES OF THEM DESTROYED BY HEALTH OFFICERS.

A SOLUTION WHICH FARMERS IN SOME SECTIONS. HAVE USED TO DESTROY THE PHYLLOXERA HAS RENDERED THE FRUIT UNFIT TO EAT.

On Wednesday a citizen took several bunches of Concord grapes to the Health Department and called attention to a green substance that clung to the stems of the bunches directly under the grapes. It was pale green and looked like paris green. The grapes were sent to Mr. Ernest Lederle, acting chemist of the board, for analysis. He found that the stuff was sulphate of copper, which is poisonous, and paris green possibly mixed with it. An officer of the sani-

An 1891 report of grapes poisoned with arsenic from the *New York Times*.

Chapter 11

ARSENIC REGULATION, THE FDA, AND OTHER REGULATORY HOAXES

Lead arsenate and calcium arsenate almost completely replaced Paris green, London purple, and the other arsenic poisons by 1900. Both became more widely used on the larger farms, with lead arsenate preferred for fruits and vegetables and calcium arsenate used on cotton. The increased use of arsenic resulted in frequent reports of poisonings in the press.

After multiple poisonings in the mid-1890s, a storm of opposition erupted in Europe and the United States protesting the excessive presence of arsenic and lead residues on farm produce. Finally, in 1903, the British Royal Commission set limits for residues of arsenic and lead both on food and in beer. Arsenic residues could not exceed 0.01 grain per pound or gallon of product. Lead residues (twice as heavy as arsenic) were limited to 0.02 grains per pound or gallon. These Royal Commission limits came to be known as the "world tolerance" for arsenic and lead in food and drink. Though they were accepted by most countries around the world, U.S. chemical corporations fought these tolerance limits for the next sixty years, and U.S. regulators and government officials failed to support the limits.[1]

Congress created the Food and Drug Administration (FDA) in 1906 to deal with the national uproar about the health and safety of food and as a muted response to the Royal Commission's international arsenic and lead tolerances. Consumers and legislators also reacted to the problems with filthy meat, dramatized in Upton Sinclair's book *The Jungle*, about the dangerous and deadly conditions in the meatpacking industry.[2]

Unfortunately, the FDA was placed under the jurisdiction of the United States Department of Agriculture until 1940, so its effectiveness in regulating the safety of the food supply, or the level of arsenic and lead residues, was literally nil. The Department of Agriculture was in favor of pesticide use, and the FDA avoided any confrontation with its lead agency, large-scale farmers, and the food-broker lobby, since all of them defended pesticides.[3]

While Congress refused to address concerns about pesticide safety, it did respond aggressively to farmer criticisms of pesticide quality. Ultimately, the complaints about arsenic from large-scale farmers forced the creation of the 1910 Pesticide Act, the first federal pesticide regulation in the United States. The lack of consistently dependable chemical products had caused farmers to grow increasingly more suspicious of the advertising claims made by pesticide salesmen. Despite the deadly nature of arsenic, numerous diluted, inconsistent, and ineffective (that is, nonlethal) products flooded the marketplace. This product inconsistency resulted in one batch of a pesticide being incredibly poisonous while another formulation might contain hardly any arsenic or lead at all.

Congress and the regulators promised that the Pesticide Act would address consumers' health and safety concerns as well as farmer complaints about shoddy ingredients and dangerous formulations. Many of the larger farmers who complained about bogus products had political connections. Consequently, the focus of the law was primarily on farmers' concerns.

Even though dozens of reports of serious food poisonings due to arsenic had appeared in newspapers and magazines on both sides of the Atlantic since the 1890s, the law failed to protect the public from the health dangers of arsenic and lead in their food. The Pesticide Act was only a truth-in-advertising law, which was passed to deal with arsenic-manufacturing rip-offs. The law merely guaranteed farmers a certain level of toxic potency in pesticide formulations—nothing more—and, as we shall see, the arsenic poisoning of consumers continued for five more decades.

In 1912 Congress established the Federal Trade Commission (FTC) in response to several thousand complaints of fraudulent advertising (many of which came from farmers, who continued to be angry about arsenic and other pesticide and fertilizer formulations). Congress promised that, finally, advertising and formulations would be regulated. Yet, as with all the laws passed to regulate industry, agriculture, and commerce, this law produced nothing more than an anemic and almost totally ineffectual commission. After a decade of agitation to get a fair-trade commission, consumers and farmers were left with few, if any, protections. The corporations were left to exploit the consumer at will.

Capitalizing on the ineffectual FDA and FTC regulatory legislation, the chemical corporations amped up their ad campaigns and entered a period of excessive and deceptive chemical advertising. These propaganda excesses dominated the farm and urban journals, in spite of the fact that thousands of arsenic and lead poisonings had occurred in America and Europe and consumer concern about residues was high.

However, through the struggle for and the creation of the FDA and FTC, the public was lulled into believing that these new regulatory agencies would be able to finally protect them from misleading ads and dangerous and useless products. So, even though the ads were just as deceptive as before and the products just as dangerous and useless, the public believed that the new regulations protected them. Sadly the public, including the farmers, was wrong—nothing had changed.

The large-scale farmers by now were hooked on arsenic and other quick-fix chemicals for their survival. At the same time, the rural magazines had become hooked on lucrative chemical advertisements for their survival. Ignoring the widespread health and safety concerns of the public, the editors and the ad makers further intensified their strategy of refusing to deal directly with farmers' or consumers' fears of poisoning. By the 1910s, the farm journals focused almost exclusively on providing chemical farm advice, addressing the farmers' most important economic fears and insecurities and ridiculing and trivializing the health and environmental concerns of both farmers and consumers.

Before the turn of the century, several chemical advertisers had begun to use a four-part sales model to promote their products. This model included magazine editorials by popular editors and writers, scientific testimonials (from government agencies and universities), farmer testimonials from influential farmers, and clever advertisements and promotions. Though more than a hundred years old, this four-part advertising model still dominates the farm journals today. The effectiveness of such long-term promotions is indisputable. Many U.S. farmers (as well as many consumers) still passionately cling to the belief that farm and home chemicals are not only necessary but also safe, in spite of considerable evidence to the contrary.

Clarence Poe, in the article reproduced below, argued that the government, in cooperation with universities, chemical corporations, progressive farmers, and journalists, had dramatically improved farming in America. Unfortunately, the false and puffed-up advertising and promotion of toxic and often useless products since the 1840s by universities, chemical corporations, wealthy farmers, and journalists was not discussed by Poe or most of the other journalists of his time.

Because of publishers, editors, publicists, scientists, large-scale farmers, and advertisers, the public was victimized by highly toxic, untested chemicals that were invented in the 1800s. The 1906 Food and Drug Administration, the 1910 Pesticide Act, and the 1912 Federal Trade Commission, which were supposed to provide protection from these poisons, refused to protect consumers or farmers or demand that the poisons be tested or standardized. Even when the FDA and FTC were first founded and acts were passed at the beginning of the twentieth century, they ignored the consumers but protected the large chemical corporations and large farmers. This pattern we will see repeated again and again throughout the book.

A typical ad page from *Better Fruit*, June 15, 1911.

THE GOVERNMENT AND THE NEW FARMER

THE MOST EFFICIENT DEPARTMENT OF AGRICULTURE IN THE WORLD, ADDING MILLIONS OF DOLLARS TO OUR NATIONAL WEALTH BY NEW METHODS OF TREATING SOILS AND SEEDS, AND SAVING MILLIONS BY THE DESTRUCTION OF PESTS—DOUBLING THE CROPS OF THE AMERICAN FARM

BY

CLARENCE H. POE

EDITOR OF "THE PROGRESSIVE FARMER"

I T is only after waiting ten thousand years that agriculture is becoming in our own time a really well-organized and scientific industry.

Its change from an industry requiring only physical strength to one requiring skill and trained intelligence means that it has now acquired a dignity which it has never had before.

And in bringing farming in America to this higher station, perhaps no other force has been of so much importance as the United States Department of Agriculture, which is but the outgrowth of the energy and enterprise of the American soil-tiller, discontented with outworn practices and providing in this branch of the government a gigantic scheme of cooperation for better things.

Clarence Poe was the editor of the *Progressive Farmer*. He wrote the introduction for the book *The Government and the New Farmer*.

EXTRA — EXTRA

WAR DECLARED

MILDEW INVADES GRAPES

EVERYWHERE—EVERYDAY—Due to the incessant rains this winter, 1938 promises to be a real "mildew year." Vineyardists everywhere are making plans to dust carefully and often this season.

Declaring war on grape mildew. From the *California Cultivator*, May 1938.

PREPARE NOW!

Delay may be costly. Remember—"an ounce of prevention is worth a pound of cure." Go to your dealer today and buy

ANCHOR or SIGNAL

100% Pure Sublimed Sulphur

SIGNAL

is a perfect blend of 75% Sublimed Flowers Sulphur and 25% Superfine Adhesive Sulphur. IT COSTS NO MORE than ordinary ground sulphur

Chapter 12

THE FIRST ADVERTISED WAR ON THE FARMS

Ever since the discovery of guano, constant propaganda by magazine editors, chemical merchants, and scientific experts had been prodding farmers to use commercial sources of nitrogen instead of animal manures, beans, peas, and grasses. The scientific experts, the editors, and the crafty ad agencies all tried to convince farmers that growing fertilizer crops took too much land out of cash production and thus was neither economical nor efficient.

The advertising clique also began to argue that not enough dairies and other animal operations existed to supply the quantity of fertilizer that would be needed for the expanding population. Even if there were enough, the propagandists claimed that the transportation costs for such antiquated and bulky fertilizers as manure, compost, and recycled wastes from the cities were too expensive. Hundreds of ads and editorials in farm journals and company brochures cited tests that claimed to prove the superiority of mixed synthetic fertilizers over natural manures and cover crops, and to illustrate that it took too much labor to apply the bulkier manures and grow fertilizer crops. Still, a large majority of U.S. farmers continued to depend mostly on manure, compost, fertilizer crops, and a few mineral nutrients for several more decades.

Nitrogen accounts for more than three-quarters of our atmospheric gases. Nitrogen is also one of the most important fertilizing elements. Beans, peas, clovers, alfalfa, and all the other legumes accumulate nitrogen from the air and transfer it naturally to the soil via their leaves, stems, and roots. Animal manures, including ours, also contain significant amounts of nitrogen as well as phosphorus, potassium, and

several micronutrients. For untold millennia farmers planted beans, peas, and grasses and applied animal manures to supply nitrogen and other nutrients for plant growth.

In the last half of the 1800s, however, the Malthusians in Europe and America (who were apoplectic about a worldwide population explosion and consequent food shortage) and many of the leading fertilizer experts worried that the planet was dangerously short on nitrogen fertilizer. All of these somewhat frantic experts pointed to the fact that the high-quality Peruvian guanos had been used up and that sodium nitrate from politically unstable Chile was the only significant supply of munition, industrial, and agricultural nitrogen.

In 1775 Antoine Laurent Lavoisier had demonstrated that air was a mixture of oxygen and another substance that he called nitrogen. In the 1890s, spurred by Lavoisier's discoveries and Liebig's theories, a number of German chemical companies, with government backing, launched a search for techniques that would enable the commercial-scale manufacture of synthetic nitrogen. The project was made a national priority because the German government and industry

This cyanide gas pesticide was a waste product of the first synthetic nitrogen gas production plant at Niagara Falls, Canada. American Cyanamid promoted cyanide as the most efficient fumigator. From *California Cultivator*, July 9, 1914.

anticipated that there would be a blockade of the Chilean nitrate deposits in the event of war. The research effort was wildly successful. German scientists and inventors discovered three processes for creating nitrogen, two of which became especially successful and very profitable.[1]

Before the turn of the twentieth century (1898), Fritz Rothe, working for Adolph Frank and Nicodem Caro, demonstrated that nitrogen passed over calcium carbide in an electric furnace heated to 1,000 degrees Centigrade produced calcium cyanamide. By 1907 plants for producing nitrogen in this manner were established in Europe and America. The Frank-Caro process produced the first marketable synthetic nitrogen fertilizer.

The Frank-Caro ammonium cyanide process was widely used in the first four decades of the twentieth century. In 1907, Frank Washburn and the tobacco magnate James Duke purchased the right to use this process in America and founded American Cyanamid. They tried to develop a site on the U.S. side of the Niagara Falls but could not obtain permits. They finally located their plant just across the border in Canada and used the electrical generating capacity of the falls to produce the heat required to liquefy atmospheric gas and capture nitrogen. In 1909 American Cyanamid began the commercial production of cyanamide for use as a fertilizer and cyanogas for use as a pesticide.[2]

The cyanamide nitrogen process also produced by-product cyanide gases. One form, cyanogas (advertised at left), was used as an important pesticide in the early 1900s. A much stronger formulation, Zyklon B (advertised below), became the chemical used as the "Final Solution" gas to exterminate Jews, Slavs, intellectuals, leftists, and Gypsies during the Third Reich. More than half of the twenty million people "exterminated" by the Nazi death machine died from this by-product of synthetic nitrogen production.

France's Henri Le Chaterlier established the temperature and pressure under which the synthesis of ammonia could be achieved from hydrogen and nitrogen in the presence of a catalyst. In 1901 he took out a patent but failed to move forward because of an explosion. In 1908 Fritz Haber carried Le Chaterlier's process further using osmium as the catalyst.

Fritz Haber worked for the chemical company Badische Analin und Soda Fabrik (BASF). After

This much more deadly waste-product cyanide gas from nitrogen production killed millions of Jews, Gypsies, Slavs, intellectuals, and Marxists in Hitler's concentration camps. From Joseph Borkin, *The Crime and Punishment of I. G. Farben* (New York: The Free Press, 1978).

Haber's initial discovery, the corporate leadership at BASF recognized that his process could be accomplished economically only on a large scale. They estimated that a company or country would have to invest in enormous industrial plant capacity and have access to inexpensive or by-product sources of feedstock energy, such as natural gas or hydrogen.

Haber collaborated with Robert Bosch on the industrialization of his nitrogen discoveries at BASF. By 1913 they had developed the technical capability to produce large quantities of synthetic nitrogen fertilizer and ultimately synthetic nitrogen bombs. By capturing waste hydrogen or natural gas from Germany's coke, gas, and coal plants, they could use it to generate the intense heat needed to turn atmospheric gases into a liquid form, after which they could separate the nitrogen from the other elements. Both Haber and Bosch later received the Nobel Prize for their breakthrough discoveries in nitrogen. Shortly after the end of World War I this process became the world's most important source of synthetic nitrogen for both fertilizer and explosives production. It continues to be the most important process today even though it requires huge amounts of energy.[3]

During the first decade of the twentieth century, German merchant and mining interests were indeed blockaded from the nitrate deposits in Chile by the British, who were using the Falkland Islands as their port of supply. Because of the blockade, both of the newly industrialized forms of synthetic nitrogen replaced Chilean nitrate in the German war machine's demand for explosives and fertilizer. Over time, both agriculture and the military came to depend almost exclusively on the new nitrogen supplies.

Haber, Bosch, synthetic nitrogen bombs, fertilizer, and newly discovered chemicals were widely praised heroes in Germany during World War I. Without their discoveries Germany would not have had enough nitrogen for food. When Bosch developed an industrial scale plant to convert synthetic ammonia to nitric acid in 1915, Germany was able to manufacture large inventories of explosives. The resulting bombs killed and wounded more U.S. soldiers than all of the other chemicals combined.[4] Just as we were proud of all our scientific breakthroughs that had military applications, so were the Germans.

An early ad for hydrocyanic acid for use as a pesticide to kill scale, which as early as the 1880s had been controlled organically with vedalia beetle predators. From *Pacific Rural Press*, July 21, 1924.

This ad is for the
synthetic nitrogen
that was developed
by Haber and BASF
and sold as the
scientific solution to
soil deficiency
problems. The
intellectual book-
farmer guy with the
glasses was Synthetic
Sam. He appeared in
hundreds of ads from
the 1920s through the
1930s. BASF was a
major marketer and
propagandizer of
synthetic nitrogen
fertilizer. This ad,
like many others,
emphasized low price.
From *Pacific Rural
Press* and the
California Cultivator,
1924. From farmer
collection.

Also important, both as war materiel and as a
fertilizer, was the BASF development of purified
sulfuric acid in 1888. In 1910 the company
industrialized the process, and almost immediately
thereafter, New Jersey Zinc and General Chemical Co.
built plants in the United States as BASF licensees.

In the 1920s and 1930s I. G. Farben screened Zyklon B
and most of the other toxic chemicals for their potential
contribution to Hitler's war and extermination effort.
Zyklon B was selected as the most effective
exterminator gas and diluted forms were highly
effective against insects and crop diseases. Fritz Haber,
who engineered the nitrogen breakthrough, directed
this screening. He selected other chemicals like Serin
and Tabun, both organophosphates, as the most
effective fast-acting nerve poisons on both humans
and insects. Chloropicrin was selected as the most
effective and economical tear gas, and a very deadly
fumigant, and so it went. By the time Germany went to
war again in the late 1930s, the Nazi government
wanted answers to the critical questions of how one
could injure, immobilize, demoralize, and exterminate
one's enemies whether they were Allied soldiers or
insect pests.

Many of the newly synthesized chemicals and waste
products were used as antipersonnel weapons for the
first time in World War I. This included chloropicrin
(tear gas and strawberry, carrot, and tree fruit
fumigation gas), mustard gas, chlorine, Zyklon
(cyanide gas), sulfuric acid, and phosgene gas. The
new weaponry of chemical warfare was used by both
sides in the war and resulted in at least 100,000 deaths
and more than a million serious injuries. After the war,
many of these chemicals were used on the farms.

During World War I, U.S. advertisers began to use the
war theme in farm chemical advertisements. "Farm for
Profit and Help the U.S. Win the War" was an eye-
catching and patriotic phrase in the ad created by

Hauser Fertilizer Co. in 1918. Around this same time, other advertisers began increasingly to employ war themes to characterize the struggle between farmers and pests. These World War I campaigns signaled the start of advertising that depicted farming as warfare and chemicals as weapons to be used in the "struggle with nature."

The German patents for nitrogen, sulfuric acid, and other chemical discoveries were no longer in force following the war, and the other chemical corporations immediately seized them. The patents, however, were almost worthless without the process knowledge that only the I. G. Farben corporations possessed. After the Great War ended, the German chemical corporations issued licenses to use processes for making nitrogen, purified sulfuric acid, and other chemicals. This need for technical expertise and process information enabled the German chemical corporations to recover quickly following the war. Another factor that helped their resurgence was the support they received from several U.S. chemical and industrial firms, including Standard Oil, DuPont, Dow, General Analine, Ford, and Allied Chemical.

Though the Frank-Caro ammonium-cyanide-nitrogen method was successful and continued to be used for several decades, most of the American synthetic-nitrogen plants built from the 1920s to the 1940s used the Haber-Bosch process, as did a majority of synthetic-nitrogen plants built in other parts of the world. At that time, there was an abundance of cheap by-product energy in all the industrial and industrializing nations. Consequently, chemical corporations promoted Haber-Bosch synthetic nitrogen as inexpensive and a modern advance over animal manure and fertilizer crops. Where waterpower was abundant, the ammonium cyanide process was often used, as it was at the dams created by the Tennessee Valley Authority and at American Cyanamid's plant at Niagara Falls in Canada.

German corporations made several profitable breakthroughs in industrial chemistry from 1850 up until the 1940s. In addition to the brilliant processes and products that they discovered, they seized other discoveries and copied them under their patent law, which was enacted in 1877. This patent law forced inventors and innovators who filed patents to develop marketable products from their discoveries within a short period of time or suffer increasing fines and, ultimately, lost patents. This prompted industry and university scientists to concentrate their efforts on developing processes and products that had a real market—and on creating products from abandoned or seized patents.

Ingenious discoveries and seized patents helped German chemical and mining corporations and cartels exercise worldwide control over the chemical and fertilizer industry for more than sixty years. Because of this control, serious U.S. shortages in fertilizer and farm chemicals resulted from German chemical and mining company blockades of the United States during World War I and their refusal to deliver products during the Ruhr War, which began in 1923.

Germany's aggressive patent laws, its predatory business approach, and a series of product blockades forced U.S. firms and the government to think strategically about all their supplies of fertilizer and chemicals. Consequently, significant domestic chemical discoveries and resource development occurred between World Wars I and II. By the late 1920s, the major commercial fertilizing elements used today—nitrogen, phosphorus, and potassium—and the

This Hauser fertilizer ad was one of the earliest uses of the farming-as-war message. Once that message began to be used it was used extensively, even excessively. From *Pacific Rural Press*, March 16, 1918.

This ad from Shell promised a 16 percent increase in yields. It failed to add that the farmers and their descendants would see a 100 percent decrease in water quality. From the University of California at Davis Shield Library Special Collections.

minor elements—sulfur and gypsum—were all being mined or synthesized in America. Even though production and mining had begun, both nitrogen and potash were in very short supply throughout the 1930s and 1940s. And it was not until some years after the end of World War II that farmers could obtain all of these fertilizers in sufficient quantity from U.S. suppliers.

With the increase in domestic fertilizer production, the advertising emphasis after World War I shifted from sodium nitrate and doctored guanos to synthetic nitrogen and mixed fertilizers. Ads for calcium nitrate, ammonium cyanide, ammonium nitrate, and anhydrous ammonia (all forms of synthetic nitrogen) began appearing regularly in farm journals by the 1930s. Increased yields, ease of application, and reduced labor costs were all promised regularly in the synthetic fertilizer ads tailored to farmers.

During this period the fertilizer merchants, especially Shell, copied the earlier German Kaliwerks strategy of advertising constantly to fit the season of a specific crop. Shell's tailor-made fertilizer ads would appear for carrots, beets, wheat, corn, potatoes, or cotton at the appropriate time of the year to affect the farmer's purchasing decisions.

The fears of Liebig and the Malthusians about nitrogen shortage finally had been laid to rest with the discovery and development of industrial processes to extract nitrogen. Advertisers and editors promised that with this single discovery hunger would disappear, farmers would make more money from higher yields, food would be plentiful and cheap, and farm labor problems would vanish. All of these promises were part of the immediately post–World War I advertising campaigns for synthetic nitrogen.

In the 1930s, the use of the war theme accelerated with the rise of war rhetoric in Europe. By the end of the decade, the war on pests and the fight to maintain or

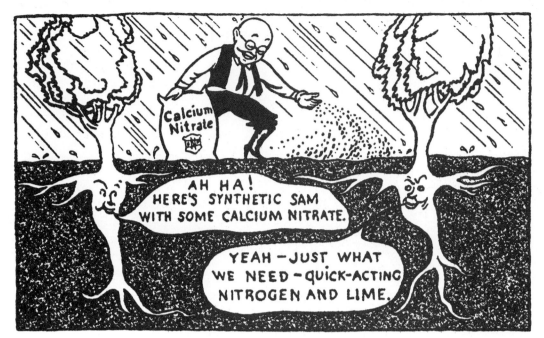

restore soil fertility were the dominant themes that advertisers used. Even the biological pesticide ads used war graphics.

Farming, always as much an art as a science, was being further reduced to mechanical military strategies. Some of the most brilliant minds on the planet were seduced into working on murder, destruction, and the Final Solution or under the dark cloud of eugenics. Others got tangled up in the drive for world domination in agriculture, the need for vastness in scale, and the industrialization of all marketable life forms, including trees, plants, seeds, and animals. The immediate consequences of these mechanistic ideas were the Dust Bowl and the Great Depression, which drove countless farmers off the land.

Farmers and farmworkers were already feeling the long-term negative effects of industrialization, warmongering, and the chemical corporations' all-out struggle with nature. Yet in spite of all the campaigns that farmers were subjected to about packaged nitrogen fertilizer, the need to modernize, and the advances in science, farmers still preferred to use animal manures and crop residues to fertilize with throughout the 1930s and 1940s.

Chapter 13

POPULIST FARMERS, THE RED SCARE, AND THE ORIGIN OF THE MODERN ORGANIC MOVEMENT

T he first years of the twentieth century were times of plenty for many U.S. farmers. Spurred by the markets that developed before and during World War I, farmers increased their yields and prices remained high until the early 1920s. These halcyon days were short-lived, however, lasting about twenty years and ending a short time after World War I. Following the war, prices were high but food supplies were also high, and the government began to manipulate prices and attempted to limit supply. As a result, farmers began to suffer as government interference distorted the real markets and caused the prices farmers were paid to fall. The United States and most of the Western world seemed to be stumbling toward another depression. Farmers began to suffer bankruptcies by the mid-1920s, well before the Wall Street crash of 1929. They responded by trying to resurrect the Grange movement and joined the IWW and other radical groups to change the nature of American culture.

While some farmers were in the midst of a twenty-year period of prosperity, the country was experiencing another avalanche of labor unrest. The Industrial Workers of the World and the Socialists organized strikes in the mines, in the factories, and on the large-scale farms. Conductors, brakemen, porters, and stewards organized strikes against several railroads and railroad manufacturers, including the Pullman factories.

Before the signing of the Treaty of Versailles in 1919, at the end of World War I, many of the largest industrial firms formed cartels and trusts similar to the model developed by Rockefeller at Standard Oil and copied by I. G. Farben. These cartels were usually alliances between, or consolidations of, chemical, petroleum, dynamite, and fertilizer corporations within one of the dominant industrial nations: France, England, Switzerland, Italy, Germany, and the United States. Over time these cartels interlocked with firms from other nations. The period from 1895 to 1945 represents an unparalleled fifty-year consolidation of and cooperation between industrial firms from all the countries listed above.

The industrial capitalists monopolized resources all over the world. Their cartels and trusts were unions of corporations, yet they opposed unions of laborers and the nationalization of resources that unions and other activists demanded. They collectively fought against civil rights, labor, and resource-protection movements with all the power that industry, government, and the press could muster.

Nonetheless, even though they were outgunned and outspent, the populist movements on the farms and in the cities remained strong throughout the first two decades of the twentieth century. But after that, working people and small-scale farmers became increasingly desperate as economic times worsened and the depression sapped people's optimism.

As farms failed again in the 1920s and rural families moved to cities and became factory workers, the rural populist movement followed them. Many factory and farm union organizers, from the1890s to the 1950s, came from small farm families whose ancestry was populist, but whose farms had gone bankrupt (including Cesar Chavez, cofounder of the United Farm Workers union). Many of the Industrial Workers of the World's most intense organizing struggles focused on large-scale farms in California and the West, where workers were exploited and cheated.[1] Again, in the 1920s and 1930s, as with the earlier populist movements, there was shared support between small farmers, farmworkers, and the factory workers who had recently lost their farms when the deluge of farm foreclosures drove tens of thousands of farmers into jobs in the city and into jobs on farms that they had previously owned.

Because of the passion generated by the loss of the family's land on a farmer's watch, it was hard to tell whether the dispossessed farmer would commit suicide, try to find a job and live in town, or become a fire-eating activist. From the late 1800s, activists streamed off the land in their times of crisis in an effort to redeem themselves by becoming a voice for change and creativity. Major civil rights breakthroughs owed their passage to decades of rural populist agitation and urban organizing. Women's suffrage, which finally passed in 1920, began to be a national issue in 1848 and was promoted by trade unions, the National Association of Colored Women, religious groups, and farm women. The eight-hour workday that was enacted in 1938 was first promoted seventy-four years earlier, in 1864, by workers, farmers, and reformers. The creation of 1930s legislation to establish marketing orders and commissions, which would help farmers with marketing and research, traces back to populist farm and labor groups advocating for assistance with just those problems.

In the first three decades of the twentieth century, most of the major farm magazines were preoccupied with both chemical agriculture and the Red Scare that they thought was behind all of the populist demands for labor protection and civil rights for farmworkers. Because of these preoccupations, some very significant advances in biological agriculture during this time were often buried on back pages of the farm journals or entirely ignored.

These breakthroughs did not receive visibility in large part because the techniques did not need to be industrialized, purchased, or packaged; they required mostly farmer input and a minimal investment in seeds, inoculants, or infrastructure development. As a consequence, there was no huge advertising budget that provoked the journals to promote biological products or ideas. Consequently, the journal publishers who were focused on ad revenues from chemical manufacturers did not give a lot of space to breakthroughs in biological farming.

While government and journalistic support was lacking for biological farming approaches, a small number of researchers and farmers rediscovered forgotten practices, developed new strategies, and wrote about organic techniques. Many of the innovations in organic farming from this time period are still being used profitably in farm communities all over the world. The work of F. H. King, William Albrecht, and Rudolph Steiner was known and appreciated by freethinking farmers and activists but largely ignored by the U.S. farm press.

In 1911, F. H. King published the book *Farmers of Forty Centuries*, which describes what we now call organic farming practices from China, Korea, and Japan. Many farmers owned King's book and used ideas gleaned from his compendium of strategies and techniques; however, it wasn't widely promoted, so its overall impact was not as great in the United States as it should have been. The book included management ideas that had inspired populist farmers in Asia for centuries and in America for the hundred years since John Taylor published *Arator*.[2]

In 1918, William Albrecht began to report on his discovery of beneficial bacteria that helped bean and pea plants (legumes) greatly increase the amount of atmospheric nitrogen they could collect. Albrecht was actually a farm boy who became the organic industry's

While major developments occurred in biological farming throughout the world, most biological promotions in the farm journals stopped or were relegated to back pages. The ads almost always included promotional text, since the editors were not promoting biological approaches any longer. The two ads in the upper right corner are from *Farm Journal*, March 1884. The Hallock's seed ad on the left is from *Farm Journal*, March 1889. (Lucerne is called alfalfa in the United States.)

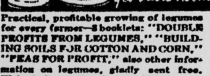
This ad was used in the 1930s by *Pacific Rural Press* for bean inoculants that William Albrecht discovered. From the University of California at Davis Shields Library Special Collections.

first scientific research guru as professor of soils and chemistry at the University of Missouri. His publications began in 1918 and continued until 1974. His remarkable body of work scientifically documented the importance of organic soil management and established that calcium was the prince of nutrients. Many innovative farmers tried Albrecht's cover crops, inoculants, and soil-health strategies because these techniques made their farms more vibrant and productive—and because his techniques made farmers money.[3]

In the early 1920s, several farmers from Hungary, Poland, and eastern Germany asked Rudolph Steiner, an Austrian philosopher, to help them deal with their problems of yield loss, plant disease, and soil degradation that resulted from their dependence on industrial agriculture's chemicals, seed varieties, and monocultural strategies. Steiner turned almost all of his energy to agriculture to try to resolve the soil-fertility and plant-health issues that plagued farmers and farm researchers after the introduction of chemical fertilizers and pesticides. Steiner combined a study of ancient farming practices with a strident emphasis on obtaining all farm inputs from the farm itself. He set forth the framework for an agriculture that views the farm as a self-sustaining system or organism. This is the opposite of Liebig's paradigm, in which the "law of the minimum" rules, and synthetic farm nutrients are purchased from chemical suppliers for the cheapest price. In Steiner's system there are few purchased inputs. In Liebig's chemical replacement system, the farmer needs to purchase and import almost everything the farm needs.

In 1923, Steiner delivered his first lecture on what has come to be known as biodynamic agriculture. The practices he outlined were part of a creative and integrated worldview merging physical and spiritual contexts. For example, while corporations and nations rushed to synthesize nitrogen, Steiner argued that farmers should use carefully composted manure for nitrogen because it mattered where nitrogen came from and how it was composted. He argued that "proper manuring" took fertilizer nutrients directly out of the sophisticated laboratory of the cows' stomachs.[4] Though creative, Steiner's innovations in the area of composting, compost inoculants, and land management have been almost completely ignored by the U.S. agribusiness community until very recently. Still, his work has been and remains important to organic, biodynamic, and biological farmers. Steiner's Waldorf schools are one of the fastest-growing educational programs in the United States. And both the biodynamic farm movement and the Waldorf schools are rapidly growing in many parts of the world.

Even during the depths of the Depression, as farm prices remained low and soil was blowing away in the Dust Bowl, the creative ideas of researchers and philosophers like King, Albrecht, and Steiner were so innovative that many farmers tried their strategies. In fact, these new ideas about agriculture might have had more immediate impact if they had followed the Great Depression instead of preceding it, since some potentially important strategies got lost in the sea of chaos and desperation that represented life on the farm during that era.

For most small and medium-sized farmers, the 1930s entailed a day-to-day struggle against bankruptcy. As the economic problems worsened around the "developed" world, populist agitation grew, and nations and states were forced to respond. The Fascist governments in Germany, Spain, and Italy responded by suppressing the unions, farmers, and civil rights struggles. More liberal capitalist nations tried to address the problems paternalistically, with one bailout program after another.

Through all the travail of the 1930s, the chemical advertisers kept right on going with their propaganda blitz. They continued to assert that chemicals made more economic sense than any other solution for fertilizer or pest control, even at a time when farmers and the common citizens were wondering where their next tank of gas or next bag of seed was coming from. In good times or bad, to these advertising hucksters, chemicals were the solution. Each new chemical product was touted to be cheap. Chemicals would solve labor problems. Chemicals would solve money problems. The advertisers inundated the farmers with commercial sales pitches and promises throughout the Depression years.

More large-scale farmers began to use the newly available nitrogen fertilizer to reduce labor problems and costs on large-scale cotton, corn, and fruit farms. The advertisers claimed high yield increases as the bait to convince other farmers to use their products. Farmers who took that bait often achieved higher yields because they switched from growing many crops together to growing mostly one. However, many achieved only short-term success with the chemicals and lost their farms as one chemical after another failed or bankrupted them when they couldn't pay their chemical bill or their mortgage because of low prices and glutted markets. Throughout the early 1930s bankruptcies increased, and the rural flight turned into an exodus, most often to the closest city.

While many large-scale American farms embraced chemicals before World War II, many others resisted them. Some of those who chose not to use chemicals adopted the ideas of the organic and biodynamic scientists and experimented on their own farms— applying farm-based biodynamic mixtures and composting fertilizers. A small group of farmers in Europe and America refused to farm with chemicals, relying instead on organic and biodynamic systems. These farmers have been responsible for keeping many ancient ideas and practices alive for the last ninety years, while at the same time developing and fine-tuning many new innovations. Thousands of farmers also continued to rigorously save their own varieties of seed, thus helping to preserve ten thousand years of local crop diversity. In recent years, many of the old-timer biological farmers and seed savers have mentored their chemical-farmer neighbors as the latter made the transition from chemical dependency to organic or biodynamic farming.

The suicide.

Famous Dr. Seuss cartoon for Flit bug spray. From Richard
Marshall, *The Tough Coughs as He Plows the Dough* (New York:
William Morrow, 1987).

Chapter 14

PESTICIDE SPRAY DEVICES, HOUSEHOLD POISONS, AND DR. SEUSS

Immediately after chemical firms began to promote pesticides to American families for house and farm use, equipment manufacturers began to produce and advertise pesticide applicators. Unbelievable contraptions appeared by the mid-1870s, invented or adapted to spread pesticide dusts and sprays. The marketing of chemicals became inextricably linked with the development of effective spray devices to apply the poison on the plant, on the pest, or under the sink.

The earliest pesticide applicator was a folded piece of paper or cardboard from which a person literally blew the poison onto the plants. This, however, was not a sufficient or efficient system, since an accidental cough or an inhalation could prove deadly.

The need to spread dust over a wider area led to the use of simple flour sifters, or, as we saw with arsenic, makeshift box dusters with milk screens nailed on the bottom. The lovable fireplace bellows was the next household item to be jerry-rigged as a pesticide duster. Modified and enlarged, the bellows could be ordered with attachments for fumigation and animal-pest control. This enabled farmers and gardeners to blow poison dust wherever and on whatever needed it.

Pump applicators supplemented this primitive arsenal in the 1880s, facilitating the spread of poisons at home and in the fields. Only slightly different from a bicycle pump, they increased efficiency and range significantly. Pumps came with a variety of attachments that could accommodate numerous poisons and different applicator needs. For the next forty years, manufacturers endlessly modified these canister pumps, especially for household use.

Fabricators and farmers developed crank and lever pumps sometime before the turn of the twentieth century. The ad makers advertised a safe and civilized lever pump, with graphics implying that one could wear a hat and tie while dusting one's crops or garden with arsenic.

Preparations for spraying were elaborate by this time, as can be seen in the photo on the following page from the University of California–Davis archives. Here the horses pull the spray rigs through the orchard, and interestingly, some of the horses wear more protection than the poison applicators. Similar two-man pumps supplied gangs of workers in the late 1800s. Workers climbed ladders or sprayed poisons from long hoses on the backs of wagons.

In addition to the stationary spray devices, there was an enormous array of sprayers—some motorized, some with hand pumps—that were pulled by horses through vegetable rows, melon patches, cotton fields, vineyards, and orchards.

This folded-paper device was fortunately short-lived. From the University of California Davis Shields Library Special Collections.

Most types of pumps that we have today were available to American farmers, nurserymen, and gardeners by the early 1900s. By the 1910s compressed-air and gas sprayers became more widely used. But pesticide sprayers were not used only by farmers; householders also used spray rigs and chemical poisons. As the U.S. population urbanized, domestic households became an increasingly important market for the chemical merchants.

Frank Presbrey in *The History and Development of Advertising* and Roland Marchand in *Advertising the American Dream*, along with several other authors, point out how advertisers linked chemical cleanliness with being American in the 1920s.[1] American interest in technology was stimulated by the press, and by advertisers who applauded, defended, and marketed each new "scientific" breakthrough in pest control. Each concoction was heralded as the product that could eradicate all household pests.

Many American cities, however, continued to have only patchwork programs of sanitation and unclean water-supply systems. Without well-designed systems

From the University of California at Davis Shields Library Special Collections. No date, though probably 1880–1890.

to deliver freshwater and carry away wastes, disease continued to be a constant reality of American cities. Filth, rats, and fear of the plague still drove buyers to get rat poisons or call the exterminator, since most people felt that rats carried almost any disease that would come along for the ride.

Preying on these real but often media-exaggerated fears, pest control and chemical advertising agencies turned out creative and dramatic campaigns that really stimulated the market. Of course, such campaigns were successful because there were always too many rats, flies, mosquitoes, and roaches. For the pest exterminators at the turn of the century, there was a greatly expanding market! As cities grew and the problems worsened, customers began to complain about the ineffectiveness and danger of the products. To deal with this consumer skepticism, advertisers increased their propaganda to convince families that they needed chemicals and spray devices that they could use at home.

Paris G

Pesticide advertising flared out of control after World War I. The United States was awash in ads, with practically every rock, barn, and flat space covered with promotions and propaganda. Amidst this visual assault, corporate advertisers desperately sought gimmicks to make their products leap out at the consumer and rise above the crowd.

As cities grew, advertisers hit on a few pivotal strategies that dramatically and permanently expanded the market demand of both city and country residents.[2] One of the most successful and innovative pesticide sales campaigns was for Flit, a fly and mosquito killer.

"THE AQUARIUS."

PATENT GREENHOUSE FUMIGATOR.

NICHOLSON'S PATENT.

Price 8s. 6d. each.

Price 8s. 6d. each.

The simplicity of this article commends it to the notice of all lovers of flowers. The necessity for fumigating delicate plants is admitted, and when the fumes of Tobacco can be blown upon them continuously for any length of time, the small insects so injurious to their growth are effectually destroyed.

When once the Tobacco is lighted, all that is required is to turn the small knob on the wheel at pleasure, and according to the rate at which the wheel is turned will the fumes of Tobacco puff out from the pipe, and can be directed by the user to any plant. The Red Spider can also be destroyed by using Flour of Sulphur, instead of Tobacco, the Flour being blown out of the pipe till these enemies of plants are covered with it.

and Powder Duster

Bellows contraptions such as the three shown here served farmers from the 1870s until after the turn of the century and proved to be a significant improvement over folded paper. From top to bottom they are Nicholson's Fumigator, the Powder Duster, and the California Beauty. From the University of California Davis Special Collections Shields Library.

The enlarged and modified bicycle pump, illustrated below was common from the early 1880s until the 1890s. By that time the bicycle pump had morphed into the canister spray rig shown at right. The canister spray pumps enabled the user to have a larger volume of poison. From the *Pacific Rural Press*, January 1885 and the back cover of *California Cultivator*, May 31, 1941.

Ads like this one for the Aquarius were common in the 1880s and 1890s. The attempt of the advertisers was clearly to imply that spraying was safe and so dignified that one could wear one's best "Sunday meeting" clothes. From the University of California at Davis Shields Library Special Collections.

"THE AQUARIUS,"

A new and most invaluable article, for which the subscribers have received Letters Patent from the United States, and it is offered to the public as the most complete and perfect hand apparatus ever invented for throwing water.

It will throw about EIGHT GALLONS of Water per minute, fifty feet high, with the power of only one hand applied. Being a most invaluable article for WASHING WINDOWS, WASHING CARRIAGES, WATERING GARDENS, SPRINKLING STREETS, THROWING on COMPOSITION, such as WHALE OIL, SOAP SUDS, TOBACCO WATER, &c.; for DESTROYING INSECTS on TREES, ROSES and other PLANTS, PUMPING WATER from the HOLDS of VESSELS, SMALL BOATS, CELLARS, &c.; WETTING SAILS, WASHING DECKS, STARTING a most valuable SPRAY or SHOWER UISHING FIRES, and for WETTING GREAT CONFLAGRATIONS might be t is so portable that it can be used where r.

peak in most unbounded terms of praise nd perfect and beautiful operation. No Barn, Hot House, Vessel or Boat, should R.

eight pounds.

½ feet of suction and 3 feet of Discharge

☞ Orders respectfully solicited,

W. & B. DOUGLAS,

d other Hydraulic Machines, Hardware, &c., &c. Middletown, Conn.

York, where the Aquarius can be seen may

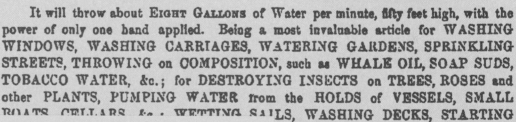

In the 1920s, America's most common household spray device was a hand pump sprayer with a pressurized canister. Each sprayer had a half-pint, pint, or quart reservoir with a bicycle-like pump attached to spray the poison. By the early 1930s, this pump became popularly known as the "Flit gun." Although Flit was the name of Standard Oil's bug spray, it also became the generic name for this type of pump, due to the popularity of the bug spray. The Flit campaign was so successful that by the mid-1930s airplane crop dusters were called "flying Flit guns."

Standard Oil, which John D. Rockefeller started as a commodities brokerage house in the 1830s, had grown to dominate the world's petroleum industry. In the late 1920s, the company needed a distinctive advertising campaign to make Flit rise above the sea of advertisements for other bug killers. Standard Oil was used to being number one in sales with its pesticide spray, and the company wanted to remain on top.

After seeing two 1927 cartoons that featured Flit guns as props, Standard Oil hired the cartoonist, Theodore Seuss Geisel, to create Flit advertisements. Geisel subsequently came to be known as Dr. Seuss. For the next fifteen years, Seuss's humorous ads, which were really commercials in the form of cartoons, appeared in thousands of weekly and hundreds of daily newspapers and magazines.[3]

The Cushman Corporation used ads to promote the use of its gasoline engine. Near right: Cushman's catalog for 1915 advertised this large sprayer, Great Western No. 1, with the heart of the sprayer being a gasoline engine. From the University of California at Davis Shields Library Special Collections. Far right: This ad for a smaller sprayer ran in several journals for years. It includes a guarantee to "double the crop and make it all first class." From *Better Fruit*, 1913.

GREAT WESTERN No. 3
Side View: machine cover closed.

The Eureka spray cart above is simpler than but similar to spray carts used today. This model or something very similar was used from the 1880s until well into the 1930s. Only the motors were changed, as gasoline became much more common after the 1920s. From the Eureka catalog. University of California at Davis Shields Library Special Collections.

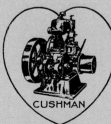

CHAMPION SPRAY PUMP.

NOTICE.

We also carry a full line of other Spray Pumps—GOULDS, STAR, EUREKA; also full line of Spray Nozzles, Spray Hose and everything connected with the Spray Pump Outfits.
SEND FOR SPECIAL CATALOGUE—Mailed Free.

WOODIN & LITTLE,

312 and 314 Market Street, Junction of Bush, San Francisco, Cal.

This excerpt of an ad in the *Pacific Rural Press*, January 1893, p. 20, shows again how pumps were towed into the orchards and used to fumigate or spray the trees.

...g Sprayer

...g and durable. Pro-
...g liquid soiling floor
... purposes, for killing
...bugs, etc.
...ipping weight 11 lbs.

...nt Sprayer

...nt.

...servoir and thoroughly
...ayer compact and ex-
... Throws fine mist
...spraying disinfectants,
...lice, water bugs, etc.
Packed 1 dozen in case. Shipping weight 10 lbs.

No. 19 Quart Glass Sprayer

Pump Chamber:
18 in. by 1½ in.
Reservoir: Quart Mason fruit jar.

Has two spray tubes. The reservoir is a quart Mason fruit jar, and may be easily replaced if broken. Easily filled. Body of sprayer is heavy tin nicely painted. Very substantial, neat and attractive.
Packed 1 dozen in case. Shipping weight 26 lbs.

No. 12 Dry Powder Duster

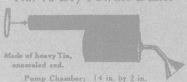

Made of heavy Tin,
enameled red.

Pump Chamber: 14 in. by 2 in.
Tank: 6½ in. by 4 in.
Capacity: 1½ quarts.

The reservoir is filled with dry powder, such as Paris Green, Arsenate of Lead, Savol, etc. Size of spray depends upon operation of plunger. Funnel and elbow are reversible, to spray up or down.
Packed 1 dozen in case. Shipping weight 14 lbs.

No. 20 Quart Glass Sprayer

Pump Chamber: 16 in. by 1¾ in.
Reservoir: Quart Mason fruit jar.

No. 20 is well made, and has an exceptionally large pump chamber, insuring plenty of air with each stroke of the handle, to produce a large, broad mist spray. Has two spray tubes. The screw cap for filling is brass. This will not corrode or rust. The reservoir is a Mason quart glass fruit jar, quickly detached if necessary and may be easily replaced if broken. Sprayer is nicely finished in red enamel as shown by illustration.

For spraying garden vegetables, house plants, shrubbery, disinfectants, etc.
Packed 1 dozen in case. Shipping weight 30 lbs.

The catalog page at right from the Smith Sprayer Co. for 1924 illustrates the evolution of the canister sprayer, mostly for household and barn use. This page also promotes the hand-crank sprayer Savage Dry Powder Duster (illustrating use from horseback) and two compressed-air sprayers. Such catalogs were advertised constantly in farm and urban journals and would be sent free for the price of postage. From the University of California at Davis Shields Library Special Collections.

Eleven
Savage Dry Powder Duster
(Continued)

The nozzle tube may be raised up or down for spraying short or tall plants or trees. Does not discharge poison in "chunks" or "gobs" but thoroughly breaks it up and dusts evenly. Feed lever adjusted from 1 to 20 pounds per acre, which is a wonderful improvement over any duster made.

Construction: Heavy sheet metal; brass; aluminum fan and housing; rubber nozzle; ball bearings throughout; nicely finished.

Weight: When empty 9½ pounds. Packed one in case. Shipping weight 14 pounds.

Hopper: Capacity 7 to 10 pounds of poison, depending on density.

Full directions for operating with each Duster.

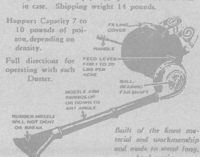

Built of the finest material and workmanship and made to stand long, hard service.

Twelve
No. 36 Lightning
COMPRESSED AIR SPRAYER

Pump inside of tank is always out of way

For spraying all garden vegetables, potatoes, plants, cotton, shrubbery, vines, tobacco plants, carbola, disinfectants, etc. The brass nozzle provided with strainer will not clog. It is fitted with brass elbow so that nozzle may be turned in any direction desired. By detaching elbow, nozzle may be screwed directly to shutoff, which throws spray straight ahead. A slight pressure of the fingers on the lever automatically throws a broad, fine mist, or coarse spray; by releasing the fingers the nozzle shuts off instantly, thereby no liquids are wasted. The carrying handle is enameled black and sprayer is attractively finished.

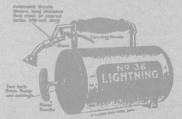

Tank: 11 in. by 7½ in. galvanized steel, double seamed; strong.
Capacity: Two gallons.
Pump: 2-in. seamless brass, screws in tank head.
Pump Castings: Heavy brass.
Nozzle: Brass, automatic, for three different sprays. Will not clog.
Elbow: Brass with union end. For spraying up or down.
Packed 1 in case. Shipping weight 7 lbs.

SMITH Nº 22 BANNER COMPRESSED AIR SPRAYER

This handle easily locks into pump-head for quickly loosening or tightening pump, also for carrying sprayer in hand. With this same handle, a few strokes of the pump charges the tank with compressed air.

Fill tank here through large opening by easily detaching pump. Pump is brass 2 in. diameter, provided with brass casting with machine cut threads for screwing into tank. Nothing to corrode, rust or wear out.

Adjustable strap with snap ends. For carrying sprayer over the shoulder.

2 feet heavy hose; detachable for cleaning. Spring wire to prevent breaking.

Golden Automatic brass spray nozzle. Throws long distance fine mist or coarse spray. Wastes no liquids.

Heavy 4-gallon galvanized steel tank, well riveted, to stand heavy pressure. Also made entirely of brass.

The exterminator-man forgets himself at the flea-circus.

MEDIAEVAL TENANT—*Darn it all, another Dragon. And just after I'd sprayed the whole castle with Flit!*

The two illustrations above are Dr. Seuss cartoons in which he used Standard Oil's bug spray Flit as a prop. Because of these cartoons, Dr. Seuss was offered a job with the oil giant. He worked for Standard Oil from 1928 until 1943 and is generally acknowledged to be responsible for greatly popularizing the use of household poisons. The rest of the illustrations are examples of Flit cartoons that Dr. Seuss created for Standard Oil. From Richard Marschall, *The Tough Coughs as He Plows the Dough* (New York: William Morrow, 1987).

An Ancient News Picture
The above photograph was found in recent excavations under the city of Rome. Noted archaeologists say it is from the Sunday Roto Section of Rome Graphic and appeared in B.C. 1073. The picture depicts that sly satyr Flit undoing the work of the unpleasant goddess Insecta.

". . . For better for worse, for richer for poorer, till Flit do us part—"

"Don't worry, Papa. Willie just swallowed a bug, and I'm having him gargle with Flit!"

At the time of his hiring, Seuss was a well-known but underpaid "screwball" cartoonist writing humorous copy and drawing cartoons for the *Judge*, a national humor magazine. With his cartoons for Standard's bug killer, Dr. Seuss turned Flit and the Flit gun into household necessities. His success, which kept Flit in the leadership position in the marketplace, also made the incredibly prolific Geisel economically comfortable and afforded him enough freedom to gestate his later cartoon masterpieces.

The Seuss taglines—"Quick Henry, the Flit!," "Swat the Fly!," "Kill the Tick!"—became nationally known slogans. Seuss helped America become friendly with poisons; we could laugh at ourselves while we went about poisoning things. In the process, the public grew comfortable with the myth that pesticides were absolutely necessary.

The Flit campaign was an advertising stroke of genius, and luck. Why luck? Because Seuss had actually considered using FlyTox or Bif for the name of the bug spray in his 1928 cartoons. The Rockefellers were twice lucky: Dr. Seuss chose to help sell Flit and several other Standard Oil products. And the public loved it!

Gradually, American householders came to depend on their Flit guns. Whether filled with Flit, Bif, Black Leaf 40, or arsenic, the home spray device had become an essential tool in the public's mind. The petroleum solvent that Seuss was selling as Flit, however, was very dangerous and probably carcinogenic in large doses, though mild when compared to the World War II chemicals that would be sprayed from Flit guns on everything from bedbugs to flies, mosquitos, and humans after the end of the war.

Considering the reverence with which Dr. Seuss is held today, it is difficult to envision him as a pivotal figure in the public acceptance of poisonous pesticides. Nevertheless, some historians feel that his campaign was largely responsible for popularizing dangerous pesticides to the American public. Adelynne Whitaker, the author of *A History of Pesticide*

Regulation in the U.S., contends that the Dr. Seuss cartoon campaign had the effect of increasing pesticide use tenfold for the nation's families.[4]

One of Seuss's later books, *The Lorax*, with its save-the-environment theme, is ironic when compared to impact of the Flit cartoons. Perhaps Dr. Seuss realized his earlier mistakes and indiscretions with Standard Oil's Flit and tried to make amends with *The Lorax*. Geisel must have known that Flit's cartoons and his World War II cartoons for DDT had an enormous impact on the public's use of pesticides and acceptance of DDT. Seuss was proud of his success as a pesticide salesman. For most of his life, however, Seuss was a reformer, a progressive, and a patriot—as his World War II cartoons for the government attacking Nazi Germany and the American right wing illustrate in later chapters.

From *California Cultivator,*
May 31, 1941.

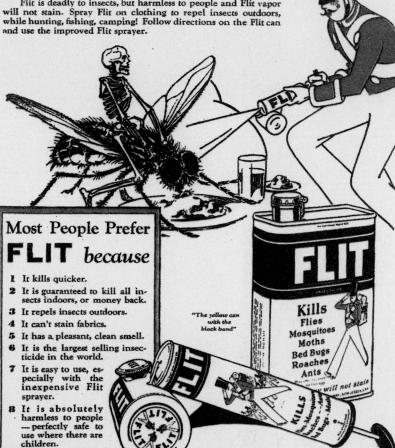

Danger!

"The fly is a little insect but a great spreader of disease," says the U. S. Public Health Service. "Thousands of people die every year from diseases transmitted by them. Remember! No filth, no flies, less disease."

Kill flies! Buy Flit at once and get rid of all flies and other disease-carrying pests—roaches, mosquitoes, bed bugs, fleas.

Flit will kill every insect in the house. It has greater killing power. Guaranteed to kill all household insects or money back.

Flit is deadly to insects, but harmless to people and Flit vapor will not stain. Spray Flit on clothing to repel insects outdoors, while hunting, fishing, camping! Follow directions on the Flit can and use the improved Flit sprayer.

Most People Prefer FLIT because

1 It kills quicker.
2 It is guaranteed to kill all insects indoors, or money back.
3 It repels insects outdoors.
4 It can't stain fabrics.
5 It has a pleasant, clean smell.
6 It is the largest selling insecticide in the world.
7 It is easy to use, especially with the inexpensive Flit sprayer.
8 It is absolutely harmless to people — perfectly safe to use where there are children.

"The yellow can with the black band"

FLIT Kills Flies Mosquitoes Moths Bed Bugs Roaches Ants

A full-page ad for Flit graced the entire back cover of *Wallace's Farmer*, May 10, 1929.

As Standard's Flit campaign soared with cartoons, other bug sprays like Bif stuck to the text-intense, scientific, war-and-fear format that Flit had used before the Dr. Seuss cartoons dominated their ads. Before long, the soldier and the text of Flit ads were replaced by Seuss's humorous cartoons.

The Ambush

In the spring of 1933 a fifteen year old girl from Billings, Montana died from eating fruit with high residues of arsenic.

This was followed in the summer of 1933 in Los Angeles, California by one of the worst episodes of arsenic poisoning in the U.S.

Dozens of people suffered from stomach poisoning. Cramps, vomiting, diarrhea, high fever, bloody urine, bloody stool, bloody vomit, cold sweats, irregular pulse and exhaustion were common symptoms.

Tests revealed that the cabbage eaten by the poison victims had thirty five times the level of arsenic allowed by the government.

Pears eaten by some of the victims had fifty-four times the limit of safety.

These examples were quoted by James Whorton in *Before Silent Spring* (Princeton: Princeton University Press, 1974). By the early 1930s almost everyone in the United States was suffering from arsenic and lead poisoning.

Chapter 15

MEDICAL HEROES, WAR HEROES, AND MORE REGULATION STRUGGLES

By 1900, chemical corporations were promising to usher labor-intensive nineteenth-century agriculture into the "age of progress and science." As we have seen, even before the outbreak of World War I, many large-scale farmers had come to believe in the chemical corporations' philosophy of "quick fixes" and used large quantities of toxic products. For some farmers, the urgency of the war effort helped accelerate as well as justify their conversion to using chemical poisons.

After World War I, as Europe recovered from the economic and structural chaos of war, poisonous products, which should have been controlled or eliminated, were either ignored or, worse, embraced by the regulators. Profit-hungry corporations and cartels marketed several pesticides as essential tools in the reconstruction of war-torn Europe. Merchants argued that these chemicals were necessary to ensure the economic vitality of American farms and industries, both of which were now shipping much larger volumes of products regularly to Europe's hungry and resource-starved markets.

As we have noted, the aggressiveness of advertisers following World War I muted the criticism of toxic chemicals. Farmers and businesses rented their barn and store walls and rooftops to soft drink, soap, tractor, automobile, and chemical advertisers. Traveling advertisers armed with paint and slogans crisscrossed the country, sometimes paying for their space, but often defacing enormous swaths of the countryside without obtaining permission or paying any ad fees.

It was once said that if one followed Lewis and Clark's tortuous route of discovery in the 1920s one would find an advertisement even in the most remote creek or inaccessible canyon.[1] In short, not only were businesses unscrupulous in their claims, but their advertisements were everywhere.

While the corporations worked to calm the public's fears, many scientists, journalists, and activists continued working without much fanfare against the uncontrolled spraying of pesticides. And there continued to be a significant and vocal opposition to the use of arsenic, lead, and other toxic chemicals. But the USDA and the FDA literally refused to address the toxicity of farm and home chemicals until the late 1930s, after the worst of the Great Depression had passed. Even then, their response was not

Find Arsenic in American Apples.
Copyright, 1925, by The New York Times Company.
Special Cable to THE NEW YORK TIMES.

LONDON, Nov. 25.—Fruit merchants were fined today for selling American apples containing arsenic to the amount of one-fifth of a grain per pound. The Public Health Department's notice was attracted by a specific case of illness which followed the eating of apples. It was stated that fruit growers in all parts of the world made a practice of spraying trees to combat insect pests, and in America, it seemed, not only the blossoms, but also the fruit, was sprayed after it had formed on the tree, to protect it from the codling moth. The season seems to have been unusually dry in America, with the result that the arsenic was not washed from the apples.

This complaint in the *New York Times* was a common item in the prominent newspapers of the 1920s and 1930s. Just as today's European consumers have rejected genetically modified and hormone-laced food from the United States, so European consumers rejected the highly poisonous U.S. food that was being exported to them even before the beginning of the twentieth century.

voluntary but forced on them by environmentalists, muckraking journalists, and public and activist outrage that had been building since the 1890s at the amount of poisonous residues on foods and the high concentration of arsenic and lead in the general environment. Yet in spite of this national and international outrage, the regulatory fight over arsenic and lead was difficult and long.

In spite of the public's awareness of the deadliness of arsenic and lead, chemical advertisers continued to promote poisonous products such as Paris green, lead arsenic, calcium arsenic, and zinc arsenic, as illustrated in this 1938 ad from *California Cultivator*.

Despite the troubles in foreign markets, arsenic advocates in the 1920s and 1930s held important positions in the United States Bureau of Agriculture and in state departments of agriculture, and support for arsenic in the U.S. Congress and farm-state legislatures continued through the 1950s. For example, Congressman Clarence Cannon from Missouri eliminated scientific testimony about the arsenic dangers from the Congressional Record in 1937. He blocked arsenic residue research at the Public Health Service and successfully demanded an increase for arsenic and lead tolerances on U.S. produce. Cannon served as chairman of the House Appropriations Committee, but he also owned commercial apple orchards.[2]

Through a variety of secretive measures, both the scientifically known facts about arsenic food residues and the dangers from lead arsenate remained hidden from consumers by public officials in Congress, the FDA, the Department of Agriculture, and the universities. The effectiveness of the arsenic defense campaign mounted by the FDA, the Department of Agriculture, large-scale growers, and the chemical corporations depended on this closed-door policy. During the 1920s and 1930s, the Department of Agriculture and the FDA actually conducted all of their meetings about and investigations into arsenic behind closed doors, under a cloud of secrecy. This was a carefully managed effort to shield the American public from the scientifically known and demonstrable health and safety risks associated with arsenic and lead residues.

In *Before Silent Spring*, James Whorton uses the quote below from the chief of the FDA's St. Louis station at a 1929 meeting of the American Pomological Society to illustrate that the FDA, the Department of Agriculture, and the big growers worked diligently to keep secret from the consumers the dangers of arsenic-spray residues on fruits and vegetables:

> What do you think would happen if the general public became acquainted with the fact that

apples were likely to be contaminated with arsenic? . . . So far, we have not given the matter any publicity, and the public as a whole has no general knowledge on the subject. We [the FDA] will be in a more enviable position, when all apples are satisfactorily cleaned to say to the public, if it gets curious, that the apples produced in this country can be eaten with perfect safety. If they were to become curious today, we would probably have to admit that we are perhaps not doing everything possible to remove excess arsenic from our fruit.

But the public knew more than the FDA wanted to think they did, as newspaper reports of poisonings illustrate. What the above quote from the St. Louis station chief shows is that FDA officials not only were aware of the dangers of arsenic but admitted that they had a policy to act only if the public demanded it.[3]

Whorton believes that government officials failed to take aggressive action because they thought that the farmers would voluntarily work harder to remove residues of arsenic and lead and adopt less toxic pest-control strategies because of their fear of being caught with high arsenic residues on their fruit. Unfortunately, in the face of inaction and secrecy by the FDA and the USDA, the large-scale farmers did little or nothing to address the residue issue. Chemical propagandists worked overtime to trivialize the problem in the magazines and newspapers, thereby guaranteeing

This painting of a barn from the early 1900s shows how ads covered walls and roofs along well-traveled roads. In this case the roof announced one cola and the wall another. The practice of legal and illegal advertising increased as increasing rates of car ownership allowed people to travel widely. Faded signs on barns and walls from this time can still be seen in rural areas. From Frank Presbrey, *The History and Development of Advertising* (Garden City, N.Y.: Doubleday, 1929).

Billboards such as these lined railroad tracks and highways near the cities. From Frank Presbrey, *The History and Development of Advertising* (Garden City, N.Y.: Doubleday, 1929).

CALIFORNIA CULTIVATOR

May 17, 1924

GAS-RESISTANT SCALE

F. G. Wyman, Manager of the Growers Fumigating and Supply Company, of Pomona, in his annual report said:

"I am not going to be unduly pessimistic but I am going to be frank. Last year I had 17 crews in the field. We covered 500,000 trees. We often used a 120 per cent schedule—and we had plenty of scale left. Not one of our growers can truthfully say that poor work was the result of poor equipment or poor workmen. We hired a night inspector. He was out all night and every night. I watched the work myself. Yet, in spite of all this, only 30 per cent of our work was good; 50 per cent was commercially clean, and 20 per cent was poor. We used to do better work than this with thinner tents and less material. There is absolutely no doubt in my mind that scale is becoming more and more resistant to cyanide gas.

"Scale extermination is without doubt our most serious problem—after 17 years of experience I cannot tell you what to do. It may yet be necessary to fumigate twice each year; or, perhaps, one fumigation and one spraying will do the trick. In orch-

---

A pesticide sprayer's complaint in *California Cultivator* about the declining effectiveness of cyanide gas on scale in fruit trees. The pesticide applicator felt scale was becoming resistant to cyanide, so he recommended that two fumigations or one fumigation and one pesticide spraying would do the trick, but even those strategies soon proved ineffective.

---

that the farmers would do nothing—because it wasn't perceived by them to be a real health and safety problem. Instead, large-scale farmers considered any alternatives to "business as usual" to be an unnecessary and expensive nuisance.

Big agriculture's defense of arsenic was based primarily on economic need. But they also argued that it was safe. History shows that this guarantee of safety was dangerously wrong. In fact, a large percent of the U.S. population began to show symptoms of both arsenic and lead poisoning before 1920. By the 1930s, well over a hundred million people in the United States suffered from mild to severe arsenic and lead poisoning.

In the middle of the Great Depression, muckraking journalists, following in the tradition of Upton Sinclair, mounted a national movement against false advertising and toxic food. The incidence of arsenic poisoning in the general population, and specifically the level of poisoning of babies and children, was one of several national issues covered by the muckrakers. A massive public education program finally exerted some impact on the arsenic lobby and the big growers. This campaign illustrated that reckless spraying of arsenic on almost all crops negatively affected the general health of the population. Still, despite criticism of arsenic's dangerous health risks in the United States, chemical corporations, large-scale farm interests, the FDA, and the USDA successfully resisted the application of the worldwide tolerances for arsenic and lead in the United States.

Though rebuffed repeatedly by government agencies and trivialized by the farm press, Arthur Kallett and F. J. Schlink, who helped found Consumer Reports and the Consumer's Union, finally alerted a much higher percentage of the general public to the problems of contamination and adulteration of foods. Their brand of public-protection muckraking journalism had a major impact on the public's perception of farm and home poisons. The books *100,000,000 Guinea Pigs* by Kallett and Schlink, *Eat, Drink, and Be Wary* by F. J. Schlink, and *40,000,000 Guinea Pig Children* by Rachael Palmer and Isadore Alpher alerted the public to the dangers of pesticide residues in food, especially arsenic and lead. They were the Rachel Carsons of the 1930s.[4]

Today there is widespread agreement on the dangers of lead in paint, solder, and water pipes. Major federal and state programs are directed at eliminating lead in the environment owing to its especially damaging effects on children. The judgment against lead today is so overwhelmingly convincing that it is hard to imagine farmers, experts, and editors ever defending its use on food crops or as a paint pigment.

Nonetheless, advocates for lead and arsenic were so powerful and well placed that a sea of hype, secrecy, and regulatory inertia swamped all arguments opposing its use. Unfortunately, this became the dominant pattern, one that is still firmly entrenched today. The company with the biggest printing press—or, actually, the largest advertising budget—emerges as the "authority," enjoys the highest product sales, and trivializes or ignores any criticism for years or even decades. Scientific experts working for the chemical corporations provide confusing scientific data, which they call "sound science," and government regulatory agencies delay restrictions on the chemicals until the public absolutely demands them

Sodium chlorate and calcium chlorate were the first widely used weed killers, having been developed in about 1898. From *California Cultivator*, June 7, 1930.

because of rising death and injury rates.

The FDA and even USDA issued warnings throughout the 1930s about pending restrictions on both arsenic and lead in the United States. This left farmers unsure about whether their produce would be seized for having higher residues than what was tolerated by the government for that year. The combination of muckraking journalism and regulatory threats were unnerving and worrisome to farmers, but as one might expect, the threat of more vigorous arsenic or pesticide restriction and suspension never really materialized into enforcement action by U.S. regulatory agencies. By the mid-1930s, many farmers realized that the useful life of most chemicals was short before serious failure or danger appeared. Recall that, within a few years after they were introduced, both Paris green and London purple became ineffective in combating certain pests they were meant to control. Shortly after their introduction, lead arsenic, carbon bisulphide, sodium chlorate, and cyanide began to exhibit ineffectiveness problems of their own.[5]

Within a few dozen insect generations (many have five or six generations per year) the insects began to tolerate the poisons. Between six and twenty years after their introduction most of the metallic and synthetic chemical pesticides experienced what has come to be

An early criticism of aerial spraying in the *Pacific Rural Press*, a major farm magazine. Here, beekeepers are complaining about the loss of beehives over a large area because of the drift and overspray of poisons from crop dusters. It also points out how pesticides drift off target and poison neighboring farms, houses, and wildlife.

known as resistance. Resistance is actually a tolerance of the insect or other pest to the poison, since tolerant individuals, who are not killed off by the chemical, come to dominate the pest population. As early as 1908, the codling moth developed resistance to lead arsenate sprays in Grand Junction, Colorado. Red scale, purple scale, and black scale developed resistance to hydrocyanic gas during the 1920s and 1930s, in several locations.

In addition to the problems with resistance, some of the new chemicals exploded, killing and injuring farmers, farmworkers, and even the pesticide applicators working for the chemical corporations. Explosions from using both sodium chlorate (the first herbicide) and carbon bisulfide caused deaths and serious injuries. Sodium and calcium chlorate caused explosions in spray rigs and crop dusters. Scientists and farmers discovered that sodium chlorate had to be mixed with another ingredient (usually urea synthetic fertilizer) to stabilize the chemical in the spray tank and prevent explosions.[6] The application of pesticides from aircraft (crop dusting) began in 1921 but did not become widespread on commercial farms until the mid-1930s. Pesticide drift complaints from crop dusting, however, occurred as early as 1935.[7]

As pests began to show resistance to arsenic and muckrakers and regulators took aim at arsenic and lead's dangers, their status plummeted from that of miraculous and heroic chemicals to that of marginally useful but highly dangerous poisons. As the 1930s drew to a close, arsenic no longer held the position as the unquestioned darling of the farm lobby. Even arsenic's most ardent supporters were constantly on the defense, but still selling a bunch of arsenic.

By the 1940s, large-scale farmers found themselves in the awkward position of defending the safety of arsenic while at the same time punishing their kids for going near the arsenic storehouse. By the 1950s, arsenic's pesticide heyday had passed, just as its

medical heyday had passed in the early 1900s. Still, arsenic remained persistent. Though it was repudiated and reviled as a medicine, arsenic's medical importance lasted into the 1930s, and doctors still prescribe miniscule amounts of arsenic as a medicine today. Though reviled as a pesticide that had poisoned nearly the entire population, lead arsenic persevered until the 1960s and arsenic is still fed to baby chicks in small doses in the giant corporate chicken farms. And arsenic is still used to defoliate cotton and to kill weeds; hundreds of thousands of pounds were still used every year during the 1990s in California, the most regulated state.

The attacks against arsenic use on the farm began long before 1900, yet the government inaction on

The ad at far left (facing page) is from *California Cultivator*, June 7, 1924. The ad at near left is from *California Cultivator*, February 2, 1924. Only the killing power and efficiency are emphasized in the ads.

protecting consumers and their direct action on defending arsenic stalled the elimination of these terrible chemicals. By the time regulators began canceling arsenic registrations, arsenic products had become much less effective, as an increasing number of pests developed resistance and orchard soils had grown increasingly toxic from excessive applications. Fortunately, its use in agriculture has decreased since the 1960s, but it is still registered for certain uses. Some of the currently used forms of arsenic are calcium arsenate, arsenic acid, cacodylic acid, DSMA, MAMA, and MSMA. All of these formulations of arsenic are used on food except MAMA, which is used on turf grass that kids play on and eat picnics on.

While most farmers in the United States no longer use arsenic, its continued presence in agricultural soils in many areas of the United States remains high because so much was used for so many years. And since arsenic is a heavy metal, it is persistent and difficult to leach or flush out of most soils, so it is stuck where it is. And we are stuck with it for probably hundreds of years.

Several other chemicals replaced arsenic and were very important pesticides in the early twentieth century. These chemicals also experienced regulatory and public relations struggles of their own. As we have shown, after 1909 cyanide products became very popular. The major U.S. innovator and advertiser was American Cyanamid. Cyanide gas was an important pesticide for more than forty years, and American Cyanamid was a dominant chemical supplier beginning in the 1910s. Cyanide products were widely promoted either as high-quality nitrogen fertilizers or as deadly pesticides. Cyano-gas has been used as a soil and grain-warehouse fumigant. Early advertisements lauded the effectiveness of this gas without defining the dangers. Other suppliers, such as the California Cyanide Company, also sold both ammonium cyanide fertilizer and cyanide gas.

A 1938 American Cyanamid ad plays on themes of patriotism, politics, and the end to the depression. This ad tries to cover as many bases as possible, although a base the ads seldom covered at this or any other time was safety. From the University of California at Davis Shields Library Special Collections.

Cyanogas is listed as a dangerous Toxicity Class I poison, meaning it can be fatal if swallowed. It is easily absorbed through the skin and is life-threatening at a concentration of only fifty parts per million.[8] EPA registration for cyanogas was discontinued by American Cyanamid in 1978 after more than three decades of public injuries, deaths, and organized opposition from consumer groups.

Sodium cyanide has been used as a rodenticide to control rabbits, rats, gophers, moles, and voles. It also was used extensively on termites in California and Florida. It too is a highly toxic chemical, being a dangerous Class I poison and very toxic in small amounts. Sodium cyanide is still in use and should be handled carefully, if one is forced to handle it at all.

Another chemical product that enjoyed widespread early use is paracide. It has been used as an insecticide since the early 1900s. Its long history of effectiveness against peach-tree borers and other difficult-to-kill insects and their pupae made the regulation and eventual elimination of paracide complicated. Paracide has a long history of efforts to regulate it because of the health damage and poisonings it can cause. It is a suspected carcinogen and mutagen and causes liver, kidney, and lung damage and anemia.[9]

Though enormously dangerous, all these chemicals remained on the shelf and available to farmers and households until public resistance to them finally overwhelmed the chemical defenders and promoters. Only after bitter battles over each individual chemical were some poisons prohibited or regulated. Chemicals have rarely been voluntarily removed from a corporation's product line, even if the corporate officers knew that the poison was deadly or caused birth defects or cancer.

State and federal regulators and state and national legislatures have refused to adequately regulate and prohibit the use of these chemicals. Their refusals, resistance, foot-dragging, and road-blocking practices continue to this day. Just as the FDA recently closed its

eyes with regard to heart attacks and deaths attributed to Vioxx and Celebrex, so have the FDA, USDA, and EPA closed their eyes when confronted with pesticide damage and death for more than one hundred years.

Only organized protests against the remaining survivors of this terrible first generation of poisons will finally end their damaging and deadly careers. We cannot depend on the government or the supposedly "sound science" from the chemical corporations. These deadly killers are still used on our parks, on our football, baseball, and soccer fields, on our cotton plants and cottonseed (80 percent of which goes into dairy feed), on corn, soy, and rice crops, and on our stored grain.

This paracide ad included no advisories about how dangerous the chemical was or the fact that its manufacturers had struggled with the government regulators. From *California Cultivator*, March 29, 1924.

PACIFIC RURAL PRESS

February 24, 1940

Doctor Cox of the State Department of Agriculture insists on washing the pears to a tolerance well below the federal limit, and our pears, stripped of their natural wax by the rigorous washing, add market losses to the cost of washing.

❧ ❧ ❧

AMONG the growing army of state scientists who criticize the stiff-necked and inconsistent attitude of the federal purefood leaders, a pioneer has been Dr. S. Marcovitch of the University of Tennessee.

His specialty has been investigations of fluorine, while Doctor Cardiff has paid most attention to lead and arsenic.

On another page in this issue you will read that the dentists of the country are finding that a proper amount of fluorine is necessary to preserve teeth from decay.

It is true that an excess amount of fluorine will discolor teeth. In a few locations in this country drinking water contains that excess amount which produces mottled teeth.

But if you haven't sufficient fluorine, your teeth will rapidly decay.

Thus theorists are gradually learning commonsense from Nature.

Nature puts lead, arsenic and fluorine in our foods, because they are needed.

And when the scientific emotionalists try to improve on Nature, they merely make themselves silly, and play into the hands of people who make money out of scrubbing fruit at the expense of growers.

Reasonable spray tolerance would not be objected to by thoughtful fruit men. A reasonable amount of washing improves the appearance of apples and pears which have been heavily sprayed.

But the pure food scientists have thrown reason to the wind, and they reap the whirlwind.

That's the whole story, and the PRP has been glad to be of service to growers these last few years in helping put the unreasonable rule makers "on the spot."

That's the only way the farmer can get his rights—fight for them.

No Spray Residue Danger Found

FEDERAL authorities who have been grudgingly and ungraciously backing up from their inconsistent position in regard to lead and arsenic spray residue on apples and pears will gain no comfort from a recent report of the National Institute of Health.

This body observed 1,231 persons over a period of 14 months, to see whether they were injured by eating sprayed fruits or by breathing poison while spraying the fruit. Their findings will be published in a Public Health bulletin.

Their report does not show injury to human health. They found some persons who had not eaten any sprayed fruits or been exposed to any spray lead and arsenic, yielding measurable amounts of lead and arsenic in their urine.

There is no mystery about this to anybody except the stiff-necked pure food scientists, who made an inconsistent spray residue tolerance and have not wanted to admit their error. Lead and arsenic are widely found in the foods we eat and the water we drink. As the PRP has reported many times a lot of our common foods and some popular baby foods contain more lead and arsenic than a farmer is permitted to have on his apples or pears.

No evidence was found that spray residue lead and arsenic affects the blood, or contributes to disease. Nor has the Food and Drug Administration ever been able to prove in court injury to health.

Agricultural scientists and spray manufacturers are busily searching for non-poisonous sprays for codling worm control, hoping to get away from the senseless and endless and costly arguments about spray residue, but up to date no efficient substitute for lead arsenate has been discovered.

Meanwhile fruit washing, required by the pure food dictators, costs the western growers many millions of dollars.

This editorial argues that lead and arsenic are totally safe in spite of irrefutable evidence that both were deadly. From *Pacific Rural Press*, March 8, 1941.

The farm journals were constantly looking for scientists who would defend pesticides. Dr. Marcovitch, of the University of Tennessee, was happy to oblige. From *Pacific Rural Press*, February 24, 1940.

Chapter 16

THE PESTICIDES OF THE 1930s

Editorials like the ones shown at left dominated the rural journals in the late 1930s and the early 1940s. The writers conclude that spray-residue analysis is distorted by the natural presence of lead, arsenic, and fluorine in a person's body. This magazine argued that these necessary and natural quantities of poisons significantly increased the level of lead and arsenic when federal investigators checked for residues in people or test animals.

The editor was so defensive about pesticides that he argued that nature supplies these toxins to our bodies because they are needed. He does not add that only miniscule amounts are produced by the body— nothing like the high concentrations used on crops that poisoned many millions of people and killed thousands in the 1920s and 1930s.

By the late 1930s, many family farmers in America were left in a state of financial desperation by the Great Depression. They were alive and mostly well fed from fields and gardens, but they were broke from low prices, defaulted contracts, failed chemicals, and poor technology. Many farmers lost their land by the early 1940s, and the corporate farmers further consolidated their acreage by buying up land from these destitute farmers, who had failed to pay their mortgage or their equipment and chemical bills during the depression years.

In the 1930s pesticide resistance, chemical muckrakers, regulatory threats, and the public rejection of produce with high residues of arsenic and lead plagued the large-scale single-crop farms, especially those of apples, grapes, citrus, stone fruit, and lettuce. These conditions especially affected growers who depended on interstate and international markets. The large-scale farmers and the farm journals continued to protect arsenic but grew increasingly anxious to find and promote any new pesticide that would reduce residues, would still kill crop pests, and would not attract too much attention from the muckrakers. Arsenic's troubles were mounting, even though the farm journals continued to deny there was any evidence that arsenic and lead had poisoned almost the entire population. The chemical advocates continued to trivialize any criticism of arsenic, lead, sodium chlorate, sodium bisulfide, and cyanide through the 1940s and 1950s.

Because of so much public criticism during the 1930s, however, many farmers tried to replace arsenic with other, less notorious, and supposedly less dangerous pesticides. One of the pesticides that farmers, the USDA, and the farm journals seized upon was fluoride, a new discovery from DuPont.

Since there were at the time no FDA-announced residue tolerances for fluoride, the growers assumed that there would be no problem using it. Fruits began

Answered Questions Also on Other Pages. Inquirer Must Give Full Name and Address. Inquiries Improperly Signed Will Not Receive Attention.

Cockroaches Are Bothersome

To the Editor: Am sending you a bug for identification. What is it and how may it be controlled?—T., Los Angeles.

Your bug proved to be an Oriental cockroach The cockroach family are sunlight haters, hiding away in dark crevices during the day and doing most of their prowling after sundown.

One very effective way of ridding premises of roaches is dusting with commercial sodium fluoride, either pure or diluted one-half with some inert substance such as powdered gypsum or flour. Most of the commercial roach powders now on the market contain sodium fluoride.

With the use of a duster of some type the sodium fluoride can be thoroughly dusted over floors and runways and into hiding places of the roaches. If any roaches are present you will soon know it for they will come dashing out of their retreats and rush about more or less blindly, showing evidence of discomfort to be eventually followed in the course of a few hours by their death. The dead or paralyzed roaches can be swept up and burned and complete extermination may be effected within 24 hours if the dusting is well done.

Sodium fluoride is deadly poisonous, however, and extreme care should be used that it is not dusted on foodstuffs.

Dusting with pyrethrum is also suggested. Pyrethrum is non-poisonous.

Another fairly good roach control method is by using traps. There are a number of home-made roach traps that can be made, but one of the simplest that we know of is to use tin bread pans, having vertical sides about three inches in height. Grease the sides with a little rancid butter and place them where the roaches are plentiful. The roaches will be able to get into the pan, but the slippery sides should prevent them from getting out.

———‹PRP›———

This editorial about sodium fluoride use on roaches praised its effectiveness but advised users that it was deadly poisonous. From the *Pacific Rural Press*, August 12, 1939.

to show up in the market with high levels of fluoride on them, and the FDA promised to act quickly to regulate it. They didn't, even though the fluoride pesticides were dangerous by-products from plutonium production. The chemical merchants acted much more quickly than the government, and the FDA set no meaningful residue limits until 1940.

Fluoride pesticides very effectively killed pests, so the chemical corporations aggressively promoted their use, and fluoride sales soared for several years before DDT and the World War II chemicals (discussed in chapter 18) became popular and captured most of the market. Other corporations besides DuPont eagerly entered the market, and ads for cryolite and other fluoride products began to appear frequently in farm magazines during the 1930s and 1940s, as the ads that follow show.[1]

ALCOA ads used a stylized farmer to testify for its brand of cryolite. Pennsylvania Salt used an Eskimo. Their ads claimed that fluoride, a uranium mining waste product, would kill pests on tomatoes, peppers, lettuce, cotton, cabbage, and cauliflowers.

Some of the "new" poisons sold to farmers in the 1930s actually had been known for decades, and a few for more than a hundred years. Carbamates and organophosphates were among the first organic chemicals to be discovered. Organophosphates were synthesized in the laboratory as early as 1822. Carbamate concoctions made from the African Calabar bean, were known to be deadly poisons in the 1700s. European chemists learned how poisonous carbamates were only in 1863, but by 1864 they had isolated the toxic component and created a synthesized copy.

It wasn't until the 1930s, however, that corporate and government scientists finally developed pesticide formulations from each of these classes of poison. In 1931, the DuPont Corporation began experimenting with carbamate pesticides with some limited success. In the 1930s, I. G. Farben conducted trials with organophosphates, first developing the antipersonnel poisons Tabun and Serin, and then formulating the pesticide TEPP in 1938.

None of these acutely toxic nerve poisons, however, were widely used even on European farms in the 1930s. Because of World War II, they remained unavailable to civilians and largely undeveloped until the 1950s. Yet between their introduction in the 1930s and the first years of widespread use in the 1950s, extensive research on a variety of carbamate and organophosphate formulations was being conducted by several corporations and cartels. I. G. Farben experimented with all the known chemicals on concentration-camp victims throughout World War II. Others also conducted trials on these two classes of nerve poison, including DuPont, Shell, Union Carbide, Basel AG (Ciba, J. R. Geigy, and Sandoz), American Cyanamid, and Rhone-Poulenc. But no major marketing efforts were conducted for either the carbamates or the organophosphate pesticides in the 1930s and 1940s.

The highly toxic bromines, however, were promoted and marketed for the first time in the 1930s. Swiss, English, and German chemists knew about the toxicity of the bromines before the 1880s and conducted extensive tests on a variety of product ideas. By 1897, America's Dow Chemical knew about the toxic character of bromines and began to research the potential value of several forms of bromine salt from Lake Michigan.

By the mid-1930s, the bromines began to be sold as pesticides to American farmers. In 1936, the California Department of Agriculture, Dow Chemical Corporation, and the farm magazines began to promote methyl bromide. By 1940, the University of California aggressively, and very effectively, promoted methyl bromide and chloropicrin blends.[2]

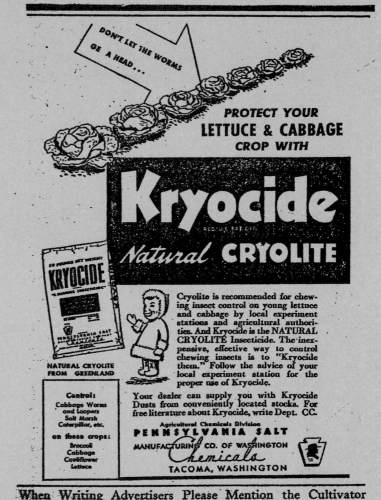

This ad for cryolite (a form of fluoride) stresses that it is a natural way to control chewing insects. Cryolite is a uranium mining by-product that is highly synthesized, highly toxic, and hardly natural.

Methyl bromide was touted as another of chemistry's "silver bullets" because of its broad-spectrum pest-killing power and its supposedly nontoxic character. Magazine editors, of course, advertised it as safe and nonpoisonous. After four or five years of vigorous promotion by magazines and scientists working for state and federal departments of agriculture and the

CRYO-SULPHUR 50-25
Controls Tomato Insects

CRYO-SULPHUR DUST 50-25 contains natural Cryolite, specifically recommended by your Governmental Experiment Station for Tomato Pin Worm. It also controls the other common chewing insects — Tomato Fruit Worm, Tomato Horn Worm and Flea Beetle. CRYO-SULPHUR DUST 50-25 also contains Sulphur, the recognized control for Tomato Russet Mite.

In Northern and Central California we also recommend the use of SULPHUR CALCIUM ARSENATE 25-75 for the control of Tomato Worm and Russet Mite.

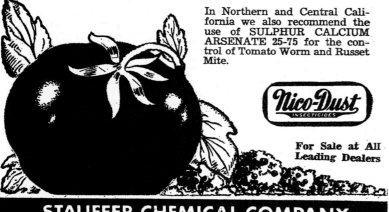

Nico-Dust
INSECTICIDES

For Sale at All
Leading Dealers

STAUFFER CHEMICAL COMPANY
Nico-Dust Manufacturing Division

636 California St., San Francisco 8 ◄ 3200 E. 26th St., Los Angeles 23

Stauffer, the longtime seller of carbon bisulfide, also sold fertilizers and a wide variety of pesticides, including sulfur, arsenic, and fluoride poisons such as cryolite, all advertised in this ad. From *California Cultivator*, June 9, 1945.

bromoxinil, dibromochloropropane (DBCP), and ethylene dibromide (EDB).

Even the most honest labels did not tell the whole story about methyl bromide. In the sixty-five years since its introduction, methyl bromide has turned out to be one of the worst. The editors, the scientists, and the agencies of agriculture were dead wrong in the 1930s about methyl bromide's safety. In all likelihood, methyl bromide will prove to be the most toxic widely used pesticide in history. It has even migrated into the upper atmosphere, where it damages the stratospheric ozone layer. Scientists have shown that methyl bromide is the most damaging ozone-depleting element, atom-for-atom, since it potentiates with chlorofluorocarbons (CFCs), thus prolonging the destructive effects of both chemicals.

In hindsight, we can say that methyl bromide is poisonous and destructive and dangerous, even though it was advertised as a safe chemical. The California Department of Food and Agriculture, which introduced this awful chemical, and the California EPA continued fighting for the use of methyl bromide well after the beginning of the new millennium—despite its terrible toxicity. And, like methyl bromide, all of the bromine chemicals cause serious problems, including environmental degradation, male infertility, water pollution, and death.

universities, farmers accepted methyl bromide, and its use escalated rapidly.

Like arsenic and the fluorine compounds, methyl bromide was an easy sell to large-scale farmers and grain elevators because of its deadly effectiveness and lack of regulation. As a result, sales were strong and increasing as America entered World War II. A host of bromine-based or highly bromated pesticides followed methyl bromide onto the farm, including

Another chemical that showed considerable promise to control a broad spectrum of pests was rotenone. This chemical was derived, without any significant alteration, from the roots of tropical plants. Since these plants were natural, rotenone enjoyed an extended life as a presumably safe pesticide. Lately, however, it has come under more careful scrutiny, and tests have shown it to be dangerous to humans. In the mid-1930s, however, rotenone looked like it would have a big "career," as an editor from the *Pacific Rural Press* predicted in February 1935.

As we noted earlier, in addition to carbamates, organophosphates, bromines, fluorines, pyrethrum powder, and rotenone, the U.S. Army, the Office of Management and Budget (OMB), and the USDA tested thousands of old and new chemicals for their toxic potential during the war years. The dominant chemical companies on the American side became participants and partners in these tests, often sharing results and processes with each other during the war. On the German side, the Nazi government and I. G. Farben continued their review of chemicals that they could use as pesticides, antipersonnel weapons, and explosives.[4]

By 1944, the ground war and the Allied bombing of German cities and industry allowed Russian, American, and British troops to win major battles and overrun industrial facilities. With each victory, more I. G. Farben chemical secrets and industrial processes became available to the military forces and the chemical corporations in France, England, and America. Near the close of the conflict in the mid-1940s, the American chemical merchants grew anxious to peddle these recently discovered or seized chemicals before the federal government had a chance to recover from the European and Pacific conflicts and began to test and regulate all of the new poisons. They theorized openly that the sooner they could get these products in the hands of farmers, the easier it would be to forestall the regulators, because they knew the corporate farmers would use their political clout (the Farm Belt and rural America still controlled Congress in the 1940s) to keep the poisons once they became dependent upon them, just as they had with arsenic, cyanide, and fluorine.

I. G. Farben factories, including those adjacent to concentration camps, were seized and the cartel disbanded by decree of the Allied Command in 1947. The longest Nuremberg sentence for any I. G. Farben criminal executive was four years. So, in spite of their active participation in the war and the labor abuse, torture, and murder of concentration-camp slaves, I. G.

● *It Gets the Invisible Pear Worms*

METHYL bromide, the amazing fumigant introduced to the United States in 1936 by D. B. Mackie, of the California Department of Agriculture, has a new victory to its credit.

More than 1500 cars of pears have been given an inexpensive bath of methyl bromide gas this year, disposing of the invisible codling moth worms, and causing the pears to ripen much more evenly and satisfactorily.

Both canners and shippers of pears are enthusiastic about the new method of treatment.

* The treatment of pears is readily done. After the pears are wrapped and packed and loaded in the refrigerator car, methyl bromide is introduced in the car, it gasifies and the air is stirred up with a fan. It is not necessary to have the gas under pressure, nor is it required that the car be calked. Any good-condition refrigerator car is a proper chamber in which to do the treating. Cost is only about $3.25 per car.

The methyl bromide gas penetrates the calyx and follows up the codling worm. Receivers of shipping pears are happy about the even ripening of the fruit. Worms, of course, hasten the ripening of infested fruit. The methyl bromide gas kills the worms and prevents this sort of experience.

Observers are very enthusiastic about the new use of methyl bromide, and call it a great discovery. They expect much wider use of the method next year.

Codling worms have been a plague of all the years to fruit growers, but this year was often called a "worst-ever" year.

Last year was a very severe year, and growers will recall a condition which has been described by D. B. Mackie and W. B. Carter in those words: "Large numbers of newly hatched larvae of the second brood that first appear at the time of early harvesting. Too small to be visible on external examination, yet noticeable to a degree on cutting, these young larvae were a source of much trouble and debate by the growers and inspectors during the 1938 season. These little worms grown in transit between the shipping point and marketing point and at the latter place show visible evidence of their presence, which facts strongly tends to depress the market."

Not only does the treatment discourage this sort of development of the worms in transit, but the pears arrive with good, uniform color, appear to ripen a bit earlier, ripen evenly, and bring a premium price, over those where there may be worm development, soft spots, uneven color and the like.

Back in 1936, when Scientist Mackie first announced his experiments with methyl bromide, it was thought of as a fumigant for nursery stock, where it is a very effective agent, easily penetrating the soil or moss around roots.

Then it demonstrated its value in fumigating potatoes, tomatoes, and other possible carriers of troublesome insects. The ability of the gas to follow up the worm into its burrow, was an improvement over other methods. The potato tuber moth yields readily to the methyl.

It is used for nuts and dried fruits.

It is the best lethal dose for the ground squirrel. 10 cubic centimeters of methyl bromide to a runway, gets the squirrel and is useful in control of bubonic plague, because it kills the fleas on the squirrel, both adult and in the egg.

Grain warehouses and mills use it.

One grain mill cleaned out its insect infestation, treating 680,000 cubic feet with 680 pounds of the methyl bromide. The material at 70 cents per pound thus cost less than $500 for a big job.

In the East methyl bromide is being widely used for the Japanese beetle and the white fringe beetle.

In Arizona they are fumigating baled hay to kill a new alfalfa weevil.

Incoming Hawaiian papayas are fumigated with methyl bromide.

It is an effective fumigant for overwintering codling moths, by gassing the fruit boxes. Sometimes 100 or more of the codling pests are found in one box.

The methyl bromide is a liquid with a low boiling point, thus it readily gasifies. Very little of it is absorbed, and its residues are inorganic, and non-poisonous.

This editorial/advisory promoted methyl bromide and claimed that its residues were nonpoisonous. Methyl bromide has been found to be one of the most toxic compounds on the planet, damaging everything from aquifers to the ozone layer. Thousands of people have been killed by methyl bromide. From *Pacific Rural Press*, September 23, 1939.

SPECIMEN LABEL

DOW

METHYL BROMIDE

Active Ingredient: Methyl Bromide ---------------------- 100%

FOR FUMIGATION
SEE MANUFACTURER'S PRINTED DIRECTIONS
Comply with Local Ordinances or Regulations

---DANGER---
**VAPOR EXTREMELY HAZARDOUS
HIGHLY VOLATILE • CAUSES BURNS**
Do not breathe vapor. • Keep away from heat.
Do not get in eyes, on skin, on clothing.

In case of contact, immediately remove all contaminated skin
covering, including shoes, clothing, adhesive or other bandages.
Do not re-use shoes or clothing until free from all contamination.
Wash skin thoroughly with soap and water, and flush eyes with
water for at least 15 minutes. If irritation or blisters develop,
get medical attention.

☠ POISON ☠
SEND FOR A DOCTOR IMMEDIATELY IN CASE OF ACCIDENT

ANTIDOTE—Place patient in fresh air, face downward, with head
slightly below level of lungs. Keep warm. Give artificial respiration
if breathing has stopped. Oxygen inhalation and stimulants of caffeine
at doctor's discretion.

NOTICE
 Seller makes no warranty of any kind, express or implied,
concerning the use of this product. Buyer assumes all risk
of use or handling, whether in accordance with directions
or not. 86-1098--Printed in U. S. A.

A 850

THE DOW CHEMICAL COMPANY
MIDLAND MICHIGAN
MIDLAND DIVISION

**While ads and editorials were effusive in their praise of and
encouragement for using methyl bromide, the labels were forced
to tell a bit more of the truth about chemical dangers, as this
1940s–1950s label illustrates. From the University of California
at Davis Shields Library Special Collections.**

Farben firms received only a slap on the wrist. And
after the breakup, each individual corporation within
the cartel retained ownership of all of its factories and
research centers and could still use its old processes to
produce chemicals for sale. However, they no longer
retained any patents on their products and they could
no longer form cartels.

Within ten years Bayer, BASF, and Hoechst had
regained much of their old power, and all the other I.
G. Farben firms have produced thousands of
poisonous products since the end of World War II.
Bayer, BASF, and Hoechst are currently among the ten
largest chemical corporations in the world, with each
continuing to exercise considerable clout in the world
of chemical research, product development, and
marketing.[5]

FIRST GRIST MILL IN NORTHERN CALIFORNIA, BUILT IN 1846 BETWEEN
CALISTOGA AND ST. HELENA, ABOUT 70 MILES NORTH OF SAN FRANCISCO

There are a number of plants that yield the drug known as rotenone chief of which are derris and cube root.

It has been widely experimented with throughout the United States last year and gave a mighty fine account of itself in most cases. In California it did a fine job of killing grape leaf hoppers, alfalfa loopers, cabbage worms, aphis and thrips.

Rotenone is both a contact and stomach poison although it has been chiefly used as a contact poison so far.

We predict a big "career" for this new insecticide.

First rotenone ad in the *Pacific Rural Press*, February 23, 1935.

Milling...a Pioneer Enterprise

East, west, north or south—an old mill is likely to be one of the oldest standing landmarks in its vicinity. After the War for American Independence and the subsequent westward movement of population, grain was planted and harvested from coast to coast.

Nearly every pioneer settlement had its grist mill to which grain was brought in wagons drawn by oxen and horse teams, to be ground into breadstuffs. Thus did commercial milling in America have its beginning. Today, volume shipments of grain come by rail, water and highway to the huge mills which serve the many branches of the food and beverage industries.

Dow has participated in the progress of the milling industry by contributing dependable grain fumigants and insecticides. As an evidence of our continuing research, we now offer Methyl Bromide with Chloropicrin added, on request. Our complete line includes spot and space fumigants, grain and food fumigants, and insecticides—including DOWKLOR, powerful new Chlordane insecticide.

> *This advertisement is No. 6 of a series covering early days in American milling. If you have a photograph of an old mill, with an historical background, possibly of interest to the milling industry, Dow would appreciate an opportunity to consider it for publication.*

The 1948 *American Miller and Processor* ad concentrates on the value to the user. Often the use of such bucolic ads confused the reader, whether farmer or consumer, and made it difficult to see such an "essential" poison in a bad light.

DOW METHYL BROMIDE
THE PENETRATING FUMIGANT

THE DOW CHEMICAL COMPANY · MIDLAND, MICHIGAN

New York · Boston · Philadelphia · Washington · Cleveland · Detroit
Chicago · St. Louis · Houston · San Francisco · Los Angeles · Seattle
Dow Chemical of Canada, Limited, Toronto, Canada

CHEMICALS INDISPENSABLE
TO INDUSTRY AND AGRICULTURE

Historical Markers

1946 ● First electronic computer built by US researchers.
● **First cases of DDT resistance in European houseflies.**

1947 ● **The Federal Fungicide, Insecticide, and Rodenticide Act (FIFRA) replaces the 1910 Pesticide Act.** The law, like its predecessor, regulates the synthetic organic pesticides and does not specifically address environmental or public health. The Agriculture Department (USDA) administers the law but does not set up a procedure for recall of products until 1969. The USDA filed criminal proceedings only once from 1947–69, although by 1969 documented pesticide accidents had risen to 50,000 per year. The USDA investigated only 60 accidents to people and 120 to animals in 1969.
● **The growth hormone DES is approved for use in cattle by the USDA and FDA, although animal tests since 1938 had shown it to be carcinogenic.**

1948 ● **First reports of DDT resistance in US houseflies.**
● **Gandhi is assassinated in India.**
● Transistor is invented in the United States.

1950 ● **Houseflies show almost 100 percent resistance to DDT on US dairies.**
● **Korean War begins.**
● Senator Joseph McCarthy's anti-Communist witch hunt begins.

1952 ● *Readers Digest* publishes an article titled "Organic Agriculture—Bunk."

1953
- **Structure of DNA is discovered.**
- TV becomes a popular form of entertainment.
- **Korean War ends.**

1954
- **French lose to the forces of Ho Chi Minh in Vietnam. The United States immediately continues the fight in Vietnam, contending that it is a civil war.**
- Segregation in US schools is declared unconstitutional.
- **Assistant Secretary of Agriculture Earl Butz claims agriculture "is now a big business" and farmers must "adapt or die."**

1958
- One hundred fatal cases of parathion poisoning are reported in India and sixty-seven in Syria.
- The **Delaney clause becomes law.** It disallows carcinogenic additives in processed food.

1959
- **Four University of California scientists advise farmers to use farming practices later known as integrated pest management (IPM), which is a major pest-control tool for modern organic farmers.**
- Antarctic Treaty is signed.
- Cuban Revolution. Rebel forces under Fidel Castro are victorious over the government of Fulgencio Batista.

1962
- **Rachel Carson publishes *Silent Spring*. It is an immediate best seller. Carson warns that excessive use of chemical pesticides is killing wildlife and damaging human health. Chemical corporations, the American Medical Association, and large-scale farmers begin an immediate and aggressive counterattack to Carson's views.**
- **The Committee for Economic Development publishes *An Adaptive Program for Agriculture*, which promotes huge corporate farms at the expense of family farms. This report guides government policy for more than a decade.**

1963
- **The President's Science Advisory Committee publishes a report vindicating Rachel Carson's research and assertions.**
- The National Agricultural Chemicals Association doubles its public relations budget.

1965
- The President's Science Advisory Committee decries the excessive use of agricultural chemicals and states: **"The corporation's convenience has been allowed to rule national policy."**
- **Cesar Chavez and Delores Huerta organize the United Farmworkers Union in Delano, California. Pesticide abuses are a major focus of their organizing work.**

1968
- *Science* magazine publishes an article linking bird population declines with reproductive failure caused by pesticide accumulations in their tissues.

1970
- **Environmental Protection Agency (EPA) is created.** Pesticide registration is moved from the USDA to the EPA.
- Campbell Soup Co. finds dangerously high levels of PCBs in chicken. Approximately 146,000 New York chickens are destroyed.

1972
- **The registration for DDT is suspended in the United States. It continues to be sold in many foreign markets, however, that export food to the United States.** More than 250 pests worldwide are resistant to DDT.
- **Jim Hightower publishes *Hard Times, Hard Tomatoes*, detailing the agribusiness bias of the land-grant college system.**
- A nongovernmental organization (NGO), the International Federation of Agricultural Movements (IFOAM), publishes the first standards for organic agriculture.
- The US EPA assumes principal authority over pesticides.

1974
- The first organic certification organizations are formed in the United States: California Certified Organic Farmers (CCOF) and Oregon Tilth. Both are NGOs.
- **First genetic engineering research is proposed.**

1975
- The Federal Insecticide, Fungicide, Rodenticide Act (FIFRA) reauthorized.
- **Economic risk assessment is imposed on the EPA, reviewed by the Secretary of Agriculture and both House and Senate agriculture committees.**

1979
- **First state organic law passed, in California.**
- **The registration for the growth hormone DES, used on US cattle, is suspended because of evidence that it is carcinogenic.**
- **Antibiotic- and hormone-treated meat from the United States is banned in Europe.**

1980
- Washington State implements the first state-run organic certification program.
- First Common Ground Conference is held in Maine by the Maine Organic Farming and Gardening Association (MOFGA), an NGO.

1981
- First Ecological Farming Conference is held in Winters, California, by the Steering Committee for Sustainable Agriculture, an NGO.

1982
- Chemical manufacturers push for support of genetic engineering to make crops resistant to herbicides. "If they wanted to reduce chemical use they would have pushed for allelopathic [weed-suppressant] characteristics," said Chilean-born University of California Professor Miguel Altieri in 1997.

1983
- Executives of the largest food-testing lab in the country are found guilty of falsifying tests determining food safety. Some 80 percent of

the testing is found to be phony. This lab worked on contract for the chemical corporations. The poisons tested remain on the market.

1984 ● **By this date 447 species of insects and mites are known to be resistant to one or more pesticides; 14 weed species are resistant to one or more herbicides.**
● **Michigan State University becomes the first land-grant college to sponsor a major conference on sustainable agriculture.**
● Milk containing dangerously high levels of chlordane/heptachlor (an organochlorine and relative of DDT) is recalled in Arkansas, Texas, Louisiana, Kansas, Missouri, and Oklahoma. Chlordane/heptachlor had been suspended in 1975 with a drop-dead date of December 31, 1980, except for subsurface treatment of termites.

1986 ● **Israel's breast cancer rate drops 30 percent in women below the age of forty-four, just eight years after the country banned organochlorine pesticides.**
● **The first national guidelines for organic production are published by the Organic Foods Production Association of North America (now called the Organic Trade Association).**

1988 ● **The US EPA finds seventy-four different pesticides in the groundwater of thirty-eight agricultural states.**

1989 ● **The conservative World Health Organization estimates that there are 1,000,000 pesticide poisonings in the world each year, resulting in 20,000 deaths.**
● **Two environmental groups, Natural Resources Defense Council and Mothers and Others for a Livable Planet, release a study implicating the apple growth regulator Alar as a cause**

of cancer. Alar is banned, and the furor over its widespread use helps stimulate a public demand for organic food. Apple growers suffer significant losses as the market for apples evaporates for several months.
● **The European Union (EU) bans the importation of US meat treated with any growth hormone.**

1990 ● There are 2,140,000 farms in the United States on about 1.1 billion acres. More than 40 percent of the land is rented. Some 600,000,000 acres is cropland.
● **Farmers spray more than 800,000,000 pounds of pesticides on US crops.**
● **Two out of five people are expected to get cancer in their lifetime.**
● The Organic Foods Production Act passes the U.S. Congress, establishing a framework for creating standards for certifying organic food and fiber products.

1992 ● **Tissue analysis demonstrates a substantial link between pesticides and breast cancer. The European Union approves the first government-enforced standards for organic production.**

1993 ● **The National Research Council warns that children are at particular risk from pesticide residues because standards were created to measure the effect of pesticides on adult males, ignoring children's lower body weight.**
● The National Academy of Sciences calls for greater regulation of pesticides and stricter tolerance standards.

1994 ● **The FDA approves genetically engineered recombinant bovine growth hormone (rBGH or rBST) after a long activist fight. The hormone was developed to increase milk production. The**

United States already has an enormous surplus of milk.
● **The genetically engineered Flavr-Savr tomato is approved for sale. It tastes terrible and no one buys it.**
● Studies link organochlorine chemicals with male reproductive problems and breast cancer. Problems include a 50 percent drop in male sperm count in forty years.
● Sales of organic food top two billion dollars. Average sales increases have exceeded 20 percent per year since 1991.
● **Monsanto and Delta and Pine Land get federal permission to grow genetically modified Roundup Ready cotton.**

1995–98 ● **More than 100 square miles of the Antarctic ice shelf drop into the ocean due to the warming of the seas. The resultant sea-level rises are already becoming catastrophic for island nations.**

1996 ● **Congress unanimously passes the Food Quality Protection Act. It repeals the Delaney Clause in exchange for a provision forcing the EPA to evaluate the combined toxic effects of all pesticide exposures, including those from water and food. This evaluation must use body weight and possible impacts on children as a baseline.**

1997 ● **The FDA approves food irradiation as a way of killing bacteria such as *E. coli* in beef.** *E. coli* results from the same dirty slaughterhouse conditions that Upton Sinclair exposed in *The Jungle* at the beginning of the twentieth century. Activists claim the process destroys nutrients and creates chemicals that may be mutagenic and carcinogenic. **A CBS News poll says 77 percent of Americans oppose irradiation.**

- **More then 600 insects and mites are resistant to one or more pesticides.**
- **Approximately 120 weeds are resistant to one or more herbicides.**
- **Approximately 115 disease organisms are resistant to pesticides.**
- **4 million acres of genetically engineered cotton, soybeans, and corn are grown in the United States.**

1998
- **The USDA finally proposes federal organic standards. These proposed standards include irradiation, genetic engineering, sludge as fertilizer, sodium chlorate as a defoliant, and allowing 20 percent of the animal diet to be not organic. The public responds with a deluge of more than 280,000 complaints.**
- **The organic food and fiber market continues to grow more than 20 percent per year.**

1999
- **In the United States, 45 percent of the cotton acreage, 45 percent of the corn acreage, and 57 percent of the soybean acreage is planted with genetically modified seed.**

2001
- The Human Genome Project finds that humans possess about 25,000 or 30,000 genes, not 140,000 as previously thought. This finding completely refutes the one gene–one protein theory advocated by DNA pioneers Watson and Crick and used by GMO corporations and US government regulators.
- **The USDA finally adopts national organic standards without including the use of sludge, irradiation, or genetic engineering as approved organic practices because of the outcry from 280,000 concerned voters.**

2003
- **Eighty-four percent of US canola acreage is planted with GMO seed.**

- **Five major weeds develop resistance to Roundup herbicide. Arsenic, paraquat, and 2, 4-D are recommended to control resistant weeds.**

2004
- **Seventy-six percent of cotton, 45 percent of corn, and 85 percent of soybean acreage in the United States is planted with genetically modified crops.** Many farmers have few options for obtaining other seed, since Monsanto and the other corporate conglomerates refuse to provide anything but genetically manipulated seed.

2005
- More than 9,000 farmers are accused by Monsanto of violating its patent rules. Most settle out of court. More than 190 farmers and businesses end up in court.

2006
- Brazil wins its WTO suit against US cotton for excess taxpayer subsidies.
- REACH is passed by the European Union. Evaluation of EU chemicals begins.
- Carbofuran becomes the only pesticide to lose its registration under FQPA.

2007
- Staphylococcus infections resistant to several antibiotics surface in various regions of the United States.
- The US House of Representatives sends a new Farm Bill to the Senate. The bill contains similar subsidy payments to those of the 2002 Farm Bill, in direct violation of the WTO's ruling on cotton.
- Wars in Iraq and Afghanistan drain most of the funding from farm programs like conservation security, equipment, sustainable practices, and water quality.
- Twelve weeds develop resistance to Roundup.

CALL TO FARMS

The Crop Corps was sponsored by the *Country Gentleman.* These promotional ads appeared from 1942 until the end of the war.

JOIN THE U.S. CROP CORPS C

FOOD FIGHTS FOR FREEDOM

Chapter 17

BOMBS AND FERTILIZERS

As our story has shown, chemical pesticide and fertilizer manufacturers had already enjoyed a long history by the time they began peddling DDT to farmers during and after World War II. They knew a thing or two about selling chemicals, but farmers were still very suspicious.

In the years that chemicals came to be used on farms—through the Civil War, the panics of 1873 and 1894, and the Great Depression of the 1930s—bankruptcy has been the rule in farming, not prosperity. Farmers knew that chemicals were not the total savior, since they had seen many fail and friends and relatives lose their farms. Unfortunately, in such desperate economic times, many farmers jumped at any cheap or free new product that guaranteed convenient fertility, high yields, the complete control of pests, or reduced labor.

It was here that the chemical corporations really demonstrated their promotional acumen and marketing skill. They had the genius and the economic wherewithal to give thousands of farmers enough of their product to conduct a test or two or three with pesticides, fertilizer, or seed. When the farmer's yield increased or pests disappeared after using the product, the farmer became convinced that the chemicals worked and would help advertise the product. When the products failed after the farmers became hooked on them, this was often enough economic loss to bankrupt farmers who had begun to believe the ads that claimed that the chemical products would help save their farm.

After World War I, synthetic nitrogen factories were built in the United States at several additional sites.

Allied Chemical and Dye Corporation (an I. G. Farben corporation) and DuPont produced 87 percent of the synthetic U.S. nitrogen before World War II. Shell-Union Oil, Dow, Hercules, Mathieson Chemical, and Pennsylvania Salt produced the other 13 percent. In 1939, the major nitrogen producers in the world, including the German nitrogen cartel (I. G. Farben), DuPont, Allied Chemical, the Chilean Nitrate Corporation, and Imperial Chemical Industries, were indicted by the U.S. government for antitrust violations of price fixing and market and supply control and were fined in 1941.

Such practices, however, were common in the fertilizer industry. In 1906, sixty-one phosphate and mixed-fertilizer companies were charged by the Justice Department with forming a cartel through a Canadian corporation. As early as 1916 the FTC published findings that profit margins on certain fertilizers were overly excessive. In 1926, thirty-seven fertilizer companies paid fines for violating antitrust laws. In 1941, sixty-nine fertilizer companies paid fines for similar violations. Unfortunately, the fines were just a slap on the wrist to the companies and amounted to $2568.00 for each defendant in 1926 and $3765.00 for each corporation in 1941. By 1958, nearly all the large potash and nitrogen producers had been parties to consent decrees with the Department of Justice, admitting their guilt. As the fines above show, the violators were not punished. A fine of a few thousand dollars for stealing millions from farmers is not punishment. And even though the government claimed that it attempted to level the competitive playing field in the nitrogen fertilizer industry, its success was limited.[1]

While the synthetic nitrogen supply in the United States was growing by leaps and bounds, it was still a tiny industry in the early 1940s. The 780,600 pounds of synthetic nitrogen produced in the United States in 1940 was only enough to fertilize a small amount of

More And More Fertilizer

Farmers used greater - quantities of fertilizer during the war than ever before, but not more than half as much as it would pay them to use "under conditions of national prosperity." This is one of the important conclusions of state production adjustment committees, as presented in a report released recently by the U. S. department of agriculture.

Published as "Cropland Use and Soil Fertility Practice in War and Peace," a section on fertilizer practices points to the sharp contrast between fertilizer use in World War I and in World War II. In the first period fertilizer consumption fell materially below the pre-war period. This was in part because of the cutting off of potash imports that had come from Germany. In World War II potash was supplied from domestic sources, as well as substantial supplies of synthetic nitrogen.

This editorial advocating more fertilizer use claims that fertilizer use is up because of the efforts between the two wars. This supposedly enabled the fertilizer merchants to supply the farmers, the troops, and Allied countries after the war. This was not true; U.S. fertilizer (especially nitrogen) merchants could not supply more than a small fraction of U.S. farmers. From *California Cultivator*, March 11, 1947.

acreage, because by this time there were more than 200 million fertilized acres in America. Though many farmers had begun to experiment with synthetic nitrogen, most did not depend on it. This was fortunate, because most of the U.S.-manufactured synthetic nitrogen and any sodium nitrate reserves were about to go to war, with most of it diverted to World War II bomb making.

Near the start of the war, Congress and President Franklin Delano Roosevelt pushed through military-preparedness bills to expand industrial capacity, including synthetic nitrogen production to make bombs and to grow more food for the war effort. The United States built eleven new nitrogen factories and financed the expansion of existing plants (there were seven before the war). By 1944, the nation's productive capacity had increased more than threefold.

This increase in production capacity in such a short period was spectacular. To the military strategists, the expanded production enabled them to combat Hitler's already significant nitrogen capability. Because of the increases, some of the larger farmers were able to obtain more synthetic nitrogen during the war than they had been led to believe they would have access to because of rationing. This was partly to encourage them to grow for the war effort at a time when farmers were growing weary of their sacrifices.

BASF's synthetic nitrogen had already proven itself a war hero to Germany in World War I, and it continued to be a boon to the German war effort in World War II. However, by World War II, the U.S. and Allied air forces also possessed the Germans' synthetic nitrogen and made use of it in bombs that destroyed city after city in Germany and Japan.[2]

During the war, farmers enlisted in a variety of "Food for Victory" campaigns and saw themselves as patriotic farmer-soldiers, which, of course, they were. Farmers lost a very large number of family members

Women Join the *"Field Artillery"*

as International Harvester Dealers

Teach Power Farming to an Army of "TRACTORETTES"

By the early 1940s many advertisements were couched in the military genre. Bombs, bullets, patriotism, and personal sacrifice all appeared as centerpieces of ads for the war effort.

What is a Tractorette?

A TRACTORETTE is a farm girl or woman who wants to help win the battle of the land, to help provide Food for Freedom. She is the farm model of the girl who is driving an ambulance or running a turret lathe in the city. Like her city sisters, she has had the benefit of special training.

> » BUY WAR BONDS
> » TURN IN YOUR SCRAP
> » SHARE YOUR CAR

Late last winter International Harvester dealers began to train this summer's Tractorettes. The dealers provided classrooms, instructors, and machines. The Harvester company furnished teaching manuals, slide films, mechanical diagrams, and service charts. The girls themselves were required to bring only two things—an earnest willingness to work and a complete disregard for grease under the fingernails or oil smudges on the nose.

They studied motors and transmissions, cooling systems, and ignition. They studied service care. They learned to drive tractors. They learned to attach the major farm implements that are used with tractors. And they were painstakingly taught *the safe way* to do everything.

'Step on it, kid; ya got gas and rubber to burn!'

U.S. JOY-RIDER

Dr. Seuss

In a climate of fear, unknowing, and personal sacrifice and deprivation the chemical corporations easily played on the nation's ragged war emotions with victorious images, powerful heroic messages, and criticism of those who refused to support the war effort. The Dr. Seuss cartoon above appeared in *PM* magazine on May 11, 1942. The ad below for Allis Chalmers, from *Wallace's Farmer*, July 11, 1942, continues the cartoon theme of ridiculing Hitler and Tojo.

to the fighting: more than 250,000 rural enlistees dead and more than 400,000 wounded.[3] In spite of the loss to the war effort of millions of men and women from the rural workforce, U.S. farmers produced most of the food and fiber that enabled the troops to win the war.

The fertilizer corporations increased their sales promotions during the war because of the increase in nitrogen and potash production and a large domestic phosphorus supply. They believed that there would be enough nitrogen and potash for war and for farming. The spectacular increases in nitrogen and potash production were advertised to the farmers as the pathway to cheap fertilizer and financial solvency after the war was won. The fertilizer companies assaulted farmers with testimonials, infomercials, and promises about the amazing yields they could get with synthetic nitrogen after the war—when they promised there would be plenty for everyone. That promise proved to be false. There was a small increase in their sales because there was an increase in production, but there still was not enough nitrogen to build bombs and supply any more than a trivial amount of U.S. acreage until well after the war.

During the war and in the immediate post-war years most Americans were respectful of anything that helped the troops destroy Hitler's Germany and Tojo's Japan, be it chemicals, gas, pesticides, or fighter planes. Americans at home had been shielded from the war but were greatly fearful and distraught over the loss and injury of so many sons. Nighttime blackouts and shortages of all commodities (including gas, sugar, eggs, and butter) were commonplace throughout the war.

Chemical merchants linked nitrogen, phosphorus, potash, and sulfur to the heroic war effort against the Japanese and Germans. Advertorials argued that without the chemical fertilizers there would probably have been a shortage of food and clothing for the soldiers. Certainly there were shortages of almost

Even the farm animals pitched in and loaded cannons with milk and food for the war effort. This ad was a joint effort between the USDA and the Chilean Nitrate Educational Bureau. From the *California Cultivator*, March 1942.

An ad for sulfur advises that the best farm defense requires the use of sulfur for the soil, for the pests, and for drying fruit. From *California Cultivator*, February 1942.

everything in the United States. Farmers suffered during the war years, and many took out loans and mortgages to stave off the threat of bankruptcy. Yet they still kept producing food and hearing promises of cheap fertilizer and good prices after the war.

While farmers struggled to stay afloat after the war, the government sold (actually gave away) the nitrogen plants to the chemical corporations. Thereafter the corporations sold the nitrogen back to the farmers and reaped most of the benefits from this wartime defense investment by the American taxpayers. This was a slap in the face to many desperate farmers, who knew that the nitrogen plants had been built with taxpayer money and expected that a significant percentage would be sold (or given) to farmer cooperatives.

The government response was that it had distributed the federal nitrogen plants mostly to firms that had not been part of the government's suit against price fixing in 1939 (settled in 1941). Ostensibly, this giveaway was designed to break up the concentration in the industry and make it more competitive. To some degree that strategy succeeded, but with only marginal effect. By 1955, eight corporations controlled 72 percent of production and five of those controlled 55 percent.

Mathieson Chemical, Spencer Chemical, Lion Oil (Monsanto), Allied Chemical and Dye, Phillips Chemical, and Hercules Powder gained the most additional nitrogen capacity from the taxpayer-built plants, but many other companies profited handsomely too, as the chart on page 166 illustrates. Six years after the war (1951), the U.S. nitrogen industry had grown 430 percent and had a 3.35-million-pound capacity to produce fertilizer and bombs. By 1955, nitrogen production had risen to 5.7 million pounds—a 730 percent increase in fifteen years.[4]

The fertilizer merchants must have known that they would be receiving the government giveaway long before the war was over, because even before the

BREAD IS JUST AS IMPORTANT AS **BULLETS**

U. S. CROP CORPS

FARM HELP IS COMING

Country Gentleman is distributing thousands of posters in small towns everywhere to help the Department of Agriculture enlist townspeople in the U. S. Crop Corps.

Above is a May 1943 ad from the *Country Gentleman* soliciting patriotic cooperation in the production of food for the war. As the war dragged on and on farmers became weary of the sacrifices, including the extra workload caused by the loss of so many farm and rural kids who went to war. But campaigns such as "Bread Is Just as Important as Bullets" and "Food Fights for Freedom" helped tie the farmers to the war effort and the troops.

fighting ended the chemical firms cranked up their war-hero ad campaigns to an unprecedented level. Editors, scientists, and the corporations promised that the fertilizers, which were part of the package of "chemicals for victory," would finally be more widely and cheaply available to all farmers. The synthetic fertilizers were advertised as one of the most important tools in the crusade to rebuild war-ravaged Europe and Asia, end worldwide hunger, and—by the way—make farmers rich.

Nitrogen plants weren't the only government giveaway at this time. A significant portion of the technology and the industrial capacity established during the war (especially for chemicals, electric devices, and farm equipment) was given to corporations to manufacture products to sell to America's farmers and householders.

Banks advised farmers how to deal with government demands on their productivity and the stresses of wartime production. This patriotic bank also offered to loan them money to "assist farmers in meeting war time problems." From *California Cultivator*, May 1942.

UNCLE SAM WANTS FOOD

DUE to the necessities of war, farmers everywhere are faced with shortages. Labor, sacks, transportation and a thousand-and-one other things are on the list of items affected.

Conditions call for many readjustments in normal operations so that crops will be harvested and protected for the use of America and her allies.

To provide farm storage and feeding facilities, many farmers—even with mortgaged or leased land—are taking advantage of Title I loans with maturities up to ten years.

To further assist farmers in meeting war time problems, Bank of America has introduced a new type of Commodity Loan to finance crops stored on the ranch under acceptable warehousing facilities.

To learn more about either of these desirable types of war time loans see your local Bank of America.

UNCLE SAM WANTS FOOD

Bank of America
NATIONAL TRUST & SAVINGS ASSOCIATION
CALIFORNIA
Member Federal
Deposit Insurance Corporation
Member Federal Reserve System

Greyhound was one of the only war advertisers in the farm journals that didn't try to sell its product. In fact, it encouraged prospective travelers to stay home unless it was really necessary! Its message in this November 28, 1942, *California Cultivator* was "War Service Comes First!"

April 18, 1942   Vol. LXXXIX, No. 8

CALIFORNIA CULTIVATOR
Los Angeles   An Illustrated Biweekly for the Golden State Farmer   San Francisco

Doing Their Share to Win the War

Covers of journals often carried a farmer-as-warrior message, as this edition from 1942 did.

Each sector of the farm economy ran emotional promotions intertwining war and production, like this advertisement for the dairy industry. This one was in the *California Cultivator* in November 1942.

Unless Really Necessary

PLEASE DON'T TRAVEL

WAR SERVICE COMES FIRST!

Don't travel unless you must . . . War service must come first . . . Ask yourself, before you travel: Is this trip necessary?

These, briefly, are admonitions and warnings of our Government, for all transportation is now vitally needed to carry our Armed Forces, workers in war plants and other essential travelers.

In the past, Americans took extra trips over the Holidays, creating an above-normal volume of travel.

But these are not normal times. Today transportation is faced with a serious shortage of equipment. Trained personnel has been depleted by the need for fighting men. Lower operating speed has become necessary in order to save rubber. Now gasoline rationing will make

it necessary for many additional essential travelers to depend on already overloaded buses.

That is why we appeal to you to postpone any pleasure trip you have planned which requires the use of public transportation. But *another* important reason is that thousands of our boys in the Armed Forces want to be home for the Holidays. Every one of them who can get leave deserves this privilege . . . we are not asking *them* not to travel. You, by giving up your trips, can add to their enjoyment and comfort. We are sure that every American will be eager and proud to make this sacrifice.

So please don't travel unless absolutely necessary, especially over the Holidays and week-ends.

GREYHOUND

VITAL TO VICTORY

The idea for this poster was originated by Steve Sacksteder, a milk inspector of the Los Angeles city department of public health, and it is being distributed by the California Dairy Council, particularly for posting on dairy barns and milk houses and wherever dairy folk gather.

Indeed, the hands that milk the cow are vital for victory. The rugged fingers of the Pacific Coast dairy farmers never tugged harder than now. Their wartime job is to meet, first of all, the needs for milk supply among the scores of military establishments which dot the west coast countryside; next, to provide their share of the 100,000,000 pounds of milk powder, 37,000,000

THESE HANDS are Vital to Victory

cases of evaporated milk, and 20,000 tons of cheese, ear-marked for lend-lease.

After that, they still must furnish the milk and other dairy foods needed by a civilian population that has grown enormously, especially in cities that are bursting with defense workers. No food is more necessary to mankind than milk, nor was the need ever greater.

Because of increased capacity, coupled with a temporary drop in the demand for nitric acid for bombs and aggressive fertilizer ad campaigns, there was a marked increase in the sale of synthetic nitrogen and triple and superphosphates soon after the war. In the short term, the success of these products seemed miraculous. And the chemical corporations advertised the products as being "miracles."

The table below illustrates the exponential growth of the fertilizer industry in California since the turn of the last century. Farmers used only 4 billion tons of fertilizer in the forty-year period from 1903 to 1944. In the fifty-year period from 1944 to 1993 California farmers applied 83 billion tons of fertilizer; nitrogen accounted for most of that growth. In the late 1930s about 3.5 percent of U.S. farm acreage was fertilized with synthetic nitrogen. By the late 1950s nearly half of the acreage was farmed with synthetic nitrogen, and by the 1990s it was more than 90 percent. By 2000 70 percent of total fertilizer use was synthetic nitrogen; most of the rest was phosphorous, potassium, and sulfur.

This was the time of the major changeover in U.S. agriculture to chemical dependency. This is the reason that many people date the beginning of chemical farming to this period after World War II. It was not the beginning, but there was a major shift in farming practices from a dependence on farm inputs to a dependence on purchased inputs, especially fertilizer.

What the long-term fertilizer ad campaigns and government promotions never revealed—or even remotely alluded to—were the real dangers of using synthetic nitrogen, mined potash, and phosphorus. While commercial synthetic fertilizers are frequently lower in acute toxicity than synthetic pesticides, they are still very toxic and regularly cause serious injury and even death.

The damage from fertilizers (especially to aquifers, groundwater basins, and soil) can be every bit as serious as the environmental and health damage from pesticides. This chemical assault on the environment, farmers, and farmworkers has created serious health problems in every agricultural area of the country. In many cases the nitrogen and phosphorous levels are so high in a water supply that death can and does result from regularly drinking the water. Pollution from nitrogen and phosphorus especially damages infants and children because of their small body size. Blue baby syndrome, which causes death in infants, is common where nitrogen pollution is high. Other ailments from contaminated water are intestinal troubles, rashes, blisters, and internal bleeding. Rural children bathe in contaminated water, they drink it plain, they drink it in teas and Kool-Aids, and they play in it. Poor farmworker children and the rural poor in general usually have no other water options. Bottled water is more expensive than gasoline.[5]

SYNTHETIC NITROGEN CAPACITY BY PRODUCER IN 1940, 1945 (BEFORE AND AFTER DISPOSAL OF GOVERNMENT-OWNED PLANTS), 1951, AND 1955 [a]

(Thousands of Short Tons of N)

Company	1940	1945 Before disposal of govt. plants	1945 After disposal of govt. plants [b]	1951	1955[c]
Allied Chemical & Dye Corp.	200.0	208.0	285.0	395.0	over 450.0f
E. I. duPont de Nemours & Co.	138.7	168.8	168.8	188.0	188.0
Shell Union Oil Co.	24.3	24.0	24.0	75.0	118.0
Hercules Powder Co.	10.0	10.0	42.0	55.0	55.0
Dow Chemical Co.	8.4	11.3	11.3	34.0	34.0
Mathieson Chemical Co.[e]	4.9	4.0	265.0d	271.0	233.0d
Pennsylvania Salt Co.	4.0	5.0	5.0	9.0	over 9.0 f
Lion Oil Co.			100.0	160.0	298.0
Commercial Solvents Co.			44.0	50.0	100.0
Spencer Chemical Co.			173.0	196.0	254.0
Phillips Chemical Co.			45.0d	125.0d	253.0d
Smith-Douglass Co.			18.0	18.0	146.0
Midland Ammonia Co.				20.0	20.0
Mississippi Chemical Co.				24.0	89.0
Hooker Electrochemical Co.					....(f)
American Cyanamid Co.					....(f)
Atlantic Refining Co.					....(f)
Consumers Cooperatives, Inc.					58.0
John Deere Co.					58.0
Delta Chemical Co.					58.0
W. R. Grace & Co.					72.0
Tennessee Valley Authority		50.0	50.0	55.0	140.0
Total Private	390.3	431.1	1,231.1d	1,675.0d	2,850.0d
Government Owned (excluding TVA)		800.0	[261.0]d	[261.0]d	[261.0]d
Total	390.3	1,231.1	1,231.1	1,675.0	2,850.0

From Jesse W. Markham, *The Fertilizer Industry*, (Nashville: Vanderbilt University Press, 1958).

Nitrogen fertilizers also cause serious injuries to farmers, farmworkers, and fertilizer applicators who come in contact with or breathe its vapors. Anhydrous ammonia (an especially toxic form of nitrogen) has blinded, injured, and killed many applicators and mixers.

In the aftermath of World War II, a new generation of scientists, led by Nobel Prize winners Norman Borlaug and Paul Muller, worked with the chemical corporations who had declared a new War on Hunger, which they called the Green Revolution. The Green Revolution encouraged farmers in many parts of the world to use synthetic nitrogen, World War II pesticides, and hybrid seeds. Borlaug and Muller argued that chemicals and "miracle" seeds were essential to reconstruct war-battered countries and feed famine victims in the undeveloped parts of Asia, Africa, and Latin America.

However noble and right-minded, the Green Revolution and the massive use of chemicals seemed, for many small-scale farmers it proved disastrous. The negative impact on millions of farmers and farm families and on their native food systems casts a dark shadow on many of the Green Revolution's claims of success. Millions in India, Latin America, and Africa have been fleeing to urban ghettos since the 1960s, after their soil fertility declined, yields flatlined, chemical prices soared, rains failed, and world market prices tanked. Farms went bankrupt trying to grow rice, grains, and cotton for export.

Claims of success due to chemical fertilizers are also often touted for farming in the United States. However, while in the short term yields rose because of fertilizer use, in the long run farm bankruptcies increased considerably because farmers were unable to pay their seed, fertilizer, and pesticide bills or their mortgages.[6] Since World War II an average of more than 650 farms per week have gone bankrupt, and big oil, big chemical, banks, insurance companies, and large-scale farmers have gobbled up the bankrupt land.

Buyers underestimated the size of Glenn County almond farmer Joe Billiou's crop. Joe Billiou, Shell, and the agricultural experts underestimated the damage to groundwater and aquifers that would occur from long-term use of synthetic nitrogen. This ad appeared in the *Pacific Rural Press* in 1938, and similar Shell ads appeared frequently throughout the late 1930s.

By 1987, fertilizer concentrations in the groundwater and aquifers of Glenn County were among the highest in the state, averaging above 45 parts per million of nitrate nitrogen in many wells. This undrinkable water can be deadly to children. Similarly damaged water systems can be found all over the rural United States. This map of high nitrogen levels is from the California Department of Food and Agriculture Fertilizer Reports, 1992.

WELL LOCATIONS WHERE NITRATE LEVELS
HAVE BEEN RECORDED AT 45 mg/l OR
GREATER DURING THE PERIOD
1975 THROUGH 1987

FIGURE 2

NITRATES ABOVE SAFE
DRINKING WATER LEVELS.
1992

GLENN COUNTY

In this ad, the happy young soldier is promoting a war-developed herbicide, Dinoseb. From *California Cultivator*, 1947.

Chapter 18

WAR TOYS

T he number of advertisements for almost all pesticides was reduced significantly in the rural magazines during World War II. Editors and publishers once again promoted beneficial insects, pyrethrum powder, Bordeaux mixture, sulfur, copper, and rotenone to make up for the lack of availability of the more highly toxic pesticides. The scarcity of supply was created mostly by demands for pesticides and other chemicals from the military and from farmers who contracted to grow for the military. Yet while the products were occasionally scarce and the number of ads reduced, the pesticide promotions continued.

An editorial called "New Insecticides," from a 1943 issue of *California Cultivator*, illustrates the vigorous advocacy in farm magazines for untested, unregulated chemicals such as DDT and DD long before they were available to farmers.

The old saying that there is nothing new under the sun, may be correct, but that doesn't mean that new uses cannot be found for some of them. Take for instance the fabulous DDT or Gesarol, which is a chemical combination that has been known for a long time but didn't have much use as far as anyone could tell. Finally, a Swiss scientist got to experimenting with it and found that it has remarkable characteristics. He couldn't patent the material because it had been made by numerous investigators so he got a patent on its use. Whether this will stick or not is probably a matter that will have to be decided by the courts. This material, which is said to be harmless to plant or animal life, is sure death to insects. Painted or sprayed on the walls of a cow barn, it will kill every fly that lights there for the next 30 to 60 days, according

to reports. It will do the same when painted on window screens, only for a longer period.

It is said to be the best treatment ever found for cooties and that soldiers' uniforms treated with the material can be sent to the cleaner several times before it loses its effect.

There is evidence that it may be of use in killing newly hatched citrus red scale for a considerable period after it is applied, which would be a big help in controlling this pest.

We will hear much more about this insecticide in the near future.

In this new deal on pest control materials the d's seem to have it as the other one is called DD and its specialty is the control of nematodes and wireworms in the soil. It is a discovery of the Shell Oil Company which has been used experimentally in the pineapple fields in Hawaii. A recently concluded experiment there showed three tons of fruit per acre on untreated check plots as compared to five times that much on treated plots.

Experiments here this summer lead to the belief that this material will be exceptionally effective in controlling both nematodes and wireworms commercially, without too much expense.

Neither material is available in commercial quantities at this time, but once their effectiveness is definitely established, production will become merely a matter of routine, particularly as soon as the military requirements ease up.[1]

Finally, in 1944, after years of propaganda and promise in the farm journals, the chemical merchants began to advertise that the war-hero pesticides and poisons that were developed in the 1930s and 1940s would soon become widely available. By 1944, with the war's end seemingly in sight, the chemical merchants began their promotional campaigns. They anticipated an immediate economic boom by turning the chemical swords they produced into plowshares. But the war stretched on much longer than anyone anticipated. Though the propaganda and the promotions for chemicals continued, the war pesticides were not widely available until nearly two years later.

The first and most famous pesticides released after World War II were DDT and DD in mid-1945. By September, advertisements for DDT began to appear in many farm magazines. As was the case for many of the chemicals introduced in the 1930s, chemists had synthesized DDT and DD much earlier. DDT was first discovered in 1874. But no one fully realized its value as a pesticide until 1939, when Paul Muller sprayed DDT on arsenic-resistant Colorado potato beetles in Europe while working for the Swiss chemical corporation J. R. Geigy.

By the time DDT was released in America, farmers had been told for four or five years that it was miraculous and safe and could kill bugs, mites, and worms that the older chemicals couldn't control any longer because of pest resistance. Corporations promised that DDT and the other powerful war-hero chemicals would replace or supplement their old medical heroes—arsenic, lead, cyanide, fluorine, sulfuric acid, sodium chlorate, and mercury. Ultimately, DDT and the war-hero chemicals did replace or complement the older poisons, until pests developed resistance and the war heroes encountered public relations problems of their own.[2]

Even more powerful than the chemical potency of DDT, however, was the sales campaign for this pesticide. When Geigy introduced DDT, the firm was the minor partner in the Swiss chemical cartel Basel AG, which included Ciba, Geigy, and Sandoz. Geigy's revenue and stock shot up after DDT. Thereafter the corporation became a major player in the cartel until the dissolution of Basel AG at the end of World War II. After the war, Geigy's stock shot up higher as it became the first corporation to promote DDT around the world. But Geigy was not the only firm to profit. Almost all chemical corporations profited handsomely from sales of DDT, since most agricultural chemical manufacturers and merchants sold several versions of it.[3]

Almost immediately after DDT became available for farm use, the soil fumigant DD came on the market. DD and the various DD-Telone formulations to follow became wildly popular. DD-Telone is still the most-used soil fumigant in the world in spite of the fact that it is a terrible toxin.

Within a few weeks after DDT and DD were released in 1945, farm magazines began to advertise the labor-saving and agronomic value of highly toxic herbicides such as 2,4-D (Tributon) and 2,4,5-T, both produced by American Chemical Paint (Amchem). Subsequently, Monsanto began manufacturing both of these chemicals and shortly thereafter made its close cousin, Agent Orange.[4]

If the public could have evaluated 2,4-D and 2,4,5-T, there might have been a chance to question their

WIN *this fight*

...and WIN *the Harvest, too!*

Growers' chances of winning the harvest are in early and proper planning for the fight against insects and fungous diseases... using the *right* ammunition and getting it "on hand" well in advance ... applying enough spray to each tree—*on time.*

Because serious losses can result from "too little—too late," the wise grower knows he must stay a step ahead of his orchard enemies at all times. In line with good battle strategy, he makes it a point to get his weapons early ... depends only on the spray materials that have shown their worth "on the prov-

ing ground"—effectiveness established by field performance year after year.

Commercial fruit growers for example, have found that they can rely on Orchard* Brand "Astringent" Lead Arsenate. They know from years of experience that the patented "astringent" feature will give quicker, better "kill" of codling moth ... and that this combined with the flake-like particles means a uniform, protective cover that cuts down worm stings and entries.

To win your fight for Harvest Victory—start right. *See your Orchard Brand dealer now.*

*Reg. U. S. Pat. Off.

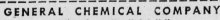

GENERAL CHEMICAL COMPANY
40 Rector Street • New York 6, N. Y.
Western Sales and Technical Service Offices:
235 MONTGOMERY STREET   1031 SOUTH BROADWAY
San Francisco 4, Cal.   Los Angeles 21, Cal.
Wenatchee, Yakima and Seattle, Wash. • Denver • Vancouver, B. C.

Pesticide manufacturers and merchandisers used military themes as much as possible. From *California Cultivator,* May 1945.

The first DDT ad in the *California Cultivator,* September 1, 1945.

RELEASED *for* HOME-FRONT USE

PENCO **DDT** INSECTICIDES

DDT has been released by the War Production Board, and Pennsylvania Salt Manufacturing Company is making this sensational insecticide for agricultural and civilian use.

Having produced DDT for the U. S. Government—for controlling the spread of louse-borne typhus in Europe and for control of insect-borne tropical diseases in the Pacific, Penn Salt is processing Penco DDT in liquid and dust forms suitable for convenient, effective use by farmers, dairymen, poultrymen and householders.

Penco DDT Insecticides are available in suitable dust and spray bases. Write us for complete information.

AGRICULTURAL CHEMICALS DIVISION

PENNSYLVANIA SALT
MANUFACTURING COMPANY
Chemicals
1000 WIDENER BUILDING, PHILADELPHIA 7, PA.

NEW YORK • CHICAGO • ST. LOUIS • PITTSBURGH • CINCINNATI
MINNEAPOLIS • WYANDOTTE • TACOMA

The second ad for DDT published in the *California Cultivator,* March 2, 1946. This General Chemical version of DDT bore the names Genicide A and Genetox S 50. To have chosen these labels at this early date now seems prophetic, since DDT is blamed in the genocidal decline and eradication of certain species.

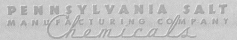

GENERAL CHEMICAL DDT PRODUCTS

FOR APPLES, PEARS, PEACHES and Other fruits and Vegetables

GENICIDE-A For Combined Control of CODLING MOTH and MITES

GENITOX S50   50% DDT for Apples, Peaches, Grapes, etc.

GENERAL CHEMICAL COMPANY
40 Rector Street • New York 6, N. Y.
Western Sales & Technical Service Offices:

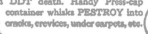

Sherwin-Williams Paint Company was one of the original marketers of Paris green, the arsenic pesticide. After World War II, Sherwin Williams found it profitable to sell DDT. Pestroy DDT was painted on screens and walls and retained its killing power for months. From the *California Cultivator*, 1946.

toxicity or danger. But the public was left out of the decision-making process on the use and safety of farm and backyard chemicals. These were heroic chemicals from the war and the Green Revolution. Questioning or criticism of the war chemicals was not encouraged or even usually dignified with a response. But some of the public did question their use and dangers soon after they were released, as we shall see shortly.

It seemed as if the chemical firms were saying "questions about safety be damned," because soon after 2,4-D and 2,4,5-T's introduction, General Petroleum (Texaco) began marketing another weed killer, 2,4,6-D or General Weed Exterminator, which was also called Dinoseb.[5] All three formulations combined chlorinated hydrocarbons and phenoxies in very toxic weed-killing mixtures. General Weed Exterminator, or Dinoseb, ultimately had all of its registrations canceled in the United States, but before that happened it was vigorously sold as cost effective, fast acting, and safer than arsenic.

These formulations were advertised to be "savior chemicals" after government restrictions were threatened for arsenic herbicides from the 1930s to the 1950s. These mixtures replaced and supplemented many uses of the first weed killers: sodium chlorate and calcium chlorate as well as arsenic herbicides. After the war, 2,4-D, 2,4,5-T, and 2,4,6-D were advertised to be effective at killing even the most persistent weeds such as Russian knapweed and morning glory.

The advertisements for these herbicides claimed they would be profitable to farmers in their wars on weeds. This resulted in huge increases in 2,4-D and 2,4,5-T use on U.S. farms. Parallel weed-killer advertisements from the same corporations aimed at military use, promising that the herbicides could kill all vegetation. This sales campaign promoted the most toxic herbicides, the same ones that came to be used extensively in Southeast Asian wars beginning in the 1950s and continuing until the mid-1970s. The

Research Aids Agriculture

By Joe Crosby

• • • Today, he said, research in agricultural chemicals is going forward with more intensified effort than at any time in the past.

In the field of insecticides, Dr. Hansberry declared DDT might well be considered the most important recent discovery, since it has started chemists thinking along entirely new and different lines. There are several compounds of DDT which will control a number of different and varied types of insects. One of these materials, 666, is used for not only controlling plant pests above the ground, but also is working out well against wireworms in the soil. • • •

The 1947 Joe Crosby editorial above promotes the use of DDT chemical relatives such as DD, which Shell called 666, "the Mark of the Devil." The 1946 ad for DD below stressed the importance of this chemical in resolving the international food shortages. It also stressed the fact that it was reasonably priced. From *California Cultivator.*

The ad for 2,4-D above appeared on April 3, 1946, in *California Cultivator.*

following World Wildlife Environment Report describes the results of using these defoliators and killer chemicals in Vietnam:

From the mid-1960s to the early 1970s the United States sent planes loaded with herbicides to Southeast Asia. They sprayed, from the air, more than 72 million litres of these deadly chemicals on to the forests and farmland of South Vietnam. Called Operation Ranch Hand, these repeated attacks had devastating ecological, economic and social consequences in the region. Scientists who studied the tragic aftermath coined a new term—ecocide. It means killing the environment. Chemical weapons were first used in World War I, when chlorine, phosgene and mustard gas were released by both sides to kill 100,000 people and injure another million. But, Operation Ranch Hand was the first massive employment of herbicides in the history of war. Three major chemical warfare agents were used—Agents Orange, White and Blue. Agent Orange and Agent White accounted for 61 per cent and 28 per cent, respectively, of the total volume dropped. They were aimed mainly at forests. Agent Blue accounted for 11 per cent and was used mainly for crop destruction missions.

The theory behind Operation Ranch Hand was that guerilla warfare would be impossible without cover, so it was decided to turn Vietnam's tropical forests into a desert. At the same time, farmland spraying destroyed food crops, making life impossible for the peasants and striking at the root of national resistance. The only animals which survived in large numbers were rodents—and they caused sudden disease epidemics. In sublethal quantities, Dioxin can result in birth defects, genetic damage and cancers. Vietnam veterans exposed to it for long periods during the war were at risk of passing on abnormalities to their children many years after they left the contaminated areas. The consequences are even worse for the Vietnamese, who had to remain behind.

The A Luoi valley is a sadly apt example of the devastation wreaked by Operation Ranch Hand. It was once a luxuriant tropical forest, teeming with wildlife. There were as many as 170 different birds and more than 30 mammals, including elephants, Javan rhinos, guar, tigers, clouded leopards, and a variety of primates. . . . Nearly 150 species of birds, all the large mammals and tens of thousands of cattle were lost. All that remains today is a desert of grass—dubbed by the Vietnamese as American grass—and a charred and stunted forest of bare tree stumps.[6]

Clearly, their "heroic" reputation did not follow these weed killers out of the Vietnam War. No one advertises 2,4-D or 2,4,5-T with warrior imagery any longer unless it is in a military defense magazine. In retrospect, Agent Orange was a terrible chemical that destroyed a generation in Vietnam and poisoned 500,000 U.S. soldiers. The millions of Vietnamese civilians and U.S. soldiers damaged from this chemical cocktail are horrible examples of what has come to be called "collateral damage." While 2,4,5-T is no longer registered in the United States, 2,4-D still is.

In the late 1940s and early 1950s the chemical corporations began pushing other highly chlorinated mixtures, the cyclodienes, which included toxaphene and the "drins" (aldrin, dieldrin, endrin, endosulfan, and endothal). These were also highly toxic and caused extensive chronic illnesses.[7]

Toward the end of World War II the Allies disclosed more results of I. G. Farben's organophosphate research. By 1947 American Cyanamid had used the cartel's discoveries to develop parathion and followed three years later with malathion. After the defeat of the Germans in the war, the patents of Bayer, Hoechst, BASF, Casselli, Siemans, AGFA, Degussa, and Kaliwerks were no longer in force, so any chemical manufacturer could make parathion, malathion, carbaryl, or any of the other organophosphate and carbamate pesticides that the German chemical corporations (I. G. Farben) had developed as early as the 1930s. But, of course, they needed I. G. Farben employees who knew the processes to be able to use the patents profitably, so many were hired by U.S. corporations, and German corporations sold the rights to use the processes and also provided consultants.

The California Environmental Protection Agency and the Department of Health have shown that parathion and the other organophosphates are responsible for an enormous injury rate among farmworkers in the state. The organophosphates are nerve poisons and are close relatives of sarin and tabun, the killer chemicals that I. G. Farben created in the 1930s. Sarin was the chemical that a Japanese cult released inside a crowded railroad station in 1998, causing several deaths and widespread panic.

In spite of the dangers of spreading nerve poisons around the countryside and the fears of the public, the advertisers ignored those concerns and focused their promotions on the ability of the organophosphates, like parathion and malathion, to control so many pests on so many crops. As the years went by, the number of

The ad above for toxaphene is from *Agricultural Chemicals*, December 1946.

This ad for aldrin, from the *Nebraska Farmer*, March 20, 1954, claims that it is not only the most effective but also the safest chemical. Effective? Yes— for a while. Safe? Never!

pests that advertisers guaranteed malathion would control kept increasing.

During the same period of time the number of people injured by malathion also kept increasing, since it was used in several federal and state eradication programs. While corporations continued to sell malathion for control of an increasing number of pests, none of them seemed to be taking account of the increasing number of injuries caused by its overly zealous use.

In 1956, Union Carbide produced the first commercially successful carbamate pesticide in America, which they named carbaryl (sold under the trade name Sevin). It too was advertised as being especially deadly but safe to handle. Like malathion it was widely used on a variety of pests. And like the organophosphates, Sevin and the other carbamates caused an enormous number of injuries.

Both the organophosphates and the carbamates were eventually made in a variety of mixtures that were designed to kill a broad spectrum of pests, including fungi, worms, plant bugs, mites, nematodes, aphids, and, of course, the Colorado potato beetle. The farm journals published editorial promotions as each new organophosphate and carbamate product was introduced and praised the high kill rate of each of the new pesticide formulations.

Among the most deadly substances known to man, these two types of chemicals have dominated U.S. agriculture since they first gained popularity in the late 1950s. By the 1960s the chemical corporations aggressively promoted these chemicals as replacements for DDT, chlordane, lindane, the drins, toxaphene, arsenic, and lead—all of which were under attack from environmentalists, regulators, and farmers who claimed they no longer worked. That promotion, coupled with the loss of many important chemical pesticides, greatly increased the use of

The earliest parathion ad found in *California Farmer*, April 8, 1950. It was advertised along with ethylene dibromide and tetraethyl pyrophosphate (TEPP)—what a trio.

organophosphates and carbamates. These two classes of chemicals now account for more than 70 percent of all restricted-use pesticides.

During the late 1960s and early 1970s farm-magazine editorials took the position that farmers were being forced to use these acutely dangerous nerve poisons to replace DDT, which was then under attack. Editors lamented that the loss of DDT required farmers to find replacements, even if they had to be the extremely highly poisonous carbamates and organophosphates. Editorials in the 1960s claimed they were being used out of desperation on American farms. As noted above, however, propaganda for these chemicals started in the late 1940s, and broad-scale use began in the1950s, long before any threats against DDT from the regulators.

Almost all formulations of carbamates and organophosphates are acutely toxic at very low dosages, and the editors and the farm public knew it. The toxic spills, injuries, and deaths from using these nerve poisons began immediately after their use became widespread. As early as 1964, 76 percent of all the systemic poisonings in California were caused by organophosphates. Consequently, it was disingenuous for the advertisers and editors to pretend that these chemicals were promoted and used only after the attacks on DDT in the late 1960s, since by 1964 they were obviously in widespread use and had already caused substantial injuries.[8] But with the loss of DDT and several arsenic formulations, any journalistic reservations against organophosphates and carbamates evaporated, and the rural journals mounted an enormous propaganda campaign to sell every pound they could.

Since the early 1950s, both of these chemicals have been involved in countless poisonings of both wildlife and human populations around the world. The most terrible and disgusting human disaster occurred in Bhopal, India, in 1984. There, methyl-iso-cyanate, the carbamate precursor for aldicarb (Temik), metam sodium, and Sevin, escaped from a poorly engineered and poorly run Union Carbide chemical plant. The escaping gases immediately killed more than 2,000 people and blinded 200,000. Subsequently, more than 15,000 additional people died as a direct result of the leak. Even today, many people continue to die in Bhopal each month from the effects of this terrible pesticide-manufacturing blunder.

It might be "smart to grow with Temik," but in 1985, Temik, the same chemical that killed, blinded, and maimed tens of thousands of people in Bhopal, India, was being applied to watermelons in California. Several people died and hundreds more became seriously ill after eating the watermelons. This incident exposed the fact that growers and pest-control advisors are less than careful or knowledgeable about which pesticides are sprayed or should be sprayed on which crops. Temik was not registered for use on the

The malathion ad above comes with a promotional from Sunland Industries. In this ad malathion is claimed to control eighty pests on forty-five crops. From *California Farmer*, November 24, 1956.

watermelons, but it was legal to use on cotton, which the watermelon farmers also grew.

Four years later, in 1989, a chemical closely related to Temik, metam sodium, was spilled into the upper Sacramento River as the result of a train wreck. Aquatic and shore life along more than forty miles of the river was destroyed for several years because of this spill, and no one knows the future impacts on human and riverine populations in the area. As of this writing, human illnesses continue in this previously beautiful but now damaged stretch of river.

The organophosphates methyl parathion, naled, guthion, and malathion have all been used on food and have drifted onto neighboring communities, lakes, and rivers. Malathion was sprayed in nine southern states over a several-year period in an effort to eradicate exotic flies such as the Mediterranean fruit fly (medfly) and the Oriental fruit fly. It is currently being sprayed in several southern and southwestern cotton-growing states in a bollworm eradication program. Naled (or Dibrom, which is "morbid" spelled backward) is one of the most toxic of the organophosphates on the market, and it too has been used in cities and towns against exotic fruit flies. Farmworkers and farmers suffer the most from the use of the organophosphates, but fish, bird, and amphibian populations suffer serious damage as well.

In spite of their dangerous and deadly properties, or maybe because of them, the organophosphates and carbamates became accepted and defended, not only by the chemical companies that manufacture them, but by chemically dependent farmers, corporation-dependent university professors, advertising-dependent farm journals, and government farm agents dependent on corporate projects and grants. Together, organophosphates and carbamates began to dominate modern U.S. agriculture.

Even though DDT was incredibly effective at killing pests, it didn't displace arsenic as the most used farm poison until the mid-1950s. This is because the magazines devoted enormous space to the defense of the old chemicals, such as arsenic, lead, cyanide, paracide, and fluorine. They published editorials that ridiculed the FDA and attempted to trivialize the environmentalists' concerns about the danger of farm chemicals. Their advocacy was to be expected, since most rural magazines had stopped asking the hard questions about pesticide safety and effectiveness long before the turn of the century.

Predictably, the farm magazines and the scientific world of agriculture treated methyl bromide, fluorine, DDT, DD, chlordane, 2,4,5-T, the drins, the organophosphates, the carbamates, and all of the second-generation chemicals in the same way they had treated arsenic and lead. They hopped on each chemical-propaganda bandwagon and defended the poison as if the world would come to an end without its availability. And they created clever and deceptive promotions that played on farmers' economic fears and ignorance of toxicology issues outside their communities.

This ad from Union Carbide for Sevin guaranteed fewer residue problems, the safeguarding of wildlife, and no soil contamination. It was claimed to be so safe that the applicators would need no protective clothing, and finally, that the use of Sevin made farmers good neighbors. From the University of California at Davis Shields Library Special Collections.

A 1996 two-page ad for aldicarb (Temik). There is no mention of the Bhopal disaster, just an emphasis on growing smart by using Temik and the potential loss in profit and manhood if a farmer doesn't use this deadly poison. From *California Farmer*, February 1996.

Post–World War II sales of the chemicals to farm families in the United States were consistently profitable because chemical corporations and magazine editors knew that very few farmers traveled broadly. The advertisers and salesmen knew that farmers didn't know much of what was going on outside their own community. They knew that the rural American public's grasp of history was short, its awareness local and often culturally bound within a relatively small region.

DDT had failed in Europe after four or five years of use. Most U.S. farm families, often parochial by choice or economic circumstance, didn't know about these DDT failures. Most farmers also didn't know there were already concerns about environmental damage from DDT by the mid-1940s, before they ever used a drop. There was no TV yet and reports from Europe on the radio were censored by the networks for the "war effort." Oldest surviving sons and farm fathers were exempt from the military draft so that they could run the farm. Because labor was scarce, they focused most of their attention on the farms, not on national or international events, even though farmers followed the progress of the war as closely as they could. As a result, most were not aware of any problems with the newly released chemicals. And the media affected a hear-no-evil, speak-no-evil, see-no-evil attitude toward the war poisons.

Consequently, a chemical could fail in one place and still remain popular and be considered effective just a few miles away. Often, failures occurred as close as a neighboring state or county—yet, just as often, no one in the adjacent area knew of the problems. The chemical corporations banked on the fact that farmers were not informed about most product failures, dangers, and spills. Most of us *still* are ignorant about the presence, and the dangers, of these chemicals in our environment and on our food.

Manufacturers have depended upon our collective ignorance and isolation and have continued to sell their chemicals and publicly defend them until long after their effectiveness has passed—and until long after they have damaged our health and our communities. The corporations put whatever spin on the failure of each chemical that they could, in order to keep the poisons on the market and derive as much revenue as possible, for as long as possible. They fought, and continue to fight, bitterly with all their time-worn themes—safety, incredible effectiveness, feeding the world, frightening brand names, labor savings, union busting, war, destruction, babies, boxers, beautiful women, violence, and the silver bullet (the techno-fix).

The adaptability of the ads and the ad makers to deal with the safety and product quality concerns of the customers for the last 160 years is capitalism at its finest. Their adaptability emerged in the 1980s after several terrible accidents and deaths involving their products. So, finally, after so much criticism about the accidents with and the danger of so many chemicals, the corporations shifted their focus and concentrated on the safety of handling and application. Then, while no civilians were looking, during the Gulf War in 1991, these "safe" chemicals (such as Deet and Roundup) morphed into battle-hardened warriors once again. In the late 1990s, advertisers shifted their tactics once more and began selling their chemicals as "environmentally friendly." Such doublethink opportunism punctuates the history of farm-chemical advertising and explains why it is so successful.

Farm magazines today still occasionally laud DDT and lament its loss. There's been noise among some politicians, here and abroad, to bring back DDT to control malaria in Africa, the argument being that banning it amounts to cynicism (even racism) by far-out Birkenstock-wearing environmentalists, who sit in their comfortable American homes and care more for the snail darter than they do for the poor suffering people of Africa.

There is truth to the indictment of spoiled Americans, but this is rhetoric designed to point fingers at well-intentioned people in order to mask the fact that DDT is still widely used in many parts of Asia, Africa, Latin America, and the Middle East. It comes home to your dining room table on a regular basis if you buy nonorganic foreign produce. Proponents argue that it is still effective on a wide variety of pests and it still kills mosquitoes in many malaria-infested regions. Environmentalists, toxicologists, and public-health advocates have argued against the use of DDT for malaria everywhere since the 1960s, but they have not been able to stop its use.

Others continue to advertise and promote arsenic herbicides and defoliants. Contemporary rural magazines still tout atrazine, Telone II, 2,4-D, and Kelthane, all of which are highly toxic and highly persistent relatives of DDT. During the Flit campaigns of the 1930s and 1940s, advertisers created a huge market for household pesticides. Over that same period, the chemical merchants expanded their targets to include businesses and government agencies. After the war the householders and landscapers filled their Flit guns with DDT instead of Flit, and consumers comfortably sprayed everything to make sure that the disease-carrying flies, mosquitoes, and bugs and the scary spiders were killed.

Like DDT, many chlorinated hydrocarbons almost immediately developed local and regional pest resistance problems and/or caused severe environmental damage to wildlife—and people. Though ineffective in one region, DDT and all the war poisons were sold aggressively in other areas, until overwhelming local evidence of ineffectiveness or regulatory cancellation finally destroyed or stopped sales. Arsenic, lead, and other first-generation chemicals were generally effective for twenty, thirty, even fifty years. But the World War II generation of chemicals began to lose their effectiveness much more quickly, often within four or five years

In promoting these World War II products, only the benefits were included. No mention was made of the real dangers to applicators and farmworkers from short- or long-term exposure. Only on the labels of the pesticide containers did any of the toxicity of the chemicals appear, but that information usually applied to acute exposures, not long-term impacts. No mention was made anywhere of the potential dangers of birth defects or cancer. And, while many of these World War II chemicals are still in use, and excellent toxicological data now exists, the real dangers from using these poisons are still not mentioned.

It is safe to argue that the period from the 1940s until the late 1990s was chemical agriculture's heyday. It was a time when the war on bugs thrived by selling record amounts of poisons and by using an enormous amount of energy to make and sell record amounts of nitrogen fertilizer. The industry was especially successful in the 1940s and 1950s in selling these pest and fertility wars with warrior imagery and as warrior strategies.

Ron Kroese, a longtime farm and rural activist, feels that the most important chemical sales messages after World War II employed the war motif, with WAR in capital letters. In his timely chapter "Agriculture's War against Nature" in the book Fatal Harvest, Kroese argues persuasively how and why this particular advertising strategy became especially successful after World War II:

> Most pesticides continue to be sold with logos and messages that convey power, dominance, and, often, violence. The overarching message they shout again and again at the farmer is that to succeed as a top producer you must do battle with nature, and you are dependent on our pesticides to win that fight.
>
> The all-out effort to achieve victory in World War II had a profound and lasting effect on American agriculture. The war established and legitimized the moral tone that continues to dominate agribusiness and government policy today. Indeed, it is scarcely an exaggeration to state that World War II didn't so much end as the guns and bombs were turned on the land. The result of this battle for production continues to be viewed in many quarters as a triumph. Yields of major crops have doubled and tripled from prewar levels, efficiency in terms of human labor has increased dramatically and U.S.-controlled transnational corporations dominate global agricultural commerce. It is increasingly obvious, however, that it was a Pyrrhic victory. We are only now beginning to fathom the real costs of our war on nature as the detrimental effects on the environment become clear and the casualties in terms of social costs continue to mount.[9]

Why then did American agriculture embrace or, perhaps more accurately, become enveloped by the military-industrial complex during World War II and the decades that followed? The most obvious answer is because the culture as a whole did. That is, as a result of the events involved in the mobilization and victory of the war, and the decision by government and industry that our country remain a global military and industrial power after the war, the United States became a national security state, a country constantly poised for war, and militarism became, and remains, an accepted aspect of our national identity. Agriculture merely played its role in that big picture. It should not be surprising then that today the United States is both the planet's number one arms merchant and the world's leading user and exporter of pesticides. Nor is it surprising that several large Pentagon contractors are also pesticide manufacturers.[10]

This photo is from *Hog Farm Management*,
September 1979.

Chapter 19

ANIMAL CONFINEMENT
AND THE PHARMACEUTICAL FARM

The small American farms that raised livestock as well as row or orchard crops and had suffered through depression and war faced even greater challenges shortly after World War II. Farmers were buffeted by the costs of changing both equipment and practices as farms became more chemically intensive and mechanized. Small farmers struggled in vain to compete with the further consolidation and expansion of the highly integrated farm monopolies, such as Continental Grain, Cargill, Archer Daniels Midland, and Richland Rice and Dunavant Enterprises and J.G. Boswell in cotton.

In their struggle to survive, many sought to copy the corporate industrial model of farming and got bigger. Others attempted to sell directly to local outlets as they always had. But following the war, an urban and suburban housing expansion took place near many cities that raised the value and the tax base on millions of acres of farmland. The increased tax base and increased noise and smell ordinances made it more difficult to farm conveniently and profitably close to cities. This cut tens of thousands of growers off from marketing their products directly to consumers.

Then, in the early 1950s, chemical researchers threw another technical dagger at family farmers. The chemical-pharmaceutical firms introduced a dramatically new technology that completely transformed the meat, poultry, and dairy industries—antibiotic farming. This miracle technology allowed farmers for the first time to confine large numbers of cows, pigs, and chickens in enclosed areas and still keep them healthy. Advertised as a revolution in animal management, this breakthrough resulted in no small part from the discovery that antibiotics, when added in subtherapeutic dosages to animal feeds, could be used to promote weight gain and prevent diseases. These management practices set in motion the next great reduction of small and diversified family farms in the United States by allowing the creation of huge, intensively managed animal-confinement operations.

In 1949 Dr. Thomas Jukes, working for Lederle Laboratories, a division of American Cyanamid (the first manufacturer of ammonium cyanide in the Americas), was studying several microorganisms to find which ones could produce quantities of B-12, a vitamin largely absent from the soybean- and corn-dominated livestock diet. One microorganism he studied was the precursor for the antibiotic chlortetracycline, also known as tetracycline. After extensive experimentation, Jukes accidentally found that feeding small amounts of this antibiotic to chicks, piglets, and calves caused them to significantly increase their weight. No one is sure what happened physiologically, but the use of antibiotics somehow suppressed harmful bacteria inside the animal's digestive tract, which enhanced growth.

Jukes and the media heavily promoted the immense possibilities of this discovery, calling it one of the most important developments of the century. In March 1950, the New York Times ran a front-page story on Jukes's findings. Word of Jukes's discoveries occupied the attention of the media for weeks. The New York Daily News even ran a political cartoon that showed Harry Truman administering a growth-promoting dose of tetracycline to a pig in an effort to increase pork for Truman's supporters.[1]

The confinement strategy for farm animals that the Jukes discovered came to dominate U.S. agricultural practices. The photo at right is from the inside of a "bacon barn" or "hog hotel," as these jails for hogs are variously called. Notice the girth restraints on the hogs, which further confine them to their prison cells.

Livestock rancher and journalist Orville Schell studied the meat industry between 1978 and 1984 for his book, Modern Meat: Antibiotics, Hormones and the Pharmaceutical Farm. Schell, who was a cattle rancher as well as an investigative journalist, realized that the use of antibiotics and hormones could be disastrous to farming, to farm animals, and to people who ate the meat. He took on issues of antibiotic abuse and endocrine disruption due to chemical farming practices long before farmers or almost anyone else in the United States realized the dangers. Schell found that by 1954, six years after the discovery of tetracycline's effect on

animal growth, U.S. farmers used 490,000 pounds (245 tons) of antibiotic feed additives in livestock feed.

In 1977, Jukes boasted that the results produced on farms were so spectacular, especially with pigs, that we could not begin to supply the demand." By 1960, 1.2 million pounds were used annually. By 1985, it was 9 million pounds. Schell felt that by 1985 it was the exception rather than the rule to find a farmer who did not use antibiotic feed additives on his livestock. By 2001 more than 19 million pounds were being used.[2]

Pharmaceutical corporations claimed that with antibiotics, the stress of the crowded conditions of confinement could be overcome. Again, the federal agencies did not challenge these claims. So, the path was paved for today's industrial bacon bins, chicken factories, and feedlots where pigs, chickens, and beef cows are raised in cramped conditions and treated with several chemical additives and drugs to keep them from getting sick in crowded cells and pens. However, Schell found that within a decade after the large-scale introduction of antibiotics into feedstocks microbiologists in Japan had discovered that their extensive use had led to the emergence of bacteria that were resistant to the antibiotics themselves.

The Japanese scientists called the resistance phenomenon "infectious drug resistance," which means that subsequent generations of bacteria could develop resistance to antibiotic drugs, then transfer that property of resistance to other bacteria as well. Within a few more years, the overmedication of animals to induce weight gain had effectively begun to eliminate any of the benefits that antibiotics offered in terms of weight gain and resistance to disease in crowded pens.

By the 1970s, hundreds of generations of bacteria succeeded in developing resistance to the drugs being fed to cattle at low dosages but over a long period of time. As more bacteria became resistant to the same antibiotics that humans depended on, the continued large-scale use of them in chicken, beef, pork, and milk cows threatened the elimination of these antibiotics for human illnesses.[3]

It soon became clear to doctors and researchers that antibiotics had to be used only as a last resort in order to preserve their capacity for protecting human health, or they would be lost to medicine. Instead of heeding the warnings of the doctors and medical scientists, however, the drug companies convinced farmers to use antibiotics like vitamins or food supplements. In 1979, Dr. Richard Novick of the New York City Public Health Research Institute told Orville Schell that "the only hope of maintaining the usefulness of antibiotics is to use them for specific purposes in a limited and carefully controlled manner, and only against organisms that are known to be sensitive to them."[4]

American Cyanamid's promotions in *The National Hog Farmer* pose these questions:

> ASK YOURSELF: Can I afford to finish my hogs with Aureomycin tetracycline during a tight year?
> THEN ASK YOURSELF: Can I afford slow growth, poor feed efficiency, cervical abscesses, bacterial enteritis, and the drag of atrophic rhinitis during any year?
>
> Aureomycin doesn't cost, it pays.

Sadly, the problems that the farmer is forced to treat in this ad are the results of confinement hog management. The pigs of farmers who use rotational grazing suffer these ailments either not at all or at

From *Hog Farm Management*, September 1979.

least not at anywhere near the same rates as continuously confined pigs.

Responding to Orville Schell's questioning in the early 1980s about the amounts of information on the use of subtherapeutic drugs in trade journals such as *The National Hog Farmer*, *Pig American*, and *Hog Farm Management*, Stanley Falkow of Stanford University said that a lot of misinformation was being disseminated:

> And I think that this misinformation has been fueled in part by people from some of the pharmaceutical houses. These drugs are the source of a lot of revenue, and of course their manufacturers want to defend them. It's hard to know if antibiotics really contribute to growth promotion, because most of the research the drug companies have done has been aimed at justifying their claims rather than understanding the actual mechanisms of what happens when these drugs are used.[5]

In addition to antibiotics, hormones also became one of modern medicine's crossover drugs for the food industry. Among them was a synthetic estrogen called DES, diethylstilbestrol. After being discovered in the 1930s, DES was prescribed to six million women between the 1940s and early 1970s to prevent miscarriages. In 1966, however, Dr. Arthur Herbst discovered that a fifteen-year-old girl had developed clear cell adenocarcinoma of the vagina. This girl would be just the first of thousands of children who were born to mothers who took DES and who subsequently suffered cancer and/or dangerous abnormalities of the reproductive organs.[6]

It was also discovered that DES could cause cattle to gain weight quickly. Pellets of DES were implanted behind the ears of steers, and doing so caused a 15 to 19 percent weight gain. Schell, as cattleman, showed that this speed-up of beef toward slaughter weight

could potentially save livestock and poultry farmers 7.7 billion pounds of feed per year. Finally, after lengthy battles, DES was banned in 1979, even though drug companies continued to manufacture and sell it and veterinarians, cattle hands, and ranchers continued to administer it to livestock.

When Schell conducted the research for his book he found the head cowboy who had blown the whistle in 1980 on Coronado Feeders (owned by Allied Mills). The disgruntled cowboy wrote a letter to company executives in Chicago disclosing that over 50,000 head

The American Cyanamid ad below and the Hoechst Chemical ad on the facing page are typical of the antibiotic ads that have flooded the journals from the 1950s to the present. Both are from *Hog Farm Management*, July 1979.

of their cattle had been illegally implanted with DES by the managers of the feedlot. His letter put a serious scare into meat consumers, feedlots, the USDA, and the FDA. But the DES scare quickly vanished off of the front pages as other crises emerged, just as the multiple meat recalls of millions of pounds of tainted beef vanished in only a few days of press in August, September, October, and November of 2007.

When the media spotlight focused on other stories in 1980, the drug and chemical companies rushed to introduce numerous alternatives for dairy, livestock, and poultry producers while still defending DES as long as possible. Though out of the headlines, the FDA discovered that more than a half million head of cattle had been treated with DES by more than 300 separate feedlot companies after the November 1, 1979, the cutoff date for its use. Forty-nine drug companies were found to be distributing the drug illegally after July 13, 1979, the cutoff date for sales. None of the ranchers or the chemical corporations were ever punished for violating the ban.

Some of the most horrific cases of human DES exposure emerged in Puerto Rico, where Dr. Saenz de Rodriguez, a doctor of pediatric medicine, treated nearly four hundred cases of premature sexual development between 1979 and 1981. Rodriguez launched a heroic fight to discern what was causing boys and girls under six years of age to develop oversized breasts and pubic hair, finally tracing the cause to hormone-contaminated chicken meat, which had been served to children at a school cafeteria.

The unfettered and unregulated use of hormones in beef and milk production has caused grave concern in public health circles about eliminating the medical value of hormones for humans. The clash since the late 1970s between health advocates such as Rodriguez, Herbst, and Schell on one side and the chemical firms on the other remains unresolved, despite profound evidence of serious health risks,

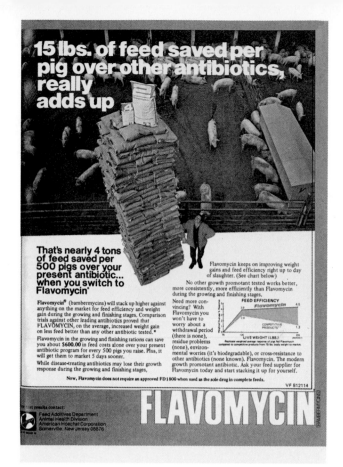

The Flavomycin ad above claims that $600 is saved on feed by using it on 500 pigs. That's only $1.20 per pig. But what is the cost to consumers in increased resistance to antibiotics?

injury, and death from the use of hormones in meat production that these scientists and investigators predicted. The meat and drug industries continue to argue that the amounts of hormones administered to animals are miniscule in comparison to dosages for humans, and therefore any fear on the public's part is largely exaggerated. On the other side, what Orville Schell wrote more than twenty years ago remains indelible today:

> . . . what is forgotten is that hormones, substances that are active even in the minutest amounts, are now mass-produced and used with an unnerving indiscrimination. Not only

are these compounds capable of bringing about the kind of imbalance of the endocrine system that I have described, but by exciting tissue cells that are sensitive to them, hormones are also able to promote carcinogenesis.[7]

Roy Hertz, M.D., who worked in the National Cancer Institute's Endocrinology Branch for twenty-eight years, emphatically told Schell that:

> . . . in the case of hormonal food additives . . . I simply don't see any reason why a person should be involuntarily exposed to them. I simply take a purist view about these kinds of food additives. Unless there is a compelling need—in the case of a famine, for instance—I don't think the actual, or even potential, risk can be justified. For drug companies to present these compounds as being without risk . . . is completely out of order, since we don't really know what their long-term cancer-causing consequences will be.[8]

Samuel Epstein, M.D., of the School of Public Health at the University of Illinois Medical Center in Chicago, also told Schell:

> One can easily appear Luddite about the use of hormones in meat animals. Even though the consequences of their use are unclear, one can still look at them by way of analogy to our experiences with other hazardous drugs and chemicals. Over the last forty years we have paid such a terrible price for refusing intelligently to assess the risks of new technologies before leaping into them. We have too often perturbed natural systems for short-term gains, but with disastrous long-term consequences. And quite apart from science, it is a terrible mistake to interfere with anything as delicately poised as the endocrine system, with so little information about where it is leading us.

In 1998, Epstein again warned the public that confidential farming industry reports revealed high residue levels of hormones in U.S. beef. Because of these high levels, the E.U banned U.S. beef shortly thereafter.[9]

With the introduction of antibiotics, confinement management, and growth hormones the raising of meat and the production of milk became highly industrialized. It also became more of a gamble for farmers, creating the need to borrow additional money for hormonal and antibiotic inputs, confinement facilities, and equipment updates. For farmers who faced increasingly competitive corporate markets, using feed additives to increase production—by even a few percentage points—became almost essential in their mind just to survive.

That is why the Hoechst ad on the preceding page trumpets the $1.20 saved on feed by using Flavomycin. Constant propaganda in drug-company pamphlets, magazine ads, and editorials led farmers to believe that antibiotics and hormones were worth using merely as an insurance policy. And that mind-set continues to dominate the animal production literature, the advertisements, and the lack of bureaucratic oversight today by the FDA, USDA, and EPA.

In March 2007, journalist Rick Weiss reported that the FDA was planning to approve a new antibiotic to treat a pneumonia-like disease in cattle, in spite of advisories from health groups and a majority of the agency's own expert advisers that the decision would be dangerous—for people. The drug, called cefquinome, is closely related to the human drug cefepime, a highly potent antibiotic that is the last defense against several serious human infections. No drug in this class has ever been approved in the United States for use in animals. Scientific critics warned that giving cefquinome to animals would speed the emergence of microbes resistant to this class of antibiotics. Those supermicrobes could then spread to people, as has happened with other drugs. In the fall of 2006, the FDA's advisory board voted to reject the request by InterVet Inc. of Millsboro, Delaware, to market the drug for cattle. Despite this vote, the FDA approved cefquinome in spring 2007 as required by a recently implemented internal "guidance document" called the Guidance for Industry 152, which determines how human health is threatened by new animal drugs.

In the Guidance for Industry 152, the agency adopted language that is even more deferential to pharmaceutical corporations than is recommended by the conservative World Health Organization. The chemical industry argued that until a direct link to human mortality from the use of drugs in animals is proven, the FDA should not prevent their use. However, the agency's own scientists are worried about creating more resistance to antibiotics in the human population

The FDA found out how hard it can be to stop the use of an antibiotic or a hormone in the mid-1990s. Again, over almost unanimous objections from health experts, the FDA then approved the marketing of two drugs, Baytril and SaraFlox. Both drugs are fluoroquinolones, important for their ability to fight both the anthrax bacterium and a food-borne bacterium, campylobacter, that causes severe diarrhea.

Doctors began finding fluorquinolone-resistant strains of campylobacter in patients with severe diarrhea. When studies showed a link to poultry, the FDA tried to ban the drug's use. Abbott Labs, the maker of SaraFlox, complied and pulled its product off the market, but Bayer fought back. It fought it for years and finally gave up in 2005, after fluorquinolone resistance had greatly increased.[10] Despite that history, the FDA is prepared to approve the use of cefquinome for use in animals as dictated by Guidance for Industry 152.

The General Accounting Office found in 2005 that 60 percent of confinement animal operations go unregulated.[11] In cases where inspections have been done, resistant bacteria have been found inside and downwind from confinement swine operations at high enough levels to cause human health problems.

Orville Schell and all the other scientists and medical doctors' worst fears have been realized. The irresponsible corporate use of the same classes of antibiotics and hormones for both humans and confinement-animal operations is now common and has compromised several valuable human antibiotics and hormones. The corporate corruption of the FDA in the protection of the public health has manifested itself once again.

The federal agencies all protected the chemical corporations and the large factory-farm corporations. They failed to support regulatory efforts that would have ensured that antibiotics and hormones would be used for humans and animals only to treat an illness. The agencies refused to support legislation against the use of antibiotics and hormones in confinement animal operations. Consequently, by 2006 there were 97,000 feedlots that produced 78 percent of all the beef slaughtered in the United States. This represented about 28 million head of beef cows. Only 2,100 corporations produced 86 percent of all the feedlot beef sold in the United States.[12] In 1994, 73 percent of U.S. pigs were raised in pastures and pens on small farms. By 2007, 95 percent were raised in large confinement hog operations. This represented 57 million pigs.[13] By 2001 there were 300 million commercial laying hens and 8.2 billion broiler chickens in the United States. All of these confined animals are housed in less than humane conditions.[14] They are all on drugs. Bon appétit!

THE EDITOR COMMENTS

This editorial defends the value of DDT, even though it was under attack (at this early date) for causing everything from nervousness to cancer and for its inability to kill barn flies. Similar editorials praising DDT and ignoring both its ineffectiveness and its danger continued into the 1980s. From *Agricultural Chemicals*, December 1951.

GAINST the too frequent reports heard about DDT's being the cause of all sorts of human ills ranging from nervousness to cancer, comes a calm but significant statement on the subject by two eminent scientists. In a joint statement issued November 19 by the Illinois State Department of Public Health and the Illinois Natural History Survey, the point was made that although both agencies have long warned against use of DDT where milk is produced or processed, the insecticide is still recommended highly for control of mosquitoes and other destructive insects.

The authors of the statement, Dr. Roland R. Cross, State Health Director and Harlow R. Mills, Chief of the Natural History Survey, in reviewing their former endorsement of the U. S. Department of Agriculture's recommendation of methoxychlor for control of flies in all phases of dairying, declared that "Our statement at that time in no manner implied that DDT might not be used with safety for other insect control." On the contrary, they went on, "To abandon the use of DDT and other valuable insecticides would subject people to an inadequate and unbalanced diet due to crop loss and disease epidemic far more serious than any we now know. We should not be concerned so much with whether DDT or other insecticides should be used," they concluded, "but should concentrate upon using the right insecticide at the right time, in the right place and in the right way."

As State Health Director, and thus responsible for the general health of the people of his state, Dr. Cross' statement regarding the safety of DDT carries a considerable amount of weight. While others of smaller responsibility and with much less authority are naming DDT as the source of great human suffering, it is refreshing to hear one more man of Dr. Cross' stature state what he knows to be true about DDT.

Chapter 20

DDT AND OTHER SECOND-GENERATION PESTICIDES UNDER ATTACK

Promotions about DDT and its prodigious effects continued for more than two decades after World War II ended. However, as early as 1946, Dr. H. Speich of Geigy Chemicals documented that DDT could no longer control houseflies in northern Sweden—a mere five years after it was first used there. Other cases of resistance quickly emerged, but few outside of the chemical industry knew of or even believed such stories, because DDT's reputation as a killer was so strong.

Most of the U.S. public was mired in disbelief about any problems with or failures of DDT. However, by 1949, 87 percent of flies in the United States were resistant to DDT. By 1950, only five years after it was released from the war effort, nearly 100 percent of the house and barn flies in the United States were resistant. Yet the chemical merchants and American magazines continued to sell and champion the use of DDT to anyone who would buy it for use on flies and a variety of other pests, especially mosquitoes. Ironically, just as the negative publicity began to build against DDT in the news and scientific papers, Geigy's Paul Muller received the Nobel Prize for medicine in 1948 for DDT's control of malaria during the war in the Pacific.

Brigadier General James Simmons boldly asserted in the *Saturday Evening Post* that "the possibilities of DDT are sufficient to stir the most sluggish imagination. In my opinion it is the War's greatest contribution to the future of the world."[1] In retrospect, we know that Brigadier General Simmons's prediction about DDT was totally misguided. But in those immediately postwar days, people implicitly trusted the heroic generals and scientists, and many were awed by them. DDT wasn't to become the greatest contributor to the future of the world, but for twenty-seven years it did become the greatest contributor to chemical corporation profits. It also became a badly misplaced component of the Green Revolution, and it continues to wreak poisonous havoc wherever it is used in the world today.

The advertising image makers sold DDT as a genocidal warrior against pests, and the farmers and homemakers believed them. DDT and the other war chemicals were so effective when first introduced that their dramatic killing power helped insulate them against attacks when they began to fail. Because of its heroic reputation, people couldn't believe that DDT didn't work, nor did they want to believe that it was dangerous to their health or the environment. Many editors sold it as a miracle chemical. Salespersons persuasively convinced farmers that they could depend on it and that it was safe to use. It is ironic that it wasn't safe to use and farmers could depend on it for only a few years.

The public's readiness to accept and even expect quick-fix solutions, however, complicated the efforts of the early critics of DDT and the other war chemicals.

Though many were awed by propaganda about its war-hero status and its kill rate, DDT was attacked early and often by environmentalists, nature lovers, and even some farmers. Within twenty years of its introduction DDT was under a siege even more serious than the attacks on arsenic in the 1930s.

The decade-long struggle between the public and the chemical corporations was DDT's greatest contribution to the world. DDT became the cause célèbre that ignited the environmental movement of the 1960s and provided a significant stimulus to the organic populism that grew up after the Vietnam War.

The final nail in the DDT coffin in the United States was the furor surrounding Rachel Carson's book, *Silent Spring*. Many think of the opposition to DDT as beginning in 1962, with the publication of this book. Actually, however, people reported hundreds of accounts of poisonings and fear of habitat

sterilization almost immediately after DDT and all the other war chemicals were released for civilian use in the mid-1940s.

While fears of DDT-related crop and livestock damage were not widespread, they were more common than either the press or the chemical merchants would have us believe. Warnings even appeared in the farm magazines, which alerted farmers that beneficial insects would be killed by DDT usage:

> DDT particularly has shown great promise in experimental work for control of tomato and cabbage worms but so far has not been registered by the state [California] department of agriculture for use on these crops. Also, in many instances, the use of DDT on these crops greatly increased aphis infestation, so it is not generally recommended at present until more experimental work has been completed.[2]

D.D.T.
Powerful Insecticide
Harmless to Humans

This photograph, taken at Jones Beach State Park in New York, was published in *National Geographic* in October 1945 in the article "Your New World of Tomorrow." Even by this date the propaganda about DDT's safety was so effective that kids played in the spray.

The experimental work was never completed. Instead, prompted by advertising, editorial promotions, and a sales-propaganda blitz, most farmers ignored or were unaware of these early cautions. Many read only the ads, editorials, and farmer testimonials and began to use massive quantities of DDT.

These earliest cautions and fears weren't confined to agriculture journals, however. James Whorton in *Before Silent Spring* quotes a 1946 article from the *New Republic* that sounds almost identical to the fears in Rachel Carson's *Silent Spring* sixteen years later:

> On May 23, 1945, the sun shone warmly on a large oak forest near the village of Mosco, Pennsylvania. Bird calls and songs rang through the woodland as the birds flew about feeding hungry young ones. But the forest was ill; its leaves were covered with millions of devouring gypsy moth caterpillars. Though birds ate vast numbers of the caterpillars and carried them to their newly hatched young, the horde was beyond their control.
>
> Early the next morning, an airplane droned over the forest dropping a fine spray of DDT in an oil solution at the rate of five pounds per acre. The effect was instantaneous. The destructive caterpillars, caught in the deadly rain, died by the thousands. On May 25, the sun rose on a forest of great silence—the silence of total death. Not a bird call broke the ominous quiet.[3]

Rachel Carson's masterpiece echoed these eloquently written fears expressed shortly after DDT's introduction in the United States. *Silent Spring* finally served to crystallize the opposition and give it a national and an international voice. Carson's book had an even greater effect on ending DDT's career than *100,000,000 Guinea Pigs* had on arsenic and lead in the 1930s. Kallett and Schlink's *100,000,000 Guinea Pigs* has been referred to as the *Uncle Tom's Cabin* of the consumer

The juxtaposition of the DDT editorial with the advertisement for Pyrocide (a biological pesticide composed of pyrethrum flowers and piperonyl butoxide) and Multicide, a DDT formulation, is ironic. Note that the editorial warns that DDT will kill beneficial insects. The solution: spray with DDT another time for the aphids. Did we just get on a pesticide treadmill? From *California Cultivator*, 1947.

movement. While it alerted consumers to the dangers of arsenic and lead, it took forty years after its publication for most of the food supply to rid itself of these killers. Come to think of it, *Uncle Tom's Cabin* hasn't rid the United States of its racial biases yet either.

In contrast, it took only ten years after *Silent Spring* was published to eliminate the public support of DDT in the United States. *Silent Spring* was the most pivotal single factor that led to the cancellation of the registration for DDT and several other dangerous pesticides in the United States. Rachel Carson's work continues to inspire today's environmental movement.

Many farmers felt that the chemical corporations gave up the fight to keep DDT on the market in the United States. Whether the corporations gave up or got beat up in the United States is largely irrelevant to the rest of the world, because the corporations continued selling arsenic and DDT in many other countries. More than 250 species of insects are resistant to DDT, yet it still has the power to kill numerous pests, since much of the resistance to DDT is regional and the result of excessive use in certain locations. But that is the pattern with DDT. At first it works wonders and then in a few years it becomes less and less effective, so people use more. After about five years it often becomes useless.[4]

In spite of the dangers of DDT, and its short term effectiveness, many argue for its continued use because it can kill many pests, especially malaria bearing mosquitos. Consequently, chemical corporations continue to sell large quantities of DDT in their war against bugs.

Like DDT and the other chlorinated hydrocarbons, the cyclodienes (including toxaphene, aldrin, dieldrin, endrin, endosulfan, and endothal) experienced serious resistance problems, which meant to the farmer that the pesticide he depended on did not kill the pests anymore and that he had wasted his pest control budget. The failure to kill pests certainly accounts for some early registration cancellations in the United States, when projected sales wouldn't justify defending it. But as one chemical from this deadly family failed another was trotted out to replace its lost cousin.

In the 1960s, dicofol, or Kelthane, another chlorinated hydrocarbon relative of DDT, appeared on the market to poison spider mites. As the chorus of criticism mounted against DDT and the cyclodienes, which had been used for control of spider mites, the use of Kelthane grew. Kelthane is still used widely in the United States for mite control, in spite of serious resistance problems in several areas and in several crops, especially cotton and beans.[5]

In the early 1990s Kelthane was shown to contain a certain percentage (5 to 15 percent) of DDT in several formulations. This discovery was prompted by tests that showed that the groundwater in Stanislaus County (in California's San Joaquin Valley) had the highest levels of DDT in the state for several years in a row, which prompted many to wonder about the persistence of DDT long after it was supposedly suspended in 1972.

The editorial below is by Dick Beeler, chief editor, who feels that the cancellation of DDT's registration was a result of farce and fraud by environmentalists and pseudoscientists. From *Agrichemical Age*, July 1972.

DDT Post Mortem

Mr. Ruckelshaus' 40-page DDT decision sounded like it was written by Carson, Commoner & Erlich. It almost completely ignored the facts and scientific opinions in the 9,000 pages of testimony given at the exhaustive hearings which started August 17, 1971 and were culminated by the remarkably lucid recommendations by the hearing officer, Edmund Sweeney, on April 25, 1972.

It is easy to say in retrospect that no one should have been surprised. From Mr. Ruckelshaus' statements and attitudes on the DDT matter, it is evident that his mind was made up long before the hearings or Mr. Sweeney's report. Even the popular press was able to predict the 40-page decision with great detail and accuracy, and surely no one is so naive as to believe that between April 25 and June 15, when the ruling was issued, Mr. Ruckelshaus had time to study completely Mr. Sweeney's recommendations and the 9,000 pages of testimony on which they were based.

In short, the entire process smells badly of farce and fraud. The fraud, however, is no worse than the original one perpetrated by the environmental mystics and pseudo scientists who hoodwinked the public and its political medicine men on DDT.

Perhaps the greatest fraud of all is the one those same cultists and politicans have pulled on themselves, for the big losers in the DDT battle are the very object of their oft professed affection: the consumer, the common man, the underprivileged and the oppressed.

The western Stanislaus area is a large, dry, bean-growing region, and Kelthane had become a favorite miticide to use on beans after DDT was outlawed. Five to 15 percent DDT is a high dose rate for DDT. Thus, even though DDT's label was suspended and the product removed from the market, it remained on the shelf as a supposedly "inert" ingredient in Kelthane. The chemical corporations claimed that the DDT was merely a mistaken contaminant caused by the manufacturing process—and no fines were imposed. Even when the regulatory process works to eliminate dangerous and deadly poisons from the shelves, they sneak back in under cover, in one guise or another. And the regulatory, legislative, and legal processes seldom punish corporations for the deaths, illnesses, injuries, or pollution caused by their chemicals.

Many highly poisonous chemicals that we thought were almost buried, like 2,4-D, DD (Telone), and arsenic, are still widely used. We looked at the pesticide use data from California, the only state that compiles actual pesticide use reports from farmers and applicators. We found that an average of more than 500,000 pounds of 2,4-D per year was used throughout the 1990s in California on 500,000 to 1,000,000 acres. From 2000 to 2005, an average of 1,805,533 pounds were used each year, more than triple the use in the 1990s. And in spite of a one-hundred-year fight against arsenic, more than 318,000 pounds of arsenic were used on California farms in 1999.

DD/Telone (1,3-dichloropropene) use in California had dropped to around 13,500 pounds in 1991 and 2,122 pounds in 1994. This was in response to consumer and community fears about its use near schools, since it is an air pollutant that causes cancer. But the chemical corporations fought for it and the California EPA capitulated. By 1999, Telone use in the state had shot up to 3,261,667 pounds. By 2003, use was up to 9,355,205 pounds. So much for worrying about the kids!

Dieldrin registration was canceled in 1971 and endrin's in 1979; however, aldrin hung on until 1987. But like DDT all three of the drins were still peddled in foreign markets long after they were cancelled in the United States.

Even when chemicals have been proven to cause cancer or to be otherwise damaging or deadly, like the trio of chemicals advertised above, the chemical corporations and the farm journals have fought back. The farm journal editors continued to deny year after year that their advertisers' chemicals were lethal or dangerous. One of the classic cases of this cynical, head-in-the-sand mentality began with arsenic on apples from Washington State and continued with several other dangerous chemicals on apples, including DDT, fluoride, and Alar.

Arsenic residues were discovered on apples early in the century and reported widely by 1925, as we have already shown. In 1938, Washington State apple growers proposed washing off arsenic and lead with hydrochloric acid in an effort to avoid having their arsenic-laced fruit rejected in the marketplace. In 1940, the industry fought for (and received) a special tolerance increase, which enabled them to have 2.5 times as much arsenic and lead residue on their fruit as the regulations stipulated.

Residues from arsenic and lead had been proven to cause a host of health problems by 1933. So, well before 1938, when they were washing the apples with

acids, the large-scale apple growers knew that arsenic injured, killed, caused cancer, and debilitated millions of U.S. citizens, including kids. In the late 1930s apple growers used fluorine-based chemicals to avoid being "red tagged" (rejected) at the produce docks for having excessive arsenic and lead on their apples. In the 1940s, 1950s, and 1960s the apple growers used DDT and fought for it to the very end, when the regulators were finally forced by public resistance to cancel its registration.

Then, in the 1970s and early 1980s, large-scale apple growers were backed into another corner for using Alar, a highly toxic growth regulator. Alar turned out to cause cancer. The growers wanted to keep using it to increase the size of their apples, put more pronounced tips on the apple bottoms, and ripen the fruit all at the same time. The apple growers fought against the cancellation of Alar by the U.S. Environmental Protection Agency and won in 1986. Several states, including New York, California, and Massachusetts, subsequently passed their own restrictions on Alar after the EPA capitulated to the chemical corporations and the big apple growers in Washington State.

In 1989 the dangers of Alar as a potentially cancer-causing substance were revealed on the CBS news show *60 Minutes*. The chemical industry and the large apple growers fought back and tried to prove that they would never use a cancer-causing substance on apples intended for kids. The editorial at the end of the chapter appeared more than twenty years after Alar had been shown to be a cancer-causing chemical, yet the Washington State Horticultural Association continued to protect Alar and defended its long-term use.

For almost one hundred years, the Washington State apple growers fought against the regulation of arsenic, DDT, fluorine, Alar, and other poisons. Washington growers have long used and long

The ads for Kelthane continued, groundwater became more contaminated, and public resistance grew. This is a portion of an ad from *California-Arizona Cotton*, May 1996, that praised Kelthane and attested that 98 percent of mites were killed in trials, so there were no resistance problems with Kelthane. The concern of the ads is effectiveness, not dangers.

Find Arsenic In American Apples.

LONDON, Nov. 25.—Fruit merchants were fined today for selling American apples containing arsenic to the amount of one-fifth of a grain per pound. The Public Health Department's notice was attracted by a specific case of illness which followed the eating of apples. It was stated that fruit growers in all parts of the world made a practice of spraying trees to combat insect pests, and in America, it seemed, not only the blossoms, but also the fruit, was sprayed after it had formed on the tree, to protect it from the codling moth. The season seems to have been unusually dry in America, with the result that the arsenic was not washed from the apples.

Removing Spray Residue

According to Dr. Alvin J. Cox, chief of the division of chemistry, state department of agriculture, ten pounds of salt and two gallons of hydrochloric acid to 100 gallons of water, makes an efficient dip for removing arsenate of lead from apples and pears. The salt, he states, makes the equal removal of the lead and arsenic, and helps to prevent browning of the fruit.

Article at top left is from the *New York Times*, 1925. Article at bottom left is from *California Cultivator*, November 1938. Article at right is from *California Cultivator*, September 21, 1940.

Residue Tolerances Raised

For a number of years Dr. Ira Cardiff, Yakima, Wash., has maintained that the government's requirements on the removal of lead arsenate spray residue from deciduous fruits, were too strict and that public health would not suffer if they were eased up a bit. The higher tolerance would greatly lessen the expense of preparing fruit for market and so help to keep the growers solvent. Because the regulations have been very hard to meet, many growers have not applied as many sprays as they should and the pest control problem has become more difficult each year.

Dr. Cardiff fought almost alone at the start but was later joined by influential organizations which pressed the subject so vigorously that the government made extensive tests to see whether its rulings were right or not. Recently on recommendation of the United States Public Health Service, Federal Security Administrator Paul V. McNutt announced that the tolerance had been changed from 0.01 grain of arsenic trioxide per pound of fruit to 0.025 grain per pound, and from 0.025 grain lead per pound to 0.05 grain. These new tolerances are restricted to lead arsenate residue on apples and pears and do not apply to other foods such as vegetables.

It will still be necessary to treat fruit with a cold acid bath and to wash it thoroughly but it should require not as vigorous treatment as in the past.

Following the lead of the federal department, the director of agriculture for California announced a similar regulation to apply within the state.

"most of the 50,000 pesticide products registered for use today have not been fully tested or evaluated," the General Accounting Office reported in 1986. Investigators from this congressional watchdog agency estimate that it will take EPA another decade to finish the task. Top EPA officials agree.

Alar, known by the chemical name daminozide, is one of these older pesticides. First registered in 1963, it came up for reassessment in 1984. The registration files on a chemical like this are supposed to contain the results of dozens of chemical, toxicological and environmental tests. Frequently, however, the database has gaps in it because the pesticide was never fully tested or the tests were never properly analyzed, experts say.

This was the case with Alar, a growth regulator that is classified as a pesticide under the Federal Insecticide, Fungicide and Rodenticide Act. Alar is used to keep a dozen or more fruit and nut crops hanging on the trees or vines longer, giving the fruit more time to ripen and color.

The chemical, applied as a spray, is taken up systemically by the plants and permeates the fruit, giving it longer shelf life. Nearly a million pounds a year were being applied before the Alar controversy broke in 1985. Most of it was used on red apples.

Traces Remain

EPA experts said recent residue tests showed that once Alar was sprayed on a crop, traces of the pesticide and a breakdown product called UDMH (unsymmetrical dimethyl hydrazine) remained in virtually every sample, even after the fruit had been processed into applesauce, juices, cherry pie filling or peanut butter. Both Alar and UDMH are listed as "possible carcinogens" by EPA scientists, based on their reassessment of the test data.

By 1986, Alar had been shown to be a possible cancer-causing chemical, and this put the apple growers in crisis since many states passed tougher regulations than the federal regulators, and the states that enacted more stringent regulations were huge markets. From *Modesto Bee*, May 22, 1986.

protected cancer-causing chemicals on apples. Alar was only the more recent cancer-causing chemical used by Washington State apple growers. Currently the consumer advocates and the environmental organizations are fighting against the enormous use of organophosphates and carbamates. Because they are so dangerously toxic, organophosphates and carbamates have been and continue to be scrutinized under the provisions of the Food Quality Protection Act (FQPA), though that process is stalled. Almost all are suspected carcinogens and cause multiple birth defects.

Registrations for parathion have been suspended in the United States for all food uses except cotton and wheat. Only carbofuran had its registration cancelled by FQPA in 2006.

State and federal regulators have promised severe restrictions or registration cancellations for most of the carbamates and organophosphates currently used in agriculture. How soon they act to remove these dangerous poisons depends on the level of public outrage and on the administration of George W. Bush and the president that follows him. At present, the regulators don't seem to be in a big hurry to get rid of them. In the 1990s in California, pesticide use reports from farmers recorded an average of more than 14,700,000 pounds on 11,200,000 acres.

When the public learned about the dangers of arsenic, DDT, fluorine, methyl bromide, DD, organophosphates, carbamates, or Alar the manufacturers all claimed that the results were flawed and the chemicals themselves were harmless. The chemical firms conducted their faux fight with overly compliant regulators in public, but behind the scenes their collusion with the same regulators kept these poisons on the market for as long as possible.

The public thought they were rid of most of these chemicals after passing numerous laws to analyze them and eliminate at least the cancer- and birth-defect-causing chemicals. As can be seen, the elimination of these chemicals through pesticide, food and drug, or EPA laws and regulations has been a joke. As a result, the U.S. public, farmers, and farmworkers, as well as communities around the world, continue to eat food grown with toxic waste and industrial byproducts. Consequently, each spring seems more polluted as we stumble ever closer to Rachel Carson's silent spring.

Thwarting The Alar Scare

The Washington apple industry was rocked by unfounded outrage, as WSHA leaders wondered how people could think they'd risk giving kids cancer.

THE Washington apple industry irrevocably changed one Sunday night in February, 1989. Forty million Americans viewed an episode of CBS' "60 Minutes" that evening, which aired a report stating that consuming apples can cause cancer. Specifically, the story centered on a National Resources Defense Council report on how Alar (daminozide), a plant growth regulator, caused cancerous tumors in mice. People threw out apples and dumped apple juice. In the end, Alar was discontinued in 1990, a year after the broadcast, by Uniroyal Chemical Co., Inc., and Washington growers lost $130 million, according to a report by Desmond O'Rourke, then a Washington State University agricultural economist.

The painful irony is that it didn't have to happen. Nearly a decade later, *New York Times* health columnist Jane Brody wrote, "subsequent tests by the National Cancer Institute and the Environmental Protection Agency failed to show that Alar caused cancer." In

> No farmer would ever use a product that would give kids cancer. That's just ridiculous."
>
> — Fred Valentine (1989)

addition, Brody termed the Alar scare "a cautionary tale that should help you realize why it is unwise to leap before you look closely at what any new study actually means."

Fred Valentine could have told you that at the time the TV show aired. Watching the telecast that night, the then-president of the Washington State Horticultural Association (WSHA) was horrified by what he was seeing and hearing. "No farmer would ever use a product that would give kids cancer," he says. "That's just ridiculous!"

"Terribly Unfair"

unfair reporting as farming. As an apple grower, I could tell you scores of stories about friends and neighbors who have lost their farms as a direct result of that one broadcast on '60 Minutes.' It was terribly unfair."

In his role with the WSHA, Valentine found himself often flying back to Washington, DC, to work with USDA and FDA on the issue. What made his task so difficult is that Alar was an extremely effective product.

ments in the media in recent years, news reporters may be more inclined to check their facts. And it pushed the growers of Washington to produce a better product, says Valentine. "We had to come up with a quality control program because we didn't have Alar. We had to have a quality rule, because for a lot of growers, if you're not forced to do something, you'll find a way around it. We fought hard about these things, but they were stepping stones to

From *American Fruit Grower*, 2005.

Cotton Council Official Defends Use of Pesticides

Melon Probe Sparks Clash on Pesticide

Farmers Claim Firm Mislabeled Chemical Tied to Tainted Fruit

By NANCY SKELTON,
Times Staff Writer

Angry farmers Saturday blamed the Union Carbide Co. for improperly labeling a pesticide that health officials say tainted watermelons and may have caused as many as 270 Californians to become ill.

Farmers said labeling on the pesticide aldicarb states that the insect killer will leave the soil within 100 days, yet traces were found in fields as late as six years after the last application.

Pesticide Battle Lines

Only Unanimity Is in Criticism of Law, EPA

By RONALD B. TAYLOR, *Times Staff Writer*

The newspaper ad from one California supermarket chain promoted produce that had "No Detected Pesticide Residues." Another advertised: "Apples Without Alar," touting fruit that was free of a specific farm chemical that may cause cancer.

LAB Certified GROWN WITHOUT ALAR

Seal of Approval

Under a new plan to certify fresh fruits and vegetables with "quality and safety" seals, California's produce industry is hoping to regain consumer confidence.

Mississippi Fish Kill, *California Farmer*, April, 1964;
"Melon Probe Sparks Clash on Pesticide," *LA Times*,
July 7, 1985; "Pesticide Battle Lines," *LA Times*, May 22, 1988;
"Seal of Approval," *California Farmer*, June 17, 1989.

Chapter 21

TWENTY-FIRST-CENTURY POPULISM: ORGANIC FARMING AND THE ANTI-FACTORY-FARM MOVEMENT

A t the same time that poisonous fertilizers, pesticides, hormones, and antibiotics were being sold to American farmers and homeowners, the organic farm movement struggled to stay alive and continue its slow growth. During the late 1930s and the 1940s, theories about and strategies for organic and biological farming advanced dramatically because of the work of several creative thinkers and innovative farmers.

Especially important breakthroughs in the biological control of pests took place during this otherwise fractious period. One promising strategy was the use of "trap crops" that were more attractive to pests than the crop they were designed to protect. A classic test of this strategy was conducted by H. D. Lange on artichokes in an effort to trap the plume moth. Lange found that the plume moth, an American pest, liked native thistles better than artichokes or cardoons, both of which are thistles introduced from Eurasia. Lange felt that by growing native thistles (*Cirsium vulgare* and *C. edule*) near artichoke and cardoon fields, farmers could control the plume moth without the use of any sprays.[1] By contrast, until the mid-1990s most California artichoke farmers sprayed highly toxic insecticides at least twenty-six times a year for control of the plume moth.

Trials conducted on the Knoll farm in Brentwood, California, in the 1990s demonstrated the efficacy of H. D. Lange's approach. In three years using *Cersium edulae* and *Cersium vulgare* as the trap crop, a total of only twenty-four artichokes were damaged by the plume moth.[2]

Another critical piece of on-farm research came from Sir Albert Howard and his wife Gabriella, who spent

nearly thirty years in India from 1905 until the 1930s. They were sent to India by the British government to teach Indians how to farm with industrial chemicals. After several years, the Howards concluded that chemical agriculture was having devastating effects in India. Their study of Indian farmers allowed them to develop a systematic analysis of the indigenous farmers' long-term, nonchemical fertility and cropping strategies compared with Europe's chemical-intensive programs.

After more than two decades of experiments, the Howards determined that Indian organic-growing methods, which made use of cover crops and composts, were the most effective soil-health and production practices. Indian strategies resulted in balanced soils, even after thousands of years of use. These balanced, fertile soils enabled farmers to conduct nontoxic pest control. The Howards concluded that because the soil was alive and healthy, the plants in turn were healthier and thus less interesting to insects and diseases than plants grown with chemical inputs and "modern, industrial methods."

After returning to England, Sir Howard wrote *An Agricultural Testament*, which was published in 1940. This practical book enabled many farmers to copy the

EDITORIALS

Organic Oregon

In what may be the ultimate in consumer foolishness, the Oregon Department of Agriculture has announced it will establish legal standards for "organic" and "natural" foods, "so that proper consumer protection regulations can be established by the department."

Coming as it does at a time when food scientists and authorities throughout the country have been trying to explain the organic fraud to the public, the Oregon project is out of order. Its also a waste of the Oregon taxpayers' money, and worse than that, it amounts to government endorsement of a counterfeit cause

The Genuine Fake

If you as a consumer, have a desire to purchase a fake or a fraud of one kind or another, should your government guarantee your right to do so?

More than that, is your government obligated to prosecute one who, knowing of your propensity for fraud, tricks you into buying the genuine in place of the fake?

Remembering that "your government" is all the rest of us, is it right for you to take our time and money to underwrite such ridiculous exercises as making sure you are cheated when you want to be cheated? And must we penalize the man who breaks his promise to cheat you?

These questions may sound absurd to you, but they are no worse than the double negative type of "consumer protection" on "organic foods" that is being adopted by several states, including California and Oregon. It is also being proposed in the US Congress.

The "organic" movement is a fraud. At best its food can only equal that which is produced by standard scientific agriculture. In fact it is often much worse than standard food because of the haphazard plant nutrition prescribed by the organic cult's dogma, the lack of protection from pests, and the pathogenic conditions involved.

Yet "organic" food consistently carries exorbitant prices that have made its proponents wealthy, and they claim for it a wide variety of obscure advantages which amount to nothing more than superstitions and witchcraft. The federal legislation to which we referred would strengthen the myth. It would make this mysterious and expensive food available to minority groups, the poor and underprivileged, who are said to suffer illness, embarrassment, and anxiety for the want of it, by subsidizing "organic" farms and food processing. . .

. . .The sure mark of a fraudulent movement is an absence of precise definitions for its concepts and terminology. Amazingly enough, members of the organic cult are unable to agree on a definition of "organic food.". . .

. . .This is not as remarkable as it might first appear. The dictionary quite simply defines "organic" as "derived from living organisms." Actually, all foods, except for supplementary minerals, such as salt, are organic and any attempt to differentiate one from the other by calling it organic is a deception. . .

Dick Beeler

Dick Beeler editorial excerpts calling "organic" a fraud. Top: from *Agrichemical Age*, September 1972. Bottom: from *Agrichemical Age*, November 1972. Many farm journal and chemical magazine editors attacked organic farming in a similar manner.

Howards' ideas and experiment with organic techniques in a farm-scale setting.[3] The Howards' contributions and analyses were widely applauded by growers and were extremely important for the development of organic farming. Unfortunately, the Howards' work was largely ignored by mainstream agriculture in the United States. Instead, DDT, lindane, DD, 2,4-D, and the other war-hero chemicals got almost all the attention from the farm journals, the government, and the scientists.

The Howards did get Lady Eve Balfour's attention, though; using concepts she learned from the Howards, she helped start the Soil Association (England's organic certification organization) and wrote *The Living Soil*. The Howards also got American publisher J. I. Rodale's attention, and Rodale bought a farm in Emmaus, Pennsylvania, that became the American center for organic agriculture for the next forty years. In 1942, two years after Howard's testament was published, Rodale was publishing *Organic Gardening and Farming* and promoting the research results of Albrecht, Steiner, and the Howards.

The Rodales conducted their outreach all over the world, forming alliances in Japan, Europe, Latin America, the Soviet Union, and Africa. At the same time that the university scientists were studying spray rates and nozzle sizes for applying toxic chemicals, the Rodale Institute was studying pest and predator interactions, cover crops and composts, organic disease management, and pest-resistant varieties.

J. I. Rodale published several books on organic growing methods, including *The Encyclopedia of Organic Gardening*. The Rodale books and magazines whetted the appetites of farmers, gardeners, and consumers for information on biological farming in the United States. Though ridiculed and ignored by the proponents of big agriculture, the Rodales have championed organic

Treated   Control

GREEN-MANURE EXPERIMENT, SHAHJAHANPUR, 1928-9

Some of Howard's points about yield and quality are punctuated with pictures and graphs like this one showing sugarcane growth after using cover crops on the plot on the left. Plate IX from *An Agricultural Testament*, by Sir Albert Howard (New York: Oxford University Press, 1943).

farming since the 1940s, and their work and influence continue to this day.

Gabrielle and Albert Howard, J. I. Rodale, Rudolph Steiner, and William Albrecht all showed how chemical solutions on farms were beginning to be counterproductive. They unanimously argued that "pseudoscientific" methods and strategies that focused exclusively on chemical solutions resulted from compartmentalized farm analyses that failed to grasp the "whole systems" approach to agriculture.

Albert Howard felt that, because of scientifically reductionist approaches employed by chemists, the farm researchers usually concentrated their analyses on isolated elements of the experimental plot (pests, fertilization, irrigation, harvest, and so on). As a result, farmers came to address problems on their farms in more compartmentalized ways. Farmers, for example, stopped seeing pest control and soil fertility as interdependent factors. In fact, they were bombarded by chemical corporation propaganda that sought to depict them as separate and that tried to get farmers to buy two or three separate products to address what is essentially one single problem: soil fertility balance.

In the 1930s, the quantum leap in knowledge about soil fertility, beneficial insects, and trap cropping was complemented by a considerable increase in the supply of biological insecticides such as rotenone, sabadilla dust, and pyrethrum powder.[4] Safer and increasingly more effective biological pest controls became more widely available, and many farmers made use of the "softer" pesticides and other nontoxic strategies. But their price was high, relative to that of the chemical poisons. Most people used the biologicals only for fleas on animals, garden insecticides, and other household pests because of their high cost.

By the mid-1940s, the sales campaigns for war chemicals began endlessly cranking away. When necessary, prices of the new poisons were lowered or they were given away to farmers to stimulate interest and sales. These practices of cheap prices and giveaway samples addicted the farmers to these poisons and helped deter thousands from using biological controls. This is because after a couple years of chemical use, the beneficial insect population declines, and it takes a few years to reestablish a large enough population of predators or parasites to control pests again.

Since the 1880s, the Division of Biological Control at the University of California had been committed to studying the nontoxic control of pests, even though its school of agriculture served as a major promoter of poisonous pesticides and fertilizers. Many breakthroughs in organic and conventional agriculture resulted from the long years of research and farmer outreach by the Division of Biological Control. Their successes include trap-crop and cover-crop research, worldwide searches for beneficial insects, farm-habitat enhancement, and biologically based integrated pest management (IPM). After the early 1950s, the division waged a constant battle for survival within the School of Agriculture and Natural Resources.

Entomologist Everett Dietrick remembers the time period between the 1950s and 1960s as one of both excitement and turmoil. "While I was a graduate student at UC Berkeley in 1947," Dietrick remembered,

> I was offered a research position as an assistant to Dr. Paul de Bach in Biological Control of California red scale. I jumped at the chance to take the research position instead of pursuing my Ph.D., because there were so few positions in this science! You had to work in the field to learn it. Those were the days when biological control was still a state-wide department with its own separate budget and facilities. With the introduction of agrochemicals, the universities put chemists in charge of the biology departments. The rule of the day then became

The cover of the November #10, 1956 *California Farmer*. The cover promoted beneficial insects. However, it became increasingly difficult to keep these friendly bugs in the field when so many toxic chemicals were being advertised, promoted, and used. From the University of California at Davis Shields Library Special Collections.

California Farmer

WITH WHICH HAS BEEN CONSOLIDATED THE

PACIFIC RURAL PRESS AND CALIFORNIA CULTIVATOR

5c PER COPY
BI-WEEKLY

L. 205, No. 10—102nd Year   November 10, 1956   SAN FRANCISCO—LOS ANGELES

Oriliis nymph

Lacewing adult

Lacewing larva   Large hover fly larva

Drius adult

Friendly Bugs

Damsel bug adult

Convergent lady beetle adult

WINTER

PEST CONTROL

ISSUE

Convergent lady beetle larva

Big-eyed bug adult

These are some of the natural predators of the Alfalfa Aphid. See page 422 for detailed story.

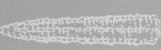

"the only good bug is a dead bug." Fortunately we had the funds and the autonomy to pursue a different approach.[5]

Dietrick worked at the University of California while there was still an active biological-control department, which included brilliant agro-ecologists such as Robert van den Bosch, Ken Hagan, and Karl Huffaker. Some folks call that a radical period. Radical? That depended upon which school you belonged to, according to Dietrick:

> Our discussions in the department of biological control usually concerned what was happening downstairs where the chemicals were being used. Talks centered around the destruction of natural enemies from the overuse of pesticides. It was all very predictable. Resistance was the issue of the day. I was studying red scale on citrus trees. And red scale was resistant to cyanide. There's a basic rule at work with pest controls: the more you hit nature with a pesticide, the harder it will come back at you. If you kill 95% of a given species the remaining 5% will come back to haunt you. Resistance is the way nature fights back.[6]

Working at the university research station in Riverside, California, Dietrick saw 1960 as a time of turmoil.

> Robert van den Bosch was my project leader. We'd moved away from citrus, avocados and tropicals to vegetable field crops. This was very important because up until that time biological control was thought to work only on tree crops or perennials. Van den Bosch had put in 10 years of very exciting work. Politics within the university system made him leave. The university was clearly endorsing chemicals more than biological controls. Van den Bosch

was transferred to Berkeley. Slinger was taken off the project. I was left without a leader and a project.[7]

In the 1970s, the National Science Foundation and the Environmental Protection Agency supported a six-crop biological IPM program that came to be known as the Huffaker Project. Directed by Karl Huffaker, from the Division of Biological Control at the University of California, this project was the largest biological IPM program ever attempted in the United States and was supported by the EPA but not the USDA. Unfortunately, the United States Department of Agriculture was philosophically opposed to the Huffaker Project and to its potential competition for research funds with chemical and genetic engineering. Consequently, the USDA failed to fund the project or participate significantly at any level. Nonetheless, with the support of the U.S. Environmental Protection Agency, the project demonstrated that biological IPM was a powerful, cost-effective tool in controlling pests in several crops.[8]

This long history of success in biological control and organics/IPM seems to have had no impact on the university administrators in the United States. In the late 1990s, the University of California School of Agriculture and Natural Resources waged a final attack on the Division of Biological Control and purged biological control from the agricultural curriculum. At a time when organic and sustainable agriculture was gaining adherents all around the world, the School of Agriculture and Natural Resources diverted most of its resources and intellectual capital into chemical agriculture and genetic manipulation. Now there are genetic manipulation centers at both UC–Berkeley and UC–Davis campuses that are supported by Syngenta and Monsanto, and very little support for biological IPM or organic agriculture research. Peanuts for clean food research, billions for chemical and genetically manipulated food.

Still, from the 1980s until the present, the organic movement and organic markets grew with hardly any government or university support. Farmers all around the world and in the United States cooperated with each other and leapfrogged over many problems. As the strength of the organic movement increased around the world, so did international antagonism to the American- and European-style factory farms and to chemicals and seeds from the Green Revolution. For more than fifty years, the belief system in U.S. agriculture rested on the presumption that bigger was better. Recently, however, research has shown that small and medium-sized farms are more efficient than large-scale farms, and that both animals and the land are cared for better on smaller-scale farms.[9]

By 1995, organic farms had attained yields equivalent or nearly equivalent to yields of conventional farms on an acre-for-acre basis in most crops. However, because small and medium-sized organic farms generally grew more than just one or two main crops, and because small farmers often harvested a crop more than once, their caloric yield (overall food production) often considerably exceeded that of large-scale conventional farms, which picked tomatoes, broccoli, cabbage, cotton, and other crops just once and plowed under many plants whose crop matured slightly later.

The latest organic movement, which began in the 1940s, suffered through fifty years of attack and derision, similar to what Deke Dietrick endured at the University of California. In spite of the attacks, organic farming has flourished in recent years and today it represents the fastest-growing sector in U.S. and international agriculture.

GENETICALLY-MODIFIED

RATATOUILLE
Provençale

Teal

This tongue-in-cheek drawing shows a tomato infused with rat genes. From *Ecologist*, November/December 2005.

Chapter 22

DESIGNER GENES: A LEAP OF FAITH

The chemical discoveries and research breakthroughs of the last two hundred years had all run their course by the late 1970s. Hundreds of pests and bugs had developed a resistance to one chemical or another, and dozens of pesticides were in perpetual regulatory trouble or under attack from consumers, farmworkers, and environmentalists. With precious few new chemicals in the research pipelines, the chemical-pharmaceutical corporations promised to design a completely new pest-control system. Using research breakthroughs in genetics, they said they would take the appropriate genes from wherever they could find them and use them to produce disease-resistant plants, fertilizer plants, and pest-resistant plants.

Throughout the 1970s and 1980s, the chemical-corporation propagandists seized upon each of the regulatory decisions against chemical poisons as opportunities to promote the necessity of genetically modified (GM) crops and animals. They argued that farmers in the United States were losing many more pesticides to regulation than competing farmers in the underdeveloped world. They promised to supply U.S. farmers with genetically modified organisms (GMOs) that would replace the need for the chemicals lost to regulation. They argued that only with GM crops could U.S. farmers continue to compete in the world commodity markets.

The corporations announced that a new age of supercrops was on the horizon, made possible by the emerging science of genetic engineering. Gene splicing would allow scientists to bypass evolutionary processes and transcend molecular and species boundaries. Traits from one plant could be transferred to another, regardless of whether they were related or not.

Early propaganda emphasized the altruistic potential of genetic engineering. Farmers were told that these discoveries would significantly diminish the need for farm chemicals and reduce the criticism and constant oversight from environmentalists, health experts, and government regulators. The corporations promised that farmers would make more money, increase production, feed the world's rapidly expanding population, and open up new markets. By the time that these products were available the years of endless bankruptcies and constant advertisements and editorials prodded many U.S. farmers into making another extraordinary leap of faith. But first let's take a look at how we got to this place in genetics and pest control where farmers are forced to choose whether or not to leap onto the genetic engineering bandwagon. Hopefully the overview below will provide some answers.

The study of plant and animal genetics was pioneered more than ten thousand years ago by farmers in the

Near East. Emmer wheat, bulgur wheat, rye, barley, spelt, and several minor grains were crossed with related plants and selected for their productivity, taste, and storability. Farmers also developed sesame, garbanzos, peas, fava beans, and countless varieties of grapes, apricots, peaches, plums, almonds, walnuts, and olives, as well as artichokes, melons, broccoli, mustards, kale, and a staggering list of other minor crops.

The Near Eastern farmers also domesticated and bred several types of animals and greatly improved them through genetic crosses with related animals. Some of the animals they domesticated pulled the world's first wagons, carts, sleds, and plows. The Middle Eastern plant and animal breeders, along with fellow farmer-breeders from Europe, Africa, India, China, and Southeast Asia, discovered and refined hundreds of other vegetables and grains, countless important fruits and nuts, and almost all of the domestic animals in use today.

The knowledge of genetics in the Americas can be traced back more than eight thousand years to plant breeders from indigenous tribes and nations. Several inventive centers of plant diversity flourished in Mexico, in the Andes, in the U.S. Southwest, and along the Mississippi River, where farmers deliberately crossed slightly and distantly related plants to modify corn, potatoes, tomatoes, peppers, beans, cotton, chocolate, peanuts, sunflowers, blueberries, blackberries, cranberries, quinoa, amaranth, guavas, sweet potatoes, tapioca, hemp, tobacco, and hundreds of other minor crops.

Native farmers crossed thousands of different plants in all the diverse areas of America, with the resulting successful crosses becoming well adapted to arid or high-rainfall climates, high, medium, and low altitudes, and temperate, subtropical, and tropical latitudes. The pre-Columbian genetic experimenters from Middle and South America discovered and developed more than 50 percent of the food plants that the world currently consumes. Among animals, though, only llamas, alpacas, vicuñas, turkeys, and guinea hens were domesticated and creatively bred in the Americas.

Applying knowledge accumulated from both of these traditions, a Catholic monk, Gregor Mendel, advanced the scientific knowledge of genetics in the mid-nineteenth century. Based on his work with fruit flies and peas in a monastery garden, Mendel proposed theories of genetic inheritance and discovered ratios of inherited characteristics in successive generations of fruit flies and peas. His theories resulted from extensive field and laboratory trials over a thirty-year period. Unfortunately, a botanist named Nägeli at the University of Munich suppressed Mendel's research for religious reasons for thirty-five years following his death. After the turn of the twentieth century, however, scientists began conducting genetic and eugenic experiments based on Mendel's theories, some well thought out, others racist and punitive.

Because of the extensive experimentation, genetic knowledge advanced rapidly in the 1930s and even during the war years. Consequently, by 1953 James Watson, Francis Crick, and Maurice Wilkins were able to discover the molecular structure of DNA. By the late 1970s researchers were using Mendel's ideas on inheritance, and what came to be called the Watson and Crick DNA model, to map genes, splice and transfer genetic material, and conduct elaborate engineering schemes on plants and animals.

By the 1980s, a generation of molecular biologists had been trained in genetic science and eagerly took positions in government and chemical-corporation laboratories, where experiments with genetic manipulation had already started. In the mid-1980s, companies began aggressively to develop (or say they were developing) genetically engineered products. This was about the same time that many corporations were losing revenue and market share as many of their most profitable farm chemicals lost their U.S. registration or reduced their product lines because of environmental, health, and resistance concerns. Among them were DDT, 2,4,5-T, dieldrin, endrin, aldrin, most of the arsenic pesticides, and several bromines.

The chemical corporations that invested heavily in genetically modified organisms needed to make promises to large-scale farmers to keep their place in the market, since their long-term advertising strategy was based on convincing growers that they consistently manufactured some of the most advanced pest-control and fertilizer products. Whether by design or because of the excess zeal of the publishers, the press got out in front of the research. Stories about genetic manipulation successes appeared before there were any real successes, or even real products.

Such predictions and claims of success, as well as reports on the inevitability of genetic manipulation, appeared in the press as a done deal before the citizens of the United States had a chance to consider its value. There was no public debate; there was no vote as to whether or not such research (funded with the taxpayers' subsidies and tax write-offs) was a good idea. Nor was there widespread public understanding at the time of how different the transgenic crossing of species was from the breeding experiments conducted by "classical" genetic breeders, in the tradition of Mendel and Native American, Middle Eastern, Chinese, Hindu, and Japanese farmers over the last ten thousand years.

In the traditional model of genetic improvement that has been conducted for ten thousand years, both closely and distantly related plants and animals are bred to produce more desirable traits. In transgenic or genetically manipulated (GM) breeding, genes are taken from unrelated species of plants or animals and literally "shot" into the genes of the target organism (with a device called a gene gun). Ostensibly, this use

of genes from, for example, rats, snakes, leeches, or insects is designed to produce more desirable traits in vegetables, fruit, grain, or animals. Genes to modify apples or corn on the cob could theoretically come from any other life form: pigs, rats, bacteria, a beauty queen, or a poisonous weed.

Instead of explaining how different the genetic manipulation of unrelated species really was from traditional breeding, the chemical corporations and their advertising agencies created ad campaigns that trumpeted every achievement and denied any failures. The campaigns claimed that genetically manipulated crops would be the answer to consumer and environmentalist assaults on farm chemicals. Ads and editorials promised that GMOs would save labor, save the environment, produce higher yields of more nutritious food, create heroic medicines, reduce or eliminate chemical use, and, of course, erase world hunger. And, in another familiar refrain, the campaigns promised that farmers would be the main beneficiaries of this new technology.

It wasn't just the farm press that promoted genetic manipulation. Biotech articles in the 1970s, 1980s, and 1990s and almost all the local and national magazines and the major daily newspapers—including the *Wall Street Journal*, the *New York Times*, the *Washington Post*, the *Los Angeles Times*, and the *St. Louis Post Dispatch*—heavily endorsed genetically manipulated crops, hormones, and medicines. The journal editors, corporate advertisers, government agencies, and scientific experts badgered farmers (and other consumers) about the promised benefits of genetic engineering for almost twenty years before a single GM product was sold.

A 1981 quote from the farm magazine *Seed World* is typical of the promises that the genetic engineers made early and often: "The goal of researchers at the . . . laboratory [of Agrigenetics in Madison, Wisconsin] is to produce better-yielding and disease-resistant crops through bioengineering—creating new kinds of superior plant varieties through genetic engineering."[1] An aura of both necessity and inevitable success permeated the promotional campaign.

While the genetic manipulators were chortling over their marketing successes, scientists were worrying about the unsound science and government enabling that had dominated the genetic manipulation debate. Well before 2000, independent genetic researchers had shown that the absolute aura of biotech success based on sound science that the corporations and the journals had promoted since the 1980s was based on an overly simplistic and faulty genetic model. Chemical-corporation and regulatory proponents of that model argued that, by splicing genes from an unrelated plant into humans, cotton, corn, soybeans, canola, or pigs they could alter those organisms one gene at a time and affect only one trait. This was based on the theory that human beings possessed about 140,000 genes, and that each gene was responsible for only a single trait. They argued that plants had fewer genes and fewer traits than humans, maybe 25,000, and about the same number of traits as genes.

In 2001, however, results from the Human Genome Project proved that humans have only about 25,000 genes. Yet we still have around 140,000 traits. So, instead of one gene affecting only one trait, we now know that most of the genes in humans and other animals, plants, insects, viruses, and bacteria influence or control several distinct traits. Single genes were originally believed by Watson and Crick to encode only one protein and to affect only one trait. Yet, in an extreme example, single genes have been found to encode up to 38,016 proteins in the fruit fly.[2]

Since scientists can now use real genetic data, instead of a theory, they have concluded that by splicing herbicide resistance into, for example, a plant, we probably will influence several different traits—not just

resistance to an herbicide. The GMO corporations and the regulators continue to claim, however, that the "one-gene, one-trait" model is sound science, even though the real world of science has proven that it isn't.

Scientists who are worried about the collateral damage from genetic manipulation have also documented how genetically altered crops are usually infused with other potentially dangerous materials at the time of the transgenic invasion. These usually include:

- a bacterium or virus that infects the organism so that the invading gene can become established—such as tumefaciens or E. coli;
- a marker virus to determine if the invasion is successful—such as cauliflower mosaic virus; and
- an antibiotic to heal any wounds the invasion causes—such as streptomycin.

This untested cocktail of antibiotics, bacteria, and viruses along with the transgenic gene is shot into the seeds or into the animal embryo. No one knows what the long-term effects might be from invading the genes of plants and animals with all these foreign bacteria, viruses, and antibiotics along with transgenic genes. No one!

So here we are, 150 years after Mendel's elegant scientific breakthroughs, stuck with lots of GMO products in the field, on the market, and in our diets without knowing all the potential risks to the food chain, our health, or the environment. How did this progress so rapidly?

Most of these changes were brought about through the government elimination of regulatory roadblocks and administrative foot-dragging, but let us explore how. The first Reagan administration put GMO regulations on a fast track to enable the corporations to get products on the shelf as soon as possible. To accomplish this, the administration made a determination, without any scientific review, that GMO crops were "substantially equivalent" to nongenetically engineered crops. Consequently, the government determined that no special testing procedures were necessary. Thereafter, genetically manipulated crops were treated and regulated the same as non-GMO crops.

Additional stimulus to the rapid development of the GMO industry came when the Reagan administration permitted the patenting of life forms and organisms for the first time in human history. This enabled the genetic modifiers and seed giants to patent everything they created in the lab and in the field and many crops and trees that other tribes, nations, and plant breeders had already discovered and developed over the millennia but had failed to patent—because in many cases there were neither patent laws to allow it nor a consciousness to patent and own life forms.

During the elder Bush's administration, Monsanto's aggressive outreach to the government and the Bush-Quayle obsession with deregulation allowed the chemical corporations to continue their self-regulation of GMO technology. Bush-Quayle regulators also followed the recommendations of the chemical industry and accepted that one gene affected only one trait and subsequently regulated (that is, didn't regulate) products according to this belief.

Ever since the earliest days of the Reagan admin-istration in 1980 there was little mandatory oversight of GMO products from any federal regulatory agency, whether the EPA, the USDA, or the FDA. Unfortunately, it was no different in the Democratic administration of Clinton and Gore. Monsanto, Novartis, DuPont, and Hoechst literally got everything they wanted from the Clinton-era regulators. And unlimited, unregulated product promotions and product research continue to this day under Bush-Cheney. The regulators still ask few if any questions at their elegant luncheons with corporate officials.

With little or no regulation or even government oversight, the GMO revolution shifted into high gear in the late 1980s and never looked back, and never had to, as there were no regulators in their rear-view mirror. Microbiologists at Monsanto, DuPont, Syngenta, Pioneer, Hoechst, Aventis, Mycogen, Genentech, the USDA, and all the other gene labs created dozens of new transgenic plants and animals by transferring the genes from one species of plant, animal, bacteria, or virus into unrelated species.

An aura of urgency drove the corporate researchers to clone sheep, monkeys, pigs, and dogs; to create monster salmon and to insert flounder genes into tomatoes to ward off frost; to complete a sheep-goat cross called the "geep"; to create artificial-growth hormones for milk cows; to increase the density of pulpwood trees; to merge pharmaceutical production and agriculture to create "pharming"; to clone pigs to grow replacement organs for human transplants; and on and on. Many of the GMO varieties created by these firms bypassed even the normal procedures of plant testing by state and federal agencies and yet ultimately received government approval.[3] Each new technique got patented with hardly any questioning, and currently there seems to be a rush to patent the genetic makeup of all biology.

Our long-term antiregulatory climate has allowed corporations to find the fastest route to profits without considering the consequences, or whose property was being patented, or what might be the negative results from hastily introduced products. As a result, hun-dreds of lawsuits have been filed against the chemical/seed corporations by foreign governments for filing illegal patents on their native crops and by farmers who claim that they got hurt financially when the gene giants' products failed.

Monsanto, unperturbed by any criticisms or lawsuits, forged ahead with aggressive acquisitions of crops and seed corporations. Its sales tactics were (are) also aggressive, as were its ads and promises. It promised that it would create better crops, ones that would address the needs of farmers and consumers, and at the same time safeguard the environment. The company agitated for immediate regulatory permission to introduce genetically altered cotton, corn, soybeans, and canola, crops that were manipulated to withstand direct sprayings of Monsanto's own Roundup (glyphosate) weed killer.

Monsanto developed many cotton varieties with another leading cottonseed supplier, Delta and Pine Land Company. Without any testing required, the companies jointly gained government approval for some GMO strains as early as 1994. Soon thereafter, advertisements touted their genetically altered Roundup Ready plants as the most successful product introduction in farming history. Monsanto began to promise that farmers could limit their herbicide purchases to just one "safe" killer and control the weeds in one application. The ads claimed that this would provide economic savings to farmers at a time of substandard prices for all commodities. And they promised to make cotton the nation's first crop in which genetically altered varieties predominated.[4]

There were lots of promises from Monsanto and the other genetic manipulators, but very little transparency about their tests or about difficulties growers should anticipate. In one case regional USDA agents in Stoneville, Mississippi, requested permission to conduct field trials so that they could advise growers of any problems or the need of any special growing practice advisories. Agents were denied the right to test even a pound of seed before it was sold. Officials from Monsanto, who owned Stoneville Seed, told USDA's Bill Meredith that there was not enough seed for government trials. Suspiciously, by the next year, Monsanto, Stoneville, and Delta and Pine Land sold about twenty million pounds of seed, enough to plant a million acres of cotton. Yet no seed had been available the previous year for government testing.[5]

Well before Monsanto released Roundup Ready cotton, farmers were barraged with performance promises, but without government or other scientific trials, farmers had to accept the products as a leap of faith and hope. Most of the promotions reached farmers through the farm journals. Today, the average farmer in cotton country still gets as much as 90 percent of his or her information through advertisements and advice from the people who make or sell the products. Magazines understand that commodity-crop farmers consider their pesticide dealer and writers in the farm magazines to be among their most trusted advisors; consequently, there is a continuous but narrow flow of information to U.S. commodity-crop growers.[6] And that information is carefully controlled in the magazines.

While they managed to suppress most of the early criticism at home, attacks from Europe in the 1980s and 1990s panicked the corporations and the farm media and spurred them to head off any further journalistic dissent in the United States. A surprisingly frank editorial by Mindy Laff in *Seed World* magazine recounts how she and several editorial colleagues were invited to a September 1997 European conference on biotechnology in Belgium, all expenses paid by the GMO corporations.

Laff and editors from other magazines met with top European journalists concerned with seeds and biotech to learn what might be applicable to address North American farmers' concerns about the environmental, food safety, and consumer backlash against biotech products from the United States and the long-term effects of biotechnology. She summarized the European journalists' recommendations as follows:

> *Toon van der Stok of Ooogst*, a Dutch agricultural/horticultural publication, recommended that more information regarding the advantages, rather than the disadvantages, of genetically modified organisms should be

provided. . . . Robin McKie of the United Kingdom said that, as annoying as this might be, it is important to repeat your message over and over until the public gets it right.

Laff concluded that

> The main point that came out of both meetings and was evident throughout the week-long tour, however, is that the best method towards communicating the benefits of biotechnology to the public will originate as representatives from the areas of science, public agencies/government, and industry work together. By appeasing consumer interests through effective communication . . . success will be implemented in a beneficial, profitable way for all those working in the biotech industry.[7]

Mindy Laff clearly articulated the long-successful strategy of introducing and advertising toxic chemicals, dangerous technologies, and untested products to farmers and consumers. Thereafter, the corporations shaped the dialogue and the U.S. farm and seed magazines and newspapers dutifully promoted a think-positive "hear no evil, see no evil, speak no evil" strategy.

With mostly good news, stock prices of Genentech, Monsanto, Pioneer, DuPont, and other GMO manufacturers soared in the bull market for tech stocks from 1990 to 2001. Affected by the propaganda as well as affecting the propaganda, Wall Street wildly supported genetic engineering. The stockholders and stockbrokers saw genetic modification as a potential huge return on investment for a wide range of genetically altered products. This investor reaction provided the genetic corporations with more capital to experiment further with the world's food and fiber supply.

Until 2003, advertisements and editorials for genetically modified crops and drugs focused on discussions of potential profit, labor, price of production, and the introduction of new vocabularies, new slogans, and new advertising campaigns. Only after several weed and insect pests started showing resistance in 2003 and 2004 did any mild criticism in U.S. farm and seed journals finally surface once again (more about that later).

When the genetic revolution began to be advertised in the 1970s, the chemical corporations invented the word *biotechnology* and used it in throughout the 1990s to promote genetic manipulation as a benign blend of biology and high technology. Corporations developed their biotech advertising strategy to make it appear that GMO intervention was clean, simple, and invisible—since it was biological. Of course, there is nothing biological, clean, simple, or even natural about GMO technology.

The word *biotechnology* is a case of deliberate linguistic deception. But, deceptive or not, the term *biotechnology* helped confuse the U.S. public that was unwittingly subsidizing it with their taxes and stock purchases. The linguistic smokescreen of biotech helped dispel concerns in the United States about inherent dangers from genetic manipulation for more than twenty years, while much of the rest of the world was alarmed, even outraged, about arrogant U.S. quack scientists toying with the genetic makeup of the planet's food supply.

By about 1999, as more people around the world became concerned about genetic manipulation, the term *biotechnology* and biotech products finally began to be attacked, first in Europe, and then by both consumer-protection groups and environmentalists in the United States. In response, Charlotte Sine, an editor of *Farm Chemicals*, decided that biotech needed a linguistic overhaul and an image remake in order to redefine the tone of the debate and alter negative public concepts of biotechnology. In *Farm Chemical* editorials Sine argued that terms such as *biotech* and even *genetically modified* were sending the wrong messages. She proposed that products of the genetic revolution should be referred to as "Genetically Enhanced." Sine's vocabulary shift was designed to allow corporations and farm journals to promote these products more positively, since Genetically Enhanced (Sine's caps) would be seen as improving the value, attractiveness, or quality of the crop.[8]

Sine's parent corporation, Meister Publications, decided that it too needed a linguistic overhaul for the new millennium. In 2001, the company changed the name of its flagship publication from *Farm Chemicals* to *CropLife*. Similar linguistic overhauls occurred when the agrochemical giants began to buy up the seed corporations. They stopped calling themselves chemical corporations and began referring to their firms as "life science" corporations. They advertised that they could now provide farmers with a complete system of farming and a line of farm products from fertilizers to seeds to pesticides and harvest chemicals. Many in the farm movement had come to realize that these corporations were death science firms, so having them ask the public to see them as life science corporations was extremely difficult to accept.

The *Farm Chemical/CropLife* articles by Charlotte Sine were promotions, not editorials. The following excerpt is from one of my favorite "infomercials" by Sine, in which she illustrates her promotional fervor. In 1999 Sine alerted GMO advocates that they would face their most difficult tests in the year 2000. She wrote about the urgency as follows:

> Remember the TV show *Mission Impossible*? It always started with a taped message to the lead characters describing the mission and instructing them that the tape would self-destruct in five seconds. The missions always seemed impossible, but somehow, they were always successfully completed because all the "good guys" pulled together.
>
> I think we're facing a mission impossible right now, and there's no way, no how this mission will be successful without the total commitment of every company, every dealership, every commodity group, every spokesperson for agriculture.
>
> Our mission is to create—and financially support—a communications and educational program on genetically enhanced seed. If this is not done, this technology, which holds so much promise for mankind by enhancing the nutritional benefits of crops and providing new tools to fight diseases will never realize an nth of its potential.[9]

Sine never discusses GMO's problems or failures in any of her promotionals—nor does she seem inclined to discuss the real "mission impossible," which is, How are we going to put serious GMO mistakes back in their Pandora's box? She also implies that anyone who is against GMOs is a bad guy, since the "good guys" are for Genetic Enhancement.

For more than twenty years Sine's editorials usually have been reflective of where the industry is headed. Seven months following her Mission Impossible editorial, the GMO corporations put together another major advertising campaign concentrating on familiar themes such as enhancement, heroic medicines, pulling together, and salvation from hunger. Sine wasted no time is applauding them for their decision to try and brainwash the public about genetic manipulation.

Though the advertisers and the publishers would like to have us think that all is well in genetic-manipulation land, the problems and failures in farm country emerged in the very first season that GMOs were grown, and serious problems have continued ever since. In 1995, after being influenced by Monsanto's slick sales effort, Mississippi cotton grower Rodney

Garrison became a willing poster boy for the Roundup Ready cotton campaign. The new strain of cotton he helped advertise was genetically engineered to withstand spraying of Monsanto's weed killer, Roundup. Garrison and many other farmers were told that they could spray Roundup on GMO cotton plants with no damage to the plant. Beneath a corporate brochure, a grinning Rodney Garrison, a professional farmer for more than three decades, proclaimed the new strain to be "as revolutionary to the cotton industry as the cotton picker."

One year after his poster-boy testimonials, however, Rodney Garrison found himself standing in the middle of a mutant cotton field. Some of the bolls, which normally contain the locks of fluffy fiber, were misshapen. Plants that should normally have six cotton bolls on each branch were lucky to carry one. For some reason, the other bolls withered and fell off. Many farmers felt that the reason was that Monsanto had conducted only five years of research and development with Roundup and Bollgard cottons, whereas it usually takes ten to twenty years to develop a successful strain of seeds.

Bollgard is a Monsanto-patented seed with a bacterial insecticide, *Baccillus thuringensis* (Bt), shot into its genetic makeup. It was only after widespread planting of Bollgard cotton that researchers noticed that the plants produced too little insecticide to keep the bollworms from feasting on the bolls. Monsanto claimed they were surprised. So were the farmers.

Unfortunately for the farmers, such terrible surprises were not even remotely alluded to in Monsanto's, DuPont's, and the USDA's promotional and advertising blitzes on commodity-crop farmers. Instead, the ads promised that farmers would be surprised with a "technology so advanced," a "technology you can count on." Cotton consultants told a different story than the advertisements, as this quote from a 1996 *Cotton Grower* shows:

Monsanto had promised 95% control of the bollworm under all but extraordinary cases. Monsanto representatives got up at the meetings and virtually said, "Your days of spraying budworm and bollworm are over with if you plant this Bt cotton." A lot of my growers, especially the ones who had to spray, felt like Monsanto really dropped the ball on that, said Danny Bennett, a consultant in Central Georgia. Monsanto blamed the failures on extraordinary bollworm pressures. Bennett found that over a ten-year period, the bollworm pressure was only light to moderate in 1996.[10]

In these early years, Monsanto dangled a "Simple, Effective, Profitable" biotechnology carrot in front of the farmers, with the promise that their herbicide-tolerant plants would greatly reduce costs for other herbicides and increase yields. But before long Monsanto wielded this carrot like a nightstick. Monsanto required all growers to sign contracts to buy seed only from Monsanto, to use only Monsanto's Roundup Ultra, and to forfeit the right to save their own seed.[11] Several farmers were prosecuted as early as 1996 because GMO seeds sprouted and grew from a previous crop they had grown on their farms. This had a chilling effect on farmers.[12]

For the most part, in the mid-1990s the chemical corporations effectively silenced the U.S. media and tied farmers up with contracts and lawsuits. Still, the *New York Times* reported—in a rarely negative piece—that more than 18,000 acres of Bollgard cotton had failed in Texas alone in 1997 owing to bollworm infestations, which the plants were theoretically designed to repel.[13]

Information began to leak out that disillusioned farmers who had been drawn in by Monsanto's and Delta and Pine Lands' advertisements faced millions of dollars in losses. Liabilities from genetic experiments such as those conducted on farmers like the trusting Mr. Garrison were far-reaching, not only affecting a farmer's performance in a single season, but damaging relationships with creditors for years to come. Farmers lost their land as a result of these early GMO experiments, just as others had lost farms when arsenic, DDT, and cyanide gas failed as the pests developed resistance. After a season of heavy crop losses, one farmer told a *New York Times* reporter, "I might have to go to my banker wearing knee pads."[14]

After the 1997 season ended, eight states reported crop failures of genetically modified cotton. Several Roundup Ready cotton varieties were taken off the market because of heavy crop losses. In an effort to get compensated, numerous farmers filed lawsuits that challenged Monsanto's rosy claims about GMO crops. To date, thousands of farmers have filed individual and class action suits against Monsanto. Almost all have found it expensive to challenge such a deep-pocketed adversary.

A year after the farmers began their class action suits, Monsanto budgeted $10 million per year for legal expenses to fight the thousands of lawsuits and to prosecute farmers who violated their contracts. The corporation hired the Pinkerton and other detective agencies to track down any farmers who were using pirated seed. The company aggressively filed lawsuits against farmers who they claimed had kept GMO seed and replanted or sold it. Monsanto also began to prosecute growers and small-scale dealers of seed in an effort to put even the smallest competitors out of business.

As the GMO acreage expanded in cotton, rapeseed, soybeans, and corn, a seemingly insurmountable problem faced Monsanto and the other GMO firms. Many farmers, rural communities, and legal experts felt that the GMO corporations or the contract farmers would be responsible for drift of pollen to adjacent fields or the spillage of seed on another farmer's property in transit from the GMO farm to the grain elevator or cotton gin.

Because of patent-law protections, however, Monsanto turned its own trespasses into illegal uses of its patented property. It got the courts (especially in St. Louis—home to Monsanto's corporate headquarters) to rule against growers whose fields had become contaminated with Monsanto's GMO plants. Instead of Monsanto being prosecuted for spillage, drift, and trespass, the farmers who were the victims of Monsanto's trespass were prosecuted and convicted for illegally possessing Monsanto's patented plants or seeds.

In an especially vicious attack, Monsanto accused Percy Schmeiser, a farmer from Saskatchewan, of growing Roundup-protected rapeseed (canola). Schmeiser had developed many rapeseed varieties and had sold his seed varieties to local farmers for more than fifty years. He formerly served as the mayor of his town and as a member of Parliament in the Canadian government. After a crippling accident he recovered and went on to climb Mount Everest twice. Schmeiser seemed to embody credibility and true grit.

However, neither his credibility nor his grit stopped the Canadian courts from deciding in favor of Monsanto, even though Monsanto admitted that its pollen drifted or its seed had blown or spilled on Schmeiser's fields and contaminated his crop. Schmeiser countersued and accused Monsanto of trespassing. Schmeiser claimed that Roundup Ready rapeseed had blown onto his farm and contaminated his crop so that it no longer had value as non-GMO seed. In Percy Schmeiser's case, he had tried to sell his seed that Monsanto's pollen drift or spilled seed had contaminated. Monsanto argued that Schmeiser knew it was contaminated and thus "enhanced" with Monsanto's patented intellectual property. It demanded that he surrender the seed and pay the company a fee for its "intellectual property rights."

Simple. Effective. Profitable.

THAT'S
ROUNDUP READY®
COTTON.

Simple. The Roundup Ready cotton system uses just one herbicide – proven Roundup UltraMAX® – and gives you a wide window of application on a broad spectrum of weeds.

Effective. Roundup UltraMAX herbicide moves weed-killing power right to the roots for unsurpassed control. And crop safety's a given – Roundup UltraMAX is proven crop safe on hundreds of millions of treated acres.

Profitable. With Roundup Ready cotton, you will increase your profit potential by reducing, or even eliminating, expensive costs like weed crews, cultivation trips and residual herbicides.

Need more proof? Just ask fellow California growers who use the Roundup Ready cotton system. They'll tell you it's the simple, effective, profitable choice.

"Simple, Effective, Profitable." A Roundup Ready ad from 1996 *California Cultivator*, 1996.

Monsanto Seed Cops Snaring Farmers

SAN FRANCISCO — Monsanto Co.'s "seed police" snared soy farmer Homan McFarling in 1999, and the company is demanding he pay it hundreds of thousands of dollars for alleged technology piracy. McFarling's sin? He saved seed from one harvest and replanted it the following season, a revered and ancient agricultural practice.

"My daddy saved seed. I saved seed," said McFarling, 62, who still grows soy on the 5,000-acre family farm in Shannon, Miss., and is fighting the agribusiness giant in court.

Saving Monsanto's seeds, genetically engineered to kill bugs and resist weed sprays, violates provisions of the company's contracts with farmers.

> Since 1997, Monsanto has filed 90 lawsuits in 25 states against 147 farmers and 39 agriculture companies.

Since 1997, Monsanto has filed similar lawsuits 90 times in 25 states against 147 farmers and 39 agriculture companies, according to a report issued Thursday by The Center for Food Safety, a biotechnology foe.

In a similar case a year ago, Tennessee farmer Kem Ralph was sued by Monsanto and sentenced to eight months in prison after he was caught lying about a truckload of cotton seed he hid for a friend.

Ralph's prison term is believed to be the first criminal prosecution linked to Monsanto's crackdown. Ralph has also been ordered to pay Monsanto more than $1.7 million.

The company itself says it annually investigates about 500 "tips" that farmers are illegally using its seeds and settles many of those cases before a lawsuit is filed.

From the *Valley News*, January 15, 2005.

Schmeiser lost on both counts and appealed. He ultimately lost on appeal to the Canadian Supreme Court. Schmeiser is one of a number of salt-of-the-earth seed growers and dealers who were targeted by Monsanto. All of them have either lost in court or settled out of court.

In fact, between 1997 and 2004, more than nine thousand U.S. farmers were investigated by Monsanto for patent infringement on proprietary GM seed, and most of those whom Monsanto pursued settled out of court. Almost two hundred farmers and farm businesses during that period were prosecuted and their cases went to court. Many farmers were forced to pay stiff penalties, including court costs, to Monsanto, totaling more than $15 million.[15] The continuous and aggressive prosecution of farmers helped prevent growers from fighting against Monsanto and the other big chemical corporations, with their huge legal defense budgets. In ten years Monsanto has spent $100 million to collect $15 million. But it has put the fear of prosecution in every commodity-crop farmer's head. This is money well spent from Monsanto's point of view.

While Monsanto was fighting its legal battles with farmers, many GMO growers in the United States, Canada, and Argentina were fighting to get loans from their bankers after suffering low yields, crop failures, bizarre crop characteristics, and failed markets. Today, farmers continue to experience low yields of GMO corn, soy, and rapeseed.[16] In areas where GMO crops have been used for several years, farmers have had to use multiple herbicide sprays to control weeds that one spray of Roundup was initially promised to control. This has greatly increased pesticide use and spraying costs as well as caused other problems.

One problem caused by increased spraying of Roundup is that invasive weeds have shown significant resistance in Roundup Ready crops. Resistance to Roundup on five major weeds had become so serious by 2003 that Monsanto and the other GMO corporations recommended using arsenic, paraquat, or 2,4-D to control the resistant weeds. With recommendations to use such dangerous chemicals to deal with resistance, it's clear that Roundup Ready and the other genetically manipulated miracles have not replaced the chemicals that the advertisements promised would be eliminated. By 2007, things got worse, as at least twelve serious weeds had developed resistance to Roundup.[17] This led to a profusion of competitors for Roundup's customers, as the 2007 advertisement from Bayer at bottom right shows.

Resistance problems have also plagued Bollgard cotton, which has Bt (*Bacillus thuringensis*) shot into it. The EPA and the corporations blame these resistance problems on the failure of farmers to adhere to the refuge requirements with Bollgard cotton, but the real issue is probably overuse of Bollgard seed in areas where worms are periodically a serious problem. Farmers must now accept mandatory refuges (land planted with non-Bollgard cotton) and a strict adherence to the allowable percentages of Bt cotton that can be planted on a given-sized piece of land in a given area. The EPA imposed a "two strikes and you're out" rule to force farmers to follow the new guidelines. Chemical corporations say they are trying to preserve Bt from showing resistance for as long as possible, even though in all their advertisements they still encourage farmers to buy it to save money, as the ad at right promises.[18]

In the ad, Monsanto claims its staff studied six hundred research plots in the South and Southeast from 1995 to 2004 and determined that farmers could make $39.35 more per acre than they could on non-Bollgard/Roundup acres. In the fine italic print at the bottom left, the ad explains that this estimate was based on the price of cotton being sixty cents per pound, even though Monsanto must have known that cotton dropped to $0.28 per pound in 1997 and stayed in the

Weed worries

If you use glyphosate, pay attention to herbicide-resistant weeds.
■ *By Rod Swoboda*

"Farmers need to understand the importance of using a management strategy to avoid problems with weeds developing resistance to glyphosate herbicide," says ISU's Mike Owen.

Articles critical of GMO crops finally began to emerge in 2003, as the article title page above shows. Weed specialist Bob Hartzler found that more than 25 percent of marestail survived Roundup sprays. From *Wallaces' Farmer*, March 2003.

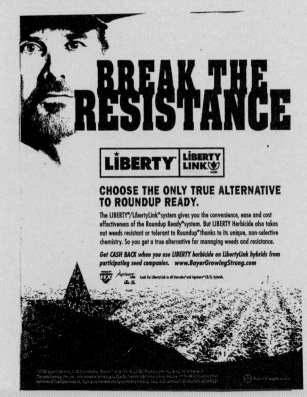

A Bayer CropScience ad from *Farm Journal*, September 2007.

$0.45 range until 2002. Even with $0.60 cotton, however, with both Bollgard and Roundup showing resistance problems, any savings using this variety of cotton could be more than used up by the cost of herbicides and pesticides to take care of the resistant pests. Is this honest advertising?

Based on historical and modern evidence, we know who will suffer from false advertising, resistance problems, and the low yields, crop failures, and lawsuits that result from genetic manipulation. In all the debacles of chemical agriculture since the mid-1800s, it was the farmers who went bankrupt from using defective and worthless fertilizers and pesticides—not the chemical corporations.

John Hester, manager of Nichols Agricultural Service in Nichols, Iowa, was asked the following question: "How has the rising controversy about genetically enhanced crops affected your region and your grower-customers?" He responded with the following:

> Farmers always say they are left holding the bag on issues that "protect the consumer" and I have to agree. The industry told them that their future is in genetically enhanced (GE) crops. Now, we tell them we aren't sure we want GE crops. That's upsetting to the farmers, and I don't blame them. We have large elevators that must meet the demands of overseas customers, and they are turning to the farmers and telling them they won't accept GE grain—instead of a premium, farmers may have difficulty finding an elevator to take their grain. Or it may increase their transportation costs. We have stuck it to the farmer, and we as an industry must do something about that.[19]

Hester wrote about these concerns in 1999, but since then he and the other farm leaders have done nothing about it. After the failure of several GMO farm products, low yields, and an increase in herbicide and

pesticide bills caused by resistance to GMO crops, who is there to compensate the farmers for these losses or their trashed credit ratings? Not Monsanto, not DuPont, not Delta and Pine Lands, not Bayer, and not the farm magazines or the USDA, FDA, and EPA that had all promoted GMOs for four decades.

Since 1998 we have seen an enormous shift toward consolidation of the seed and chemical industries. At the same time that Monsanto aggressively prosecuted growers and promoted untested products, it also began acquiring a large share of the genetically manipulated seed market as well as conventional seed suppliers. In May 1998, Monsanto bought two of the world's largest seed corporations, Dekalb and Cargill International Seed. With these acquisitions Monsanto was poised to control a majority of the nation's cotton, canola, and soybean seed markets (from 75 to 85 percent), and it promptly infused many seed varieties with GM genes.

Other chemical corporations have likewise conducted corporate takeovers during the last decade. In 1999, DuPont acquired Pioneer Seed (the largest corn-seed breeder and marketer) as well as other significant genetic seed banks. Through these acquisitions DuPont obtained control of a huge part of the corn-seed business. In 2001, the Swiss company Novartis acquired Astra-Zeneca and changed its name to Syngenta—thereby becoming the largest seed and chemical corporation in the world. However, in 2005, after Monsanto purchased the Mexican company Seminis, it became the world's largest seed corporation. In such a narrowly controlled seed market, commodity-crop farmers are left with little choice but to buy their seeds and herbicides from Monsanto, DuPont, or Syngenta and abide by their contracts.

Mississippi cotton grower Rodney Garrison was right: this was a revolution. But the revolution was not about crops that produced better yields or higher profits for farmers or even lower toxic pesticide and fertilizer use, as farmers like Garrison and consumers had been promised. Instead, the revolution was about a narrowing of the seed supply, about the vertical integration and monopolistic control of seed and chemical markets, and about the corporate control over the regulators at the EPA, the USDA, and the FDA.

The major corporations have been able to force and cajole commodity-crop farmers into taking the leap and planting transgenic corn, cotton, soybeans, and canola, even though yields are often low and failures frequent. They did this largely by limiting the farmers' seed choices and with saturation advertising

promises. The corporations were less successful, however, at cajoling international consumers in the marketplace. By 1997 it had become clear that people around the world were not convinced about the eco-friendliness and the relative harmlessness of genetically modified crops and that biotech crops had real trouble brewing in the marketplace.

Nowhere was this early antagonism toward GMOs greater than in Europe, where suspicions about high-technology farming were already intense after battles over U.S. hormone-treated beef, England's deadly encounter with mad cow disease, frequent European outbreaks of salmonella poisoning, and hoof-and-mouth disease in England and France. And suspicions about corporate factory farming continued to remain high in Europe with the outbreak of bird flu

A two-page ad from Monsanto for Bollgard combined with Roundup Ready cotton. From *Cotton Grower*, 2004. "Making higher yields a reality."

Bollgard protects more bolls to make bigger profits a reality.

Harvesting more first-position bolls delivers bigger profits.

So how do you pick more first-position bolls?

Plant Bollgard® with Roundup Ready® cotton. Bollgard works 24 hours a day, seven days a week, to protect more bolls from damaging budworms and bollworms, including the bigger, first-position bolls that deliver higher yields and profits. That's how Bollgard protection has provided an average of $39.35 more profit per acre over the last eight years.*

For bigger profits in your fields, protect more first-position bolls with Bollgard with Roundup Ready cotton. Making higher yields – and bigger profits – a reality.

*Source: Based on over 600 Monsanto research trials in Southern and Southeastern Cotton Belt locations 1995 - 2002. Advantage calculated based on incremental lint harvested at $.60/lb, and difference in technology cost versus insect control cost (number applications at estimated cost/spray). Results may vary

Bollgard
with
**Roundup Ready
Cotton**

Making Higher Yields A Reality.

in huge poultry confinement operations all over Asia, Turkey, eastern Europe, and finally England in 2006. The large-scale poultry operations in Asia that first developed bird flu were raising poultry destined mainly for Europe, so the fears of bird flu in the European Union were much higher than in the Americas.

The molecular biologists and industry spokespeople tried to ignore or trivialize the European public's fears and suspicions about high-technology foods and argued that the Biotech Century would unfold without serious environmental consequences to the planet. Editors like *California Farmer*'s Len Richardson boldly defended genetically manipulated crops and animals, in spite of the European and Asian resistance. In November 1999, Richardson called GMOs "the bullet to the future."[20]

Industry and editorial promotions such as Richardson's were met with even more scientific and consumer skepticism abroad, as well as activist outrage. Test plots and fields were destroyed by environmental groups such as Earth First, Greenpeace, Friends of the Earth, and farmer activists from the United States, France, Japan, Korea, and India. Attacks have continued every year on gene factories producing GMOs, on researchers advocating their use, on genetically manipulated crops, and on ships and trucks transporting GMO seed. Even an Iowa cornfield was painted with the banner, "Warning: Biotech Hazard." One Cargill plant in India was torn down. Cargo ships were blockaded from delivering GMO corn to Mexico. African countries even refused genetically manipulated crops as foreign aid.

The promised market euphoria for GMO food never materialized for U.S. farmers. Instead of gaining world markets as the genetic promoters promised, U.S. corn farmers alone have lost $300 million per year since the late 1990s in reduced sales to the European Union. In 2006 the EU instituted labeling and traceability requirements, which means a potential loss of an additional $4 billion per year in farm-commodity exports.[21]

Faced with a total loss of international markets, U.S. corn growers turned to ethanol, and soy growers to biodeisel. People criticize the switch of so much corn and soy acreage to fuels, but it was the biotech revolution that forced growers to look for other markets when the traditional food and fodder markets refused to accept genetically altered corn and soy, and the seed giants provided only GMO seed. And there is a glut of corn, cotton, and soy in U.S. and world markets.

Negative consumer reaction in Europe, Asia, and the United States caused FritoLay, Gerber Baby Food, and Heinz Baby Food corporations to reject genetically modified farm products in 1999/2000. The grain giant Archer Daniels Midland (ADM) apparently didn't want to be "Supermarket to the World" with genetically manipulated crops. ADM and Continental Grain notified grain elevators on September 1, 1999, that they should segregate genetically manipulated and conventional corn and soybeans because of consumer rejection of GMOs in Europe and Asia.[22]

CropLife reported that in 2003 ADM sent another letter to retailers warning against the use or purchase of GMO corn manipulated to resist rootworm because it could not be sold to the European Union:

> ADM will not be able to accept delivery of these varieties. We encourage you to carefully consider your seed purchasing to assist us in supplying products that meet the requirements of our worldwide customers.[23]

Most, if not all, of the EU countries and Japan require labeling for all genetically manipulated products. In the Italian state of Tuscany, the growth and sale of GMO crops was banned in 2003. Austria banned GMOs in 2004, and Switzerland followed suit in 2005.

Many other European and Asian consumers and farmers have also judged GMOs unacceptable and rejected or greatly restricted the marketing and growing of genetically manipulated crops.

In spite of such an overwhelming international rejection of GMOs, mandatory labeling of genetically manipulated foods is still not required in the United States. And the courts, legislatures, and regulators have even prohibited labeling in many cases, including the voluntary labeling of non-GMO products, such as dairy products that do not come from cows injected with bovine growth hormones. Any such labeling—which the corporations say implies that non-GMO products are better, purer, safer, and more natural—is considered "product disparagement." Apparently, giving U.S. consumers the choice over what they buy to feed to themselves and their kids is the ultimate sin against unwanted products.

Even U.S. consumers, however, finally began asking serious questions about genetically modified food by the spring of 2000. By 2005, eighty-three Vermont communities had voted against growing GMOs in their towns and three California counties legislatively banned the growth of genetically modified seed. With the international and national consumer backlash that continues to build against genetically manipulated crops, Len Richardson's "bullet to the future" could turn out to be another in a long series of "bullets to bankruptcy" that advertising-driven farm magazines have sold to farmers.

What are the long-term effects of this excessive dedication to genetic manipulation? As we have shown, the continuing GMO crop-and-market failures in cotton, corn, soy, and rapeseed caused economic difficulties and disaster for many of the farmers who were early adopters of the technology.[24] But microbiologists are worried about even more widespread consequences. The chemical/GMO corporations argue that the chances for catastrophe

are slim—but they do not deny that the chance for disasters exists. Depending upon one's perspective, GMO disasters may have already started to occur.

In 2002 researchers proved that genetically manipulated corn from the United States had contaminated native Mexican corn in at least eight locations. Mexico is the heartland for corn, where it was first domesticated, and where more genetic variation exists than anywhere else. Contamination of the wild races of corn puts the genetic diversity of corn at risk and severely hampers our ability to fight any future corn disease pandemics. Such a pandemic already occurred in the early 1970s, when one-quarter of the U.S. corn crop was lost to blight that affected a single susceptible gene bred into almost all commercially grown hybrid corn.

In 2007, the monarch butterfly population migrations were significantly below the normal average. Scientists feel that this could be a direct result of the increased use of Roundup and the use of other more toxic weed killers on genetically manipulated corn, soybeans, and cotton to deal with Roundup weed-resistance problems. These increased sprayings of Roundup greatly reduced or contaminated the population of milkweed, which is one of the monarch butterfly's major host plants. In one study, 23.7 percent of monarch larvae failed to reach adult stage after being exposed to genetically modified corn.[25]

Though U.S. scientists were rebuffed in their efforts to thoroughly analyze GMOs, it is fortunate that scientists in other parts of the world began to conduct independent studies on genetically manipulated crops and to check the stomach contents and the health of animals fed GM feed. Dr. Arpad Pusztai in Scotland was the first scientist to demonstrate that genetically modified potatoes damaged the spleen, thymus, kidney, and guts of young rats and retarded their growth. His findings alone proved that more independent and government testing should be mandatory. His findings and the publication of his fears also led to him being fired from the institute where he had worked for thirty-five years.[26]

Irina Ermakova, a leading researcher at the Russian Academy of Sciences, studied the differences in rats fed GMO soy, non-GMO soy, and a nonsoy diet during their pregnancy and nursing period. Thirty-six percent of the offspring from mothers who ate genetically manipulated soy were extremely small, while only 6 percent of the offspring who ate non-GMO soy or no soy were small. Worse, after three weeks twenty-five of the forty-five (55.6 percent) baby rats whose mothers ate GMO soy died, whereas only three of the thirty-three (9 percent) non-GMO-soy offspring died, and three of the forty-four (6.8 percent) nonsoy offspring died.[27]

Many scientists, researchers, and seed experts who live and work outside of genetic engineering's inner circle of promoters feel alarmed that large portions of our precious seed banks are now closed to the public. Corporate amalgamations have managed to reduce ten thousand years of seed breeding to private commercial property that can be bought and sold on the global market or discarded as irrelevant and without profit potential. Corporate efforts to convert the genetic blueprints of thousands of years of plant and animal breeding to privately owned intellectual property represents a growing danger to the world's food supply. And it puts our public property in the hands of corporations that have violated every human right, that have killed, injured, and maimed their own workers and millions of consumers and innocent bystanders.

With respect to corporations' insatiable desire to play god and clone animals, recent reports have shown that the genetic cloning of animals is very risky and plagued by fraud. Of course, the FDA and the USDA support this research. Cloning involves high rates of late abortion and early prenatal death, with failure rates of 95 to 97 percent in most mammal-cloning attempts. One scientific review stated that even the successfully cloned animals were disasters, or nearly so. Sixty-four percent of cloned cattle and 40 percent of cloned sheep exhibit some form of abnormality or birth defect such as grossly oversized calves, enlarged tongues, squashed faces, intestinal blockages, immune deficiencies, and diabetes.[28] Clearly, live births in normal animals do not express these deformities in such disgustingly high percentages. In another recent industry debacle, South Korea's most famous scientist, Professor Hwang Woo-suk, admitted in December 2005 that he had faked cloning experiments on human stem cells. Despite the concerns about deformity percentages, obvious quackery, and lying in the industry, recent reports indicate that the USDA and the EPA are comfortable with selling cloned animals and having the industry advertise the products as essentially identical to normal animals.

Corporate attempts at limiting research and journalism that might be negative to genetic manipulation have been effective. However, studies such as Pusztai's and Ermakova's are finally beginning to emerge and they illustrate that there are severe problems with GMO foods and products, as the following excerpt from Jeffrey Smith's *Seeds of Deception* shows:

> Rats fed GM corn had problems with blood cell, kidney and liver formation. Mice fed GM soy had problems with liver cell formation and pancreatic function, and the livers of rats fed with GM canola were heavier. Pigs fed GM corn on about 25 farms in North America became sterile, had false pregnancies or gave birth to bags of water. Cows fed GM corn in Germany died mysteriously. And twice the number of chickens died when fed GM corn compared to those fed with natural corn. . . .
>
> According to a report in the *Daily Express*, soon after GM soy was introduced to the UK,

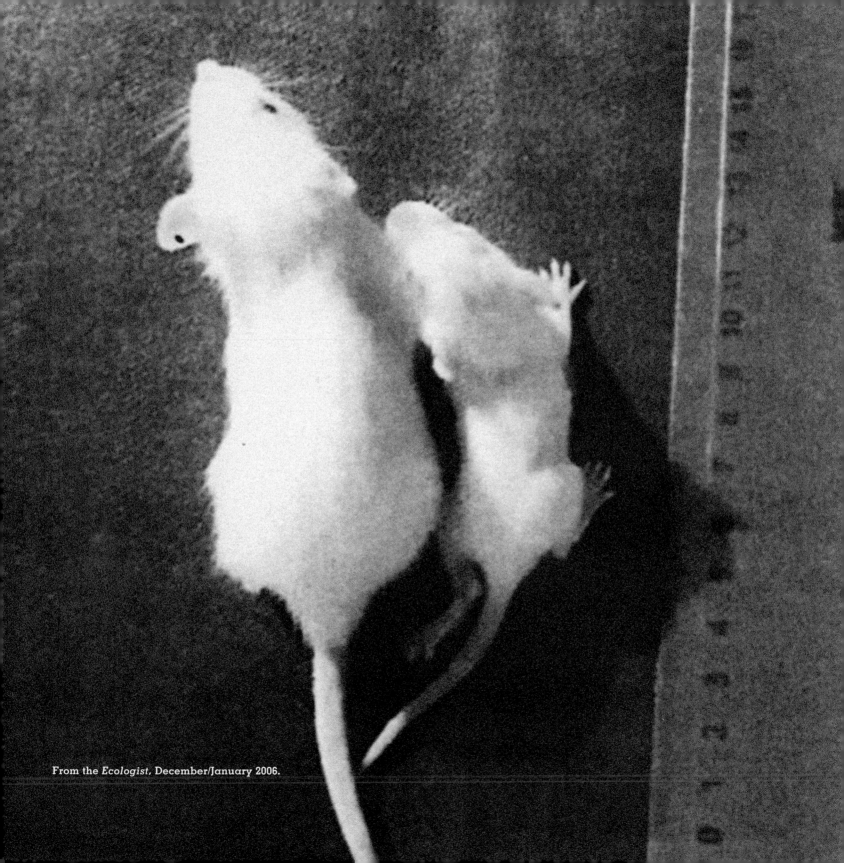

From the *Ecologist*, December/January 2006.

soy allergies skyrocketed by 50%. A gene from a Brazil nut inserted into soybeans made the soy allergenic to those who normally react to Brazil nuts. . . .

GM soy contains significantly more tripsin inhibitor, a common allergen in soy . . . data [Monsanto] had omitted from the study, and later recovered, showed that the increase was as high as seven-fold in cooked soy.

. . . Filipinos living next to a Bt cornfield developed skin, intestinal, and respiratory reactions and fever [for two years in a row]. . . . Mice fed Bt corn developed an immune response equal to that of cholera toxin, as well as abnormal and excessive cell growth in their small intestine.

A new disease was caused by a brand of the food supplement L-tryptophan, . . . created through genetic modification. . . . The disease killed about 100 Americans and caused sickness or disability in about 5,000–10,000 others.[29]

What further disasters will genetic manipulation bring? What will be the result of the secret, sensational, taxpayer-subsidized experimentation that Charlotte Sine, Mindy Laff, Len Richardson, and their colleagues are so anxious to protect and promote?

In spite of such serious problems with GMO technology, consumer and scientific pressure in the United States has not been as consistent or as strident as in Europe. This is due partly to general public ignorance and apathy and partly to the lack of grassroots environmental movements in the United States (unlike in Europe, which has very active and effective ones in several countries). But the ineffectiveness of actions in the United States also results from the gene giants acting much more aggressively than they did in the E.U.

In the United States, the corporations fought off public demands for labeling GMOs with a well-coordinated, well-financed public-relations effort. Unfortunately, the U.S. environmental movement was conflicted about whether it should try to label GMOs or work to ban them. The corporations were not conflicted; they put all their energies into the antilabeling fight because they knew that if it was labeled it was over, since polls showed that more than 75 percent of the population was opposed to genetically manipulated food. Even in states where labeling had been enacted, the corporations won legal battles to eliminate labeling requirements. These victories enabled corporations to slip GMO products into the U.S. market untested and unlabeled.

Consequently, U.S. consumers currently serve as the major guinea pigs in the consumption of transgenic products from millions of acres of unsupervised experiments with crops and animals. In the 1930s there were 100,000,000 human guinea pigs being fed arsenic, lead, sulfuric acid, and cyanide on their food. Now, there are 300,000,000 guinea pigs eating genetically altered food with nerve poisons on it.

Unlabeled genetically engineered ingredients are currently found in more than 70 percent of the processed food on U.S. market shelves. Most of the GMO commodities are dumped on unwary U.S. consumers in the form of processed foods, milk, and meat. Millions of U.S. consumers buy processed foods with GMO ingredients today without knowing that they are doing so. The advertisers claim that the GMO food is the same as any other food. They claim that transgenic food is safe, and that it is cheap. Those are all lies. Tomatoes with rat genes are not the same as heirloom tomatoes. Drs. Pusztai and Ermakova and several other scientists have shown that it is not safe. Most genetically modified food is from the major subsidized crops: corn, soy, cotton, and canola. None of these highly subsidized crops are cheap, since shoppers pay for them once at the market and once again at tax time, and since we don't know how much it is going to cost for the cleanup of chemical and genetic farming, we should expect a third payment sometime in the near future. It only seems cheap when one looks at only the purchase price.

Besides what corporations dump on U.S. consumers, the remaining GMO commodities are usually bought by US-AID and other relief agencies at elevated prices from brokers to assist famine, typhoon, hurricane, earthquake, and tsunami victims. Recently, countries have demanded that GMO grains for aid programs be ground before they are shipped so that their growers will not plant the seed and contaminate local varieties of their crops. Many traditional cultures fear the effects of GMO food.

Since U.S. shoppers are the major consumers, they should especially fear genetically manipulated foods. But because there is no labeling most consumers don't even know they are cooking with or eating GMO food.

The grain giants try to avoid mixing non-genetically modified grains with any modified ones so that they can sell to Europe and Asia. However, the reality is that grains are mixed in processing and milling, especially if they are not bound for a suspicious overseas market. Anything one eats in the United States that is not organic or not labeled as "No GMO" and has corn, soy, canola, or cottonseed ingredients is probably a GMO-contaminated product. That includes especially milk and milk products—since many milk cows in confinement dairies consume about eight pounds of GMO cottonseed and cottonseed meal and several pounds of GMO corn and soy per day and are shot up with GMO growth hormones. Milk products are probably the most genetically modified food on the grocer's shelves. And consumers are becoming wary of the safety of genetically manipulated milk

In the face of rising criticism, scientific suspicion, and failed products, the chemical-pharmaceutical-seed firms created another massive advertising campaign in 2003. Instead of going back to the lab, retesting their products, and examining their theories, the GMO corporations created even slicker ads, catchier slogans, and more passionate editorials and bought more TV slots. Their focus in this campaign was on using life-saving and hunger abatement arguments for the necessity of GMOs. Their ads went something like the following: What if your wife was going to have a baby, and you knew it was going to be born with a birth defect? Wouldn't you want the option of fixing your baby so it could have a better life? What if we could cure blindness in Asia with GMO "miracle rice" that is vitamin-rich? Wouldn't you want to cure millions of afflicted children?

These rhetorical questions and answers, coupled with beautifully done visuals on TV, were designed to provide hope that a genetically altered answer was just around the corner. There was no discussion of the fact that one gene could have the capacity to affect hundreds or even thousands of traits, and that the products therefore could prove dangerous or capricious or useless. There was no mention that with their GMO "miracle rice," kids would have to eat fifteen pounds per day to get the necessary vitamin A that a vitamin pill costing pennies could provide. These are GMO fantasy ads. They are more science fiction than science and offer no significant hope for people who have real afflictions, real hunger, and vitamin deficiencies.

In today's advertising campaigns directed at farmers, the genetic manipulators similarly raise expectations and hopes that they can't possibly meet. Advertisements promote GMOs as an alternative to chemical use, and yet chemicals haven't disappeared with the advent of the genetic revolution. In fact, the recommended weed killers for weeds resistant to

Roundup in Roundup Ready fields are among the worst chemicals ever used.

With U.S. commodity-crop farmers having few seed choices and literally no access to criticism of the technology, many signed contracts with Monsanto, Syngenta, and DuPont, and the percentage growing GMO seed ballooned. By 2003, 84 percent of U.S. canola acreage was planted with GMO seed. By 2004, 76 percent of cotton, 46 percent of corn, and 86 percent of soybean acreage in the United States grew genetically modified crops.[30]

The unrestricted and unregulated corporate power to transform, remake, and exploit nature in wholly new ways virtually guarantees that the genetic manipulation taking place on such a large scale will inflict its own form of damage on the earth's environment. We should not be deluded into believing that genetic engineering is biological or "natural." Chemicals can cause and have caused terrible injury, death, and widespread toxic pollution of the soil, air, and water. Genetic manipulation, though, is a different class of threat. Genetically manipulated pest controls are genetic polluters. We might be able to clean up most of the chemical pollution, but how are we going to clean up genetic pollution?

There is an axiom guiding scientists that holds that no research effort should be attempted unless it can be undone, unless any damage it causes can be repaired. It is proving difficult, costly, and time consuming to control and repair the damaging effects of chemical pollution that resulted from ignoring that axiom for all these years, and we have only just begun the cleanup.

If we continue to ignore this axiom with GMOs it may prove impossible to undo the damage caused by genetic pollution. Because they continue to ignore the axiom, the genetic-engineering firms have become another example of how human creativity and intelligence can be overpowered by hubris and greed. The current corporate strategy is focused on the immediate commercialization of GMO discoveries, and therefore the companies feel that they must sell products that have been barely tested. This rush to profit from GMO discoveries rivets us not on a glimmering future of biological discovery and prosperity but, instead, on a blind leap into a potentially darker age of agriculture than the one we endured with poisonous chemicals for the last 160 years.

All of the conflict and friction described in this chapter is a result of corporate efforts to immediately capitalize on new discoveries and control the choices farmers make for seed fertilizer and pesticides. In their impatience to play god and alter whatever they could to show off their new skills, these corporations tried and often succeeded in patenting, owning, and genetically modifying corn, soy, cotton, rice, and canola. Instead, they should have used marker-assisted genetic analyses derived from gene slicing and genome projects for plants and animals. This would have enabled them to do more efficient analyses of the genetic makeup of animals and plants.

Marker-assisted research allows geneticists to conduct classic genetic crosses in five years instead of the twenty years it currently takes to develop a new crop. Microbiologists could have done genetic engineering without employing transgenic techniques, which are considered odious and immoral to a large segment of the world's population and damaging to the purity and diversity of the world's gene pool. Ironically, this very diversity and purity of the gene pool is needed in order to use marker-assisted genetics, so any pollution of the gene pool that the genetic manipulators have already caused will make marker-assisted genetics more difficult.

As the begging letter from Monsanto's animal agriculture president to milk producers at right shows, the U.S. public has finally had an impact on rejecting genetically modified milk for their kids. With the loss in sales of bovine growth hormone because of consumer fears, Monsanto used an old tactic in this open letter to stall or stop the slide in sales. It prods farmers to join them and protect the farmers' tools against government regulators and nervous cooperatives or else every tool the farmer has will be at risk. It sounds desperate and it sounds like a loss for the gene giants, but don't count them out yet. They have a lot of friends in high places.

imagine

September 1, 2007

Dear Dairy Producer:

Recent announcements by major grocery chains and milk processors to sell only milk labeled "rbST-free" continue the trend of deceiving consumers about milk. For those of us actively involved in the dairy industry, we know that all milk is equally safe and wholesome and continue to be alarmed that many dairy co-ops have agreed to initiate involuntary programs to supply milk marketed in this manner.

This is bad for our business, your business and the future of the industry.

The American Farm Bureau's market basket survey shows consumers are paying 36% more for milk labeled rbST-free than conventional milk. Are you seeing 36% more in your milk check? In most cases, producers aren't receiving any premium. It's unfair to force dairy farmers to give up a useful management tool and not adequately compensate them.

We recognize that POSILAC may not be right for every operation, but this issue goes beyond rbST and affects every tool you use. What will be next? Your production choices should be yours alone – not marketers', not processors' and not Monsanto's.

We've heard from many producers who are frustrated, upset and don't know how to respond to this situation. Here's what you can do:

1. Contact your cooperative management and voice your concerns about:
   • Unfounded restrictions on safe, FDA-approved products.
   • Protecting the image of all milk.
   • Capturing a fair share of retail premiums for members.

2. Demand fair and adequate compensation for not being allowed to use valuable management tools. At current high prices, the value of POSILAC to producers has doubled, now significantly exceeding $1 in profits per supplemented cow per day.

We will continue to defend your right to use FDA-approved technologies, but we cannot do it alone. We need your help, your voice, to strengthen our actions and producer-led initiatives with federal and state regulatory channels. Only then, can we put a stop to this detrimental trend.

Best regards,

Kevin Holloway
President, Monsanto Animal Agriculture

P.S. Take action by sending letters to FDA, FTC and your congressmen requesting a revision to current milk labeling guidelines. It's easy. Go to **capwiz.com/voicesforchoices**.

From *Hoard's Dairyman*, September 10, 2007.

Excerpt of an ad from *California Farmer*, 1995. This ad used the baby, the historical caption, and large print to sell a pesticide. It is just as well the baby in the ad is not being raised in cotton fields today since seven of the fifteen most used chemicals on California cotton for the ten-year period surrounding this ad were probable carcinogens.

"It was fall, 1926. Your fields were full of cotton. Even the baby was close by. Today, you don't raise kids in the field. You just raise cotton."

"There was a time when you raised more than cotton in your fields."

Chapter 23

160 YEARS OF POISONOUS ADVERTISING

In chapter 11 we discussed the four-part sales model used by the fertilizer and pesticide industry to get farmers hooked on their products. Recall that the four dimensions of the model included editorials, scientific testimonials, saturation advertisements, and farmer testimonials about chemical successes on their farms. By the 1880s, the farm magazines had begun to use this strategy to promote chemicals and fertilizers. During the twentieth century they continued using it to sell these products but added antibiotics, hormones, and genetically modified seeds. The editorial promotions almost always began several years before any products were available. At the same time that the product was announced, or shortly thereafter, scientists from a university or one of the state or federal departments of agriculture would publish testimonials. Finally, as each product came to market, a profusion of clever advertisements would appear in the magazines. Within a short time after the fertilizer or pesticide was first used, testimonials from farmers would show up in the farm journals.

Advertising in America changed dramatically in the period between 1880 and 1920. In the 1880s, publishers and ad brokers dominated the industry and owned or bought up magazine space and then tried to sell it to prospective advertisers. The advertising journals *Printer's Ink*, which began in 1882, and *Agricultural Advertising*, which began in 1893, encouraged both advertising agencies and farm magazines to be more professional in their approach to how they sold products.

Many advertisers before the turn of the century were more likely to be P. T. Barnum types. The advertising companies usually composed the ads, and the brokers would merely consult on the artwork or edit the text. Before World War I, advertising brokers were replaced by or morphed into advertising agencies. They changed their business practices from controlling ad space to designing ads and campaigns and selling a complete package of services to advertisers and the journals. By the 1910s, ad agencies began to employ both artists and copywriters and emerged as full-service companies.

Ad agencies in America after about 1910 increasingly were owned by, and drew their employees from, members of the upper and upper-middle classes of society. Almost all of the early ad-agency employees were men, even though a few of the most successful magazines were run by and created for women, such as *Ladies' Home Journal*.[1] Usually the large ad agencies had one woman, or at most a few women, on staff to advise them on the appropriate way to sell to their female clientele. Nearly all of the themes of the ads as well were drawn from the upper levels of American society and reflected a male bias on most issues.

Many of the ad agencies did contract work for the government during the buildup to World War I and throughout its duration. This was a difficult assignment because the war was unpopular and the country's mind-set was isolationist at the time. However, with a variety of schemes the advertisers were able to get men to enlist in the military and to get citizens to buy war bonds and make sacrifices for a war that most people opposed.

After their experience selling the war, the advertising agencies realized that they could change the impression of the consuming public on almost any issue or product. In their minds, if they could sell an unpopular war, they could sell anything. Perhaps more importantly, several copywriters realized that they could mold public opinion. Most of the advertising campaigns in the farm magazines came from the same agencies that produced the majority of the other ads in America (but that is a subject for another book).

Prior to the twentieth century, conservationism was much more the national ethic than consumerism, so advertisers were trying to change deeply ingrained values and habits. They helped create in a conservationist culture a continuous reservoir of new consumers who aspired to be affluent and consume at least as many goods as their neighbors.

As we have noted many times in this book, after each war or economic depression U.S. farmland became more concentrated in the hands of fewer farmers as local banks folded and regional banks foreclosed on small-scale farmers. As the farms increased in size, the large-scale farmers depended even more on pesticides and synthetic fertilizers to manage their increased acreage, reduce labor costs, and avoid labor strife. The large-scale gentlemen farmers who

From *California Farmer*, 1995. The ad at right compares a beautiful woman to chemically produced vegetables and melons. The caption for this ad states: "Let's face it. Looking beautiful isn't easy. That's why you need the help of our chemicals." Continuing, the caption reads, "Sprayed preventatively they stop disease on contact and even work systemically within the plant for further protection. That means fewer blemishes and more marketable vegetables." Ridomil is the chemical being advertised as the fungicidal beauty promoter. Ad tactics that use beauty have been tried for decades. The beautiful woman from the McCormick Tractor and Equipment Co. calendar at left appeared in the 1903 *Sugar Beet Grower*.

used the "most advanced" farm products were increasingly paraded before the farmers and the general public as the ideal farmers. This gentleman-farmer icon was sold first through the rural magazines and newspapers and later on radio and television, including popular TV shows like *The Big Valley* and *Bonanza*. To the rural and national media, mega-farmers became the beacons of progress.

Farm ads, like most other ads in the time period after World War I, emphasized the importance of being modern. The ads and editorials advocated replacing horses, antiquated equipment, and outmoded practices. Magazines urged farmers to stay up with the times (and their wealthy neighbors) and purchase the most innovative device or the newest chemical on the market. Most corporations exploited the themes of war and the

farmer's struggle with nature. These were the most important sales themes from the mid-1920s until about 2000. Currently, though, the chemical-corporation advertisements that appear in the world's farm journals strike a wide-ranging pose, from dominant warrior to sensitive environmentalist. After eighty years of promoting the war with nature, the chemical merchants now spend considerable ad space claiming that their poisons and genetically manipulated products are environmentally friendly and safer, and that the companies themselves are good stewards of the earth and its resources.

For example, in the early 1990s Jon Arvik, manager of environmental affairs for Monsanto, argued convincingly that the company's Roundup weed killer was safer than other herbicides. And these arguments in

favor of Roundup's safety convinced many somewhat environmentally sensitive farmers to use it. But Roundup is neither safe nor benign; as farm writer Gene Logsdon has pointed out, being safer than other herbicides is not the same thing as being safe.[2]

Besides war and safety, advertisers also employed various other themes and subject matter, such as boom-or-bust, babies, boxers, cartoons, and beautiful women. The corporate perception of beauty has gradually changed over the years, from pleasingly plump in the 1890s to pencil-thin today. But whatever the perception, the corporations always promoted a product to change how women looked and promoted an ideal of how women and girls should look. The ad on the preceding page takes the female beauty concept to a ridiculous extreme, comparing a beautiful

215

woman with melons and vegetables that were sprayed with a fungicide.

Just as the perception of beauty changed, so too did corporate strategies for selling things shift on countless other issues and products. Advertisers sold a war that no one wanted. They sold tractors that no one needed. They sold farm and home products that were worthless. They sold patent medicines. Without a doubt, the advertising agencies demonstrated they could sell the proverbial ice to Eskimos. They still can!

In an effort to switch their image from that of warrior-polluter to sensitive steward, several of the largest chemical corporations—without any irony—have funded propaganda campaigns that lauded their efforts to clean up the environment. It strains credulity to think that Monsanto really cares about the environment or that DuPont, American Cyanamid, Shell, Standard, Dow, Syngenta, Bayer, BASF, and all the other chemical companies would voluntarily clean up pollutants that they dumped. All of these corporations have fought for decades against public demands for pesticide regulation, the cleanup of toxic spills, and environmental protection. In the long history of pesticide use, only after all possible legal appeals have been exhausted have the chemical firms finally agreed to do the right thing. Even then they cheat.

When the struggles took place over arsenic and lead in the late nineteenth and early twentieth centuries, DDT in the 1960s, and the drins in the 1970 and 1980s, the corporations and their scientists presented misleading arguments in the farm press and national media to illustrate the safety of their products. Muckrakers, consumer advocates, and farm activists were forced to spend enormous time and energy over the course of decades to refute the corporate defenders.

Each time, the misleading arguments from the chemical firms and their paid scientists effectively stalled the cancellation of many very dangerous and deadly chemicals. By claiming ignorance about the danger of their poisons, the corporate defenders held the critics at bay until their chemicals had been retested again and again. Many of the tests were conducted at public expense until laws were passed to

Even ads for pesticide masks, helmets, and dust protectors used cartoons. This helmet ad addressed openly the dangers from crop dusting, including parathion poisoning. From *Agrichemical West*, September 1967.

force the corporations to pay independent labs to perform the tests. Even that solution was flawed, as many "independent" testing laboratories were paid by the corporations to provide clean bills of health when a chemical failed to pass government regulations. The corporations stalled finishing the tests and reporting the results for as long as they could get extensions from state and federal regulators.

The corporations always acted as if they had no idea about the dangers of their products. However, most had known for a long time how toxic their pesticides and fertilizers were to farmers, farmworkers, crop dusters, and consumers. Indeed, the toxicity of these substances had been known since the 1920s, when BASF assigned Fritz Haber (the industrial developer of nitrogen fertilizer) the task of evaluating the pesticide and antipersonnel potential for all of the known chemical formulations in the world. These evaluations were continued by I. G. Farben through the 1930s until the end of World War II.[3] Not only the German but American, French, Swiss, and British chemical firms and governments conducted similar investigations to find antipersonnel, insecticidal, herbicidal, and explosive chemicals for the war effort. In short, war research led to the discovery of the poisonous nature of almost all the classes of chemicals known today. More importantly, the chemical firms developed the processes for making these terrible poisons and knew well their toxicity at an early date.

After World War II, the chemical corporations in Allied countries had access to the BASF/I. G. Farben, U.S. Office of Management and Budget, and quadripartite cartel (including France, England, Switzerland, and Germany) chemical-testing results. I. G. Farben conducted its tests on live human beings in concentration camps, not only on rats, dogs, or ferrets. So it is very likely that almost all chemists who were part of the quadripartite cartel knew that the pesticides that killed rats were not harmless to humans.

THIS WILL BE MORE FUN THAN PULLING THEIR LITTLE WINGS OFF YOURSELF.

Many ads, following Dr. Seuss again, tried to be funny about the poisons. Danitol, the chemical advertised in this excerpt from a 1995 ad in *California Farmer*, isn't that funny. It is a very toxic pesticide. Only 2.3 parts per billion is enough to kill half the trout in a stream (*Farm Chemicals Handbook*, Meister Press, 2004).

In spite of this mass of historical testing, the United States and the EU have been forced to enact several pieces of legislation to get the corporations to disclose, or even analyze, the real dangers of their chemical products. In the sixty years following World War II, as a result of federal and state mandates, the chemical companies were required to conduct tests to determine the birth defect, carcinogenic, and environmental risks of all their chemicals. These mandates included the 1947 Federal Insecticide Fungicide and Rodenticide Act, the formation of the EPA in the 1970s, California's Proposition 65 in 1983 (to identify cancer-causing chemicals and warn of their presence in public places), California's Birth Defect Prevention Act of

1985, the Federal Food Quality Protection Act of 1996, and the EU's REACH program. Though the intent of each of these laws was laudable, each statute was bargained down by corporate lawyers to a level that was more acceptable to the corporations but provided less protection to consumers and the farm community.

By 1995, as a result of legislative and legal actions, especially in California, nearly all the chemical-data gaps had been filled, and the toxicology of most currently registered pesticides in California was well known. This was a significant development given that California's agricultural economy is larger than that of most other nations. Consequently, by 1995 most pesticides and chemicals had been analyzed to satisfy California's regulations at that time.

The conclusions of the California chemical tests for cancer and birth defects were clear. Almost all of the restricted-use chemicals killed the test animals after exposure to very small doses. Many were found to be nerve destroyers at lower dose rates, or the direct cause of cancer and birth-defects at even lower rates. Almost all restricted-use chemicals damaged bird, fish, and other wildlife populations. The most toxic poisons account for at least 80 percent of pesticides used on California farms.

Still, the chemical merchants and the ad agencies continued to defend these terrible toxins and alleged them to be safe—that is, if the users followed the label instructions. The chemical corporations claimed once again that they had no idea that their products could kill or harm people, pollute water supplies, and toxify soil. In spite of this damning data on chemical toxicology that surfaced in California, even the most deadly chemicals are still legally registered there and continue to be marketed, and they continue to be approved by the California EPA, the U.S. EPA, the FDA, and the USDA.[4] As a result, they are still widely used around the world and still appear at your dinner table.

"If Capture doesn't get rid of the mites, lygus and worms in your cotton field, I'll eat them."

This excerpt from a 1995 *California Farmer* ad is even less funny, especially for the fish. A Capture concentration of less than one part per billion is enough to kill at least half of the trout in a stream (*Farm Chemicals Handbook*, Meister Publications, 1998).

The latest battles over the safety of pesticides and other chemicals have been focused on the U.S. Food Quality Protection Act of 1996 (FQPA), which was designed to evaluate all chemicals used in agriculture, and the 2006 EU REACH (Registration, Evaluation, and Authorization of Chemicals) program, designed to evaluate or reevaluate all chemicals currently on the market in the EU. These two efforts to control dangerous chemicals in the workplace, in the home, and in the food supply have been very successfully resisted by the corporations and high government officials.

The Crop Protection Association, a front group for the chemical manufacturers, has alleged that the thresholds of danger for the organophosphate and carbamate poisons under the FQPA were set too low. In other words, they argued that the chemicals in question were not as dangerous as critics said they were. Yet these are the same chemicals that the farm-journal editors complained were incredibly dangerous when farmers were forced to use them in the 1960s and 1970s as replacements for DDT, chlordane, and lindane. At that time the Crop Protection Association argued that the environmentalists would be forcing them to use these more dangerous poisons if they demanded that the EPA cancel the registration for DDT and other poisons. For the advertisers to change their minds now and suggest thirty-five years later that the chemicals are not that dangerous is ludicrous and probably criminal. The rapid killing power, or acute toxicity, of carbamates and organophosphates at extremely low dosages is their signature selling feature, and this has been the case since they were first advertised in the 1950s.[5]

In 2001 the EU REACH program proposed to conduct evaluations on all the chemicals used in the member countries and asked for input from chemical suppliers, users, and antichemical advocates. One month after they took office the Bush-Cheney administration asked for suggestions from, and conducted meetings with, the chemical industry pertaining to how they should respond to REACH. A Department of Commerce document, *EU White Paper: Strategy for a Future Chemicals Policy*, states:

> Since its presentation in February 2001, Commerce and USTR (United States Trade Representative) have been actively meeting with the U.S. chemicals industry to solicit their views and concerns. . . . Commerce and USTR have met with representatives from the Synthetic Organic Chemicals Manufacturers Association, the American Chemical Council, the American Plastics Council, ISAC 3, DuPont

Gentlemen, quit bickering about Bts.

AGREE. HAS IT ALL

Even the biological pesticides such as the Bt advertised here used time-worn themes; they even used time-worn boxers from the turn of the century. From *California Farmer*, 1995.

HERE'S YOUR CHANCE

Grower associations, crop protection chemical associations, agricultural lobbying groups — even members of Congress — have written EPA to express concern over how the agency is implementing the Food Quality Protection Act (FQPA). Industry representatives agree that while the law is strict, its mandates are achievable without undue damage to agriculture if EPA acts on factual — rather than theoretical — data.

For more information on the technical aspects of FQPA — and for the opportunity to make your opinion heard in Washington, DC — see the special report, "OPs And Carbamates At Risk: A Call To Action," beginning after p. 24.

and Dow to identify industry concerns. Officials from the U.S. mission in Brussels have also met with a number of European and U.S. chemical companies based in Europe to solicit their views on the Strategy and its impact on the industry.[6]

Another Department of Commerce briefing paper from a slightly later date stresses that in the meeting with the chemical corporations and associations, government officials "advised industry to develop an official position and strategy as soon as possible to assist in influencing the EU's draft text."[7]

The Office of EU and Regional Affairs and the Office of Chemicals within the Commerce Department assisted the chemical industry in developing a démarche, a form of diplomatic protest about the REACH proposal. By March 21, 2002, Secretary of State Colin Powell had sent a cable to thirty-six U.S. diplomatic posts in nations outside the EU stating that "U.S. industry, as well as European industry, have expressed serious concerns with the white paper on both sides of the Atlantic . . . examination of just four commercially important chemicals on the authorization list shows that $8.8 billion worth of downstream products are at risk for bans or severe restrictions under the new system." This analysis of lost sales was not conducted by either the U.S. Commerce Department or Secretary Powell's office but was generated by the American Chemical Council. Its estimate of $8.8 billion included all computers exported from the United States. The World Wildlife Fund concluded that the $8.8 billion figure was an exaggeration and could not be supported by a fair reading of the REACH proposal.[8]

On March 21, 2002, an official démarche was also sent by Powell to all EU member states. After Powell's cable and démarche, U.S. officials continued to meet with industry heavyweights. EPA officials traveled to Europe in 2002 to deliver the chemical industry's messages in conjunction with officials from the American Chemical

Council. Also that year, Charles Auer, the director of the EPA's Chemicals Control Division, along with officials from the American Chemical Council, had meetings with several high-level officials from the EU and various European countries. Auer and the ACC single-mindedly "advocated efficient voluntary measures and science-based decision making in regulating chemicals during a March 8 meeting with German officials and business reps."

Beginning in 2002, the U.S. EPA held senior-level meetings in Europe to explain the U.S. regulatory system and express concerns with REACH.[9] The EPA conducted seminars with agriculture and regulatory agencies all over Europe and in many countries of Asia, Latin America, and Africa. EPA officials delivered the message that regulatory policy should adopt a risk-benefit analysis instead of banning the most dangerous chemicals. Echoing the Bush antiregulatory policy, they encouraged voluntary corporate restrictions on the most toxic poisons.

In a meeting on April 3, 2003, Catherine Novelli, the assistant U.S. trade representative for Europe and the Mediterranean, advised the impacted industries "to come up with 'themes' for their concerns about the proposed legislation. The chemical industry had done a list of themes dealing with the EU process. Intel had also done a list of substantive themes."

The combined list of themes that the industry wanted to pursue totaled eleven. The USTR officials and the industry representatives stressed that the prominence of the messenger of the themes must be an elevated one. The implication was that it would take someone of Secretary of State Powell's stature or higher to convey the message in strong enough terms to get the necessary action. USTR's Barbara Norton stressed in an e-mail to industry officials that "the only thing that will get the EU to stop is having the heavyweights come in and say that the commission can't take this forward until a real cost-benefit analysis is done."[10]

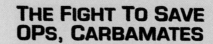

THE FIGHT TO SAVE OPS, CARBAMATES

EPA's decisions regarding use of organophosphate and carbamate insecticides may establish a precedent for all pesticide registrations.

OPS' PIECE OF THE MARKET

Organophosphate (OP) insecticides represent a sizable share of total insecticide-treated U.S. acreage. Here's a look at selected crops.

OPs' SHARE OF TOTAL INSECTICIDE-TREATED ACRES

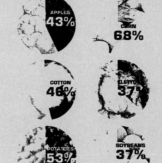

THREATENED OPs & CARBAMATES

ORGANOPHOSPHATES

Bidrin	Guthion, Sniper
Bolstar	Imidan
Counter	Lorsban
Curacron	Metasystox R
Cygon	Mocap
Dython	Monitor
D-Z-N	Nemacur
Dibrom, Legion	Orthene, Payload
Di-Syston	Penncap M
Dyfonate	Supracide
Ethion	Thimet
Ethyl Parathion	

CARBAMATES

Adios	Larvin
Drexel Carbaryl	Sevin
Furadan	Slam
Hopkins Sevin	Temik
Lannate	Vydate

-- Adapted from "Q&A Organophosphates, Carbamates and the Food Quality Protection Act" published by the American Crop Protection Association

The *Farm Chemical* ad above is from the 1998 special edition that encouraged farmers to resist by any means the implementation of the Food Quality Protection Act of 1996. These excerpts from that campaign illustrate the enormity of the market for organophosphates (OPs) and the fear of their loss to the chemical corporations.

Revealed also in the e-mail from USTR to industry groups were the administration's efforts, combined with those of U.S. chemical-industry officials, to identify countries that should be targeted and neutralized. They decided to particularly target "Germany, UK, France, Italy, Netherlands, and Ireland because they all have large production of chemicals and downstream products. In Italy, it will be important to get to [European Commission President Romano] Prodi. In addition, we need to get to the Swedes and Finns and neutralize their environmental arguments." Less than a month later, on April 29, 2003, Secretary Powell sent a second cable about the EU REACH proposal. "The cable was sent on a 'priority' basis to diplomatic posts in European Union nations. It states that it is important for posts to reiterate to the European Commission and EU Member states our general concerns before the commission reaches agreement on its formal proposal."[11]

In that cable Powell included all eleven of the themes and concerns from the chemical industry and Intel. Not surprisingly, he strongly urged cost-benefit analysis. He expressed concerns about stifling the creation of new, more effective, and safer chemicals and warned that manufacturers might halt production if the cost of testing and registration was too high. He suggested the exclusion of certain chemicals and eliminating the testing of all chemicals used in producing articles such as computers, plastic products, and textiles.[12]

In September 2003, the leaders of Great Britain, France, and Germany wrote to Romano Prodi, who was then the president of the European Commission, to voice their concerns about REACH. They requested additional analysis and an assessment of the effects on industry.[13] This letter indicates that the U.S. government and corporate efforts to sabotage the REACH process had succeeded. They got political heavyweights like Tony Blair, Jacques Chirac, Gerhard Schroeder, Romano Prodi, and Colin Powell to dilute the regulations and significantly weaken the proposal. One hundred and forty-three nongovernment environmental groups from thirty-one countries in Europe concluded:

> Unprecedented interference by the chemical producers in Europe and the U.S. has led the Commission to considerably weaken the proposal and to tip the balance away from the environmental and public health protection towards the self-interests of business.[14]

By the end of 2005, after another cable was sent from Secretary of State Powell before he resigned, and after more pressure from U.S. and other chemical-industry officials, the European Parliament voted on an even more watered-down version of REACH. Of the more than 30,000 chemicals used in Europe that have never been tested, several thousand were exempted from mandatory testing and regulation. More cost-benefit and risk-benefit analysis was included. Maybe that picture of Uncle Sam threatening us about pesticides is more accurate than any of us thought.

So, how can we protect ourselves until the regulators finally decide to regulate chemicals and GMOs? When can we know what is and what is not dangerous to us, our families, and our land? Every attempt, from the 1890s to the twenty-first century, to regulate and test chemicals has been blocked by the chemical corporations, large-scale farmers, the farm press, the universities, and the government.

It is hard to imagine that ad agencies, publishers, editors, and government leaders and regulators did not and still do not know how dangerous these chemicals are. It is unbelievable that they do not know about the toxic dangers of nerve poisons such as the organophosphates and the carbamates. It is hard to imagine that they are not aware of the dangers and damage from methyl bromide and all the other bromines. And it is dead certain they know that there are cancer and birth-defect clusters all over farm country in the United States.

Like the tobacco merchants and their advertisers, however, the chemical merchants have consistently denied (and continue to deny) any knowledge of danger and have always challenged the results when any dangers finally caught the public's eye. Since the chemical giants reap enormous profits from the organophosphates (which account for about 60 percent of total pesticide use) and the carbamates (which account for more than 10 percent of total use), the journals refuse to utter discouraging words about these products. As the struggle over FQPA and REACH shows, the U.S. government continues to defend the chemical industry and the use of these

The use of Uncle Sam as a prop to get farmers to reject provisions of a law that was passed to protect all of us, consumers and farmers alike, is an example of how bitterly and deviously the corporations fought against the loss of dangerous chemicals that they envisioned would happen with the Food Quality Protection Act of 1996. From *Farm Chemicals*, 1998 special edition.

Don't miss this opportunity to speak out for real use evaluation of OPs and Carbamates

I WANT YOUR COMMENTS

A Special Editorial Focus • OPs and Carbamates at Risk: A Call To Action • Custom published by Meister Publishing Company

dangerous poisons, both here and abroad at the highest levels of government.

In advertisements and promotional campaigns tailored for farmers or householders, the chemical manufacturers have never listed the full spectrum of dangers from using their chemicals. Most of the editorials trivialized the dangers and ridiculed the regulators. Only in ridiculously small print did the advertisers warn the customer to follow the label instructions. But chemical corporations never alerted their potential customers to the fact that their chemicals caused birth defects, cancer, or nerve damage. Certainly nowhere in any ads will you find a warning

Temik/aldicarb will occasionally increase yields, as this ad claims. This ad advises that the grower use it two times, once before planting and once after the plant is growing. The corporation promises an average increase in yield of 149 pounds of cotton lint.

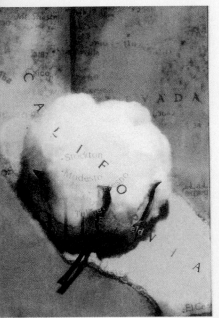

FOR A LAND WITH *Great* EXPECTATIONS, THE PRODUCT WITH GREAT *Potential.*

NOTE: Farmers and applicators should know that there are serious dangers associated with the use of and the contact and exposure to TEMIK.

The Surgeon General has determined that actual dangers from exposure to poisons like TEMIK must be displayed prominently in large type on any label or advertisement for those poisons.

DANGER!

Temik is a nerve poison, a category I poison, and an incredibly toxic chemical. Exposure to Temik has been shown to cause cancers and birth defects and to sicken or even kill farmworkers, farmers, chemical workers, and most wildlife. Temik is a cholinesterase inhibitor causing great nervous system damage. If poisoned with Temik/aldicarb head straight for the emergency room and tell the staff to apply atropine sulfate to calm nerves and stop nervous convulsions and shock.

that alerts you to these real dangers from exposure to the chemicals. Seldom, if ever, does even the fine print on the labels fully apprise the purchaser of the real dangers. And though many pesticide-regulation laws are currently on the books, the regulators have eliminated very few of the most dangerous chemicals. How much longer are they going to permit and promote the use of arsenic, fluorine, strychnine, paraquat, carbon bisulfide, methyl bromide, DD/Telone, parathion, and 2,4-D in the United States?

In the case of tobacco, the companies were finally forced to disclose the dangers and pay compensation to the smokers and secondhand-smoke sufferers. Yet pesticide drift is far more damaging than secondhand smoke. While multinational chemical companies and large-scale users argue for the absolute necessity of these chemicals, farmers and neighbors who are exposed to pesticide drift frequently contract cancer or have babies with birth defects. Pesticide drift is tantamount to chemical trespass and ought to be stopped. If bartenders can stop smokers from lighting up in bars and taverns in California, New York, Boston, Italy, and Ireland because they were contracting cancer from secondhand smoke, then surely something can be done about secondhand pesticide exposure.

Just as tobacco ads have been banned from television, pesticide and fertilizer ads should be banned from radio, TV, and magazines. But, unfortunately, as long as the corporations keep selling toxic chemicals the irresponsible farm journals seem hell-bent on running the ads to make a buck. And ad agencies seem all too eager to prostitute themselves by creating campaigns that glorify the value and trivialize the dangers of farm poisons.

There are precedents for resistance in the media to advertising campaigns that damaged the public. In 1905, *Collier's* magazine began a campaign against patent medicines that was joined by the *Ladies' Home Journal* and the *Saturday Evening Post*. By refusing to accept any more patent-medicine ads, these magazines helped eliminate a majority of quack and dangerous products from the market and added to the demand for the creation of the Food and Drug Administration.

A similarly bold move against chemical ads needs to be made by an important national publication or farm magazine today, such as *Farm Journal*, *Atlantic Monthly*, or *Newsweek*. Unfortunately, our media outlets are owned by such a small club that there is little freedom of the press, and great fear of the advertisers. So, it would be a big surprise today if one of these periodicals were to step out of line and show some courage about food poisons.

Until the press wakes up from its corporately induced coma, consumers could be a key lever in pressuring the major national magazines to make a progressive move and take a stand for public safety. A first step in that process might be convincing a magazine to advocate for a regulation that chemical corporations must advise consumers of both the short- and long-term dangers from their products, both in the advertisements and on the labels. This could be similar to the surgeon general's requirement to warn tobacco users of health dangers wherever tobacco is advertised or sold. Even drug manufacturers have been forced to state the potential side effects in their advertisements, no matter how remote the possibility of complications.

For some of the more toxic products, it would require pages of copy just to display in the ads all the dangers, and those chemicals should be immediately banned. Appropriately, the farm magazines should also evaluate the chemical and critique the advertising claims. Until the registrations of these deadly chemicals are cancelled, the sample of an advertisement layout at left, and the minimum evaluations, critiques, warnings, and advisories, would greatly help farmers and farmworkers and should be required as truth-in-advertising for ads and labels for poisonous chemical products.

"The great expectations held for DDT have been realized. During 1946, exhaustive scientific tests have shown that, when properly used, DDT kills a host of destructive insect pests, and is a benefactor of all humanity."

The great expectations held for DDT were realized early, as this ad states, but later studies have shown that its damage far exceeded its great expectations and its role as a benefactor was greatly overrated. This cartoon and the quotation below it appeared in an advertisement in the June 30, 1947 edition of *Time* magazine

Chapter 24

WHO INVITED THESE CHEMICALS TO DINNER? A LEAP OF HOPE

Serious leaps of faith were required for sick people to let themselves get bled, blistered, purged, and poisoned by physicians, surgeons, patent-medicine peddlers, and quacks. Using known and feared poisonous chemicals (medicines) on food required further leaps. Advertising and editorial campaigns allowed farmers and the general public to make those jumps—even made them seem rational, or at least economic. The promotional campaigns convinced farmers and consumers that using arsenic, lead, DDT, toxic fertilizers, nerve poisons, Alar, and thousands of other labor-saving substances was in their best interest.

None of these chemicals were tested until the chemical merchants tested them on real-life farms, farmers, farmworkers, prisoners, concentration-camp inmates, and consumers. Few knew that the highly advertised chemicals—promoted as a fast track to prosperity—were going to cause the economic and resource damage they caused. It took years of use before the deadly effects from the chemicals became too apparent to deny. As each failed in its turn, it was replaced by still deadlier chemicals, preceded by and supported with advertising and promotional hoopla.

Today, we finally have some significant parts of a chemical report card, one that results from decades of chemical use on our farms, on our foods, and on our clothes. Not surprisingly, the chemicals and the chemical corporations receive an F. Rather than being coddled by our government, the chemical corporations should be prosecuted for their reckless criminal behavior. They failed to protect us from or advise us about the danger of the chemicals they manufactured and promoted. Instead, like the tobacco companies, they hid the truth and the real dangers from us. They falsified tests. They obstructed or

employed the regulators. They and their congressional allies watered down the laws that we had passed to protect ourselves from their poisonous products. And all the while they kept encouraging us to poison ourselves, our food, and our farms.

However, instead of being prosecuted and jailed, the chemical merchants have remained wealthy in halcyon times or depression and continued promoting their dangerous products. Currently several are busily encouraging farmers to make the most enormous leap of faith yet. The same chemical corporations that polluted our water, air, and soil, tortured and killed workers in Nazi concentration camps, killed thousands in Bhopal, and caused cancer and birth-defect clusters in rural communities are trying to force everyone to jump on the genetic engineering bandwagon. In the United States they have succeeded in getting a majority of the commodity-crop farmers to jump.

None of the best genetic scientists can predict what will happen with these genetically manipulated crops: whether their genes will spread and create superresistant weeds; whether they will destroy the purity of any meaningful definition of *organic*; or whether the novel proteins they contain will damage human health as well as the environment. Until science can predict, and prevent, these kinds of threats, we should stop planting GM crops; they should be tested only in the lab to understand the long-term ramifications of their use. If the results are positive, and if worldwide food shortages truly exist— then and only then should we allow their use on a trial basis. Right now—and for the foreseeable future—we have a glut of all commodities, not a shortage.

Hunger in the world is not caused by lack of food, since there is more than plenty to feed the world's population right now. Hunger is caused by a lack of willingness on the part of commodity brokers to feed starving people. Brokers and corrupt politicians would rather let food rot in silos and take a tax loss or sell it at a loss than give it to starving people who are unable to pay. Hunger is a distribution and political problem, not an agricultural-production problem. In the midst of the greatest era of plenty in the history of civilization, millions of American children (not to mention those around the world) still go to bed hungry each night, because there is not the will to feed them, not because there is not enough food.

In this book, I have tried to draw an accurate sketch of the cultural, technical, and economic challenges farmers faced for the last two hundred years. We have used almanacs, farm journals, advertisements, editorials, and letters to the editor as a major vehicle for telling this story for several reasons. For one, the almanacs and magazines are still accessible in both university and large public libraries, and people can still read for themselves how farmers were disdainful of chemicals, advertisements, and aristocratic science from the 1760s until the 1940s. And, importantly for our tale, people can read how the farm magazines dealt with such skeptical attitudes.

Second, we wanted to showcase the graphic and editorial messages that the corporations used to influence farmers' seed, pesticide, and fertilizer choices. Even if farmers didn't read the popular agriculture books by Justus von Liebig or E. B. Voorhees, the articles, editorials, and ads in the magazines promoted the chemical theorists and chemical merchants' views.

In the advertisements and editorials the toxic peddlers tried any tactic to market a product, no matter how poisonous, dangerous, damaging, or ineffective the chemical was. As each product proved to be dangerous or useless, the corporations, scientists, and editors moved on and promoted the next chemical or set of chemical products. Each chemical or genetically modified product, in its turn, has been advertised as being necessary, both for solving the "problems" of world health and hunger and for improving the farmer's bottom line.

Today, chemical corporations continue to argue that the regulatory loss of a chemical or a genetically manipulated crop or animal will threaten the world's food supply and risk worldwide starvation. To support or supplement this "Chicken Little" tactic, they have been able to come up with advertising gimmicks that promise ease of use and increased profits to farmers, while at the same time sugarcoating the risk of poisoning and pollution. Throughout their continuous advertising campaign, chemical corporations repeatedly scared farmers with the specter of crop loss if farmers didn't use their chemicals. Chemical corporations used the same tactics to sell their dangerous products all over the world, trotting out many of the same themes and promises over and over again to try on new audiences in new languages.

Until the 1940s, the general farm public resisted chemical products and the pseudoscience theories of the corporations. The advertisers and the corporations persisted nonetheless, and their chemical advertising and propaganda campaigns, which began at the dawn of industrialization, have proven incredibly successful over the years. These promotional campaigns from the chemical industry have helped change the nature of agriculture in the Western world.

Until a mere hundred years ago (a brief period in the history of agriculture) experiential knowledge and, to a lesser degree, religion and faith were used to solve pest problems. Even today, no one knows how the Bishop of Lausanne managed to get rid of June beetles in 1497, except as an act of faith or knowledge of their habits. In the last one hundred years or so, most of the bishop's pest-control rituals have been replaced by scientific ones and faith in chemicals has replaced faith in God in the pest control business. Today, most U.S. farmers still don't know the formulas of their pesticides or how they actually work. Corporate science and the bishop both alleged that their formulas and techniques were effective. If farmers believed in them, they used them. And if they seemed to work, farmers continued to use them, no matter where they came from—bishop or alchemist.

For the last 160 years corporate chemists have claimed that their products work miracles. And indeed, some of them performed well, even miraculously, but usually only for five to ten years. After that, problems invariably emerged, though the corporations and the farm editors refused to admit product failures until they became too obvious to hide when farmers began losing their crops after applying the most recent miracle pesticide.

Corporate scientists assured farmers and consumers that their chemicals were safe, and the regulators relaxed, so much so that the chemicals are even sold in the grocery and drug stores near food and regularly sprayed in airplanes, kitchens, and hospitals and under the sink. There is now good evidence that these chemicals are not safe. We have also shown that the most toxic chemicals are the most used chemicals.

One of the most important stories told in this book is the tale of how scientists, industrialists, and merchants promoted in America farm chemicals that the public was very mistrustful of. This was science from the budding industrialists of the nineteenth and early twentieth centuries. These representatives from upper sectors of society continued to promote toxic chemicals for four generations before they really took hold. This promotion encountered great difficulty because the general population mistrusted aristocrats, alchemists, chemists, and explosive manufacturers and feared the products that these experimenters concocted.

Baron Justus von Liebig and a legion of scientists that he trained promoted his theories in the *Cultivator*. His soil-analysis system misled countless farmers, and his rejection of the humus theory confused them even more. As a result he became a laughingstock in farm magazines from the mid-1850s until the mid-1880s.

Though agricultural chemistry and "scientific" systems of agriculture suffered early setbacks, they continued to be vigorously promoted by Liebig's American students at both Harvard and Yale. Aristocratic industrialist donations created the Lawrence School of Science at Harvard in 1848 and the Sheffield Scientific School in New Haven in 1860. Both schools promoted scientific agriculture and tried to copy or improve upon the Liebig laboratory at Geissen, Germany. In spite of Liebig's bogus ideas, both the aristocrats and the merchants conducted a persistent assault from the editorial pulpit of farm journals and the lecterns of powerful universities on the necessity of more, not less, science and better, not fewer, chemicals to cure the errors of Liebig and his followers (especially Mapes).

Another story that we have told is how resistant farmers and consumers initially were to fertilizers and pesticides that chemical salesmen promoted. There was significant farmer resistance to quack "book farmers" throughout the nineteenth and early twentieth centuries. It wasn't just the farmers, however, who were skeptical, fooled, or scared. Consumer skepticism and anti-pesticide activism obviously was a factor in Standard Oil's decision to hire Dr. Seuss.

The brilliant cartoonist and gag writer was able to make people feel comfortable and even laugh about spraying a dangerous poison in their houses and barns from 1928 until 1943. And this was a very critical breakthrough for the chemical firms. Because of ad campaigns like those conducted for Flit, householders began to use chemicals in their search for ways to combat pests resulting from poor sanitation and sewage systems. Despite Dr. Seuss's efforts and the corporate advertising blitz for toxic chemicals, public resistance to chemicals, both fertilizers and pesticides, still remained substantial until after World War II, and a large percentage of American farmers continued to be suspicious about fertilizer and pesticide quality.

Before and after the war, Dr. Seuss concentrated much of his energy on attacking Hitler, the Republicans, isolationists, and German sympathizers, such as Charles Lindbergh.

After the war, advertisements, articles, editorials, and testimonials constantly appeared in the journals to address the farmers' fears of poisoning, soil sterilization, crop defoliation, livestock illness, water pollution, and worthless pesticides and fertilizer mixes. It was in the 1950s that the chemical manufacturers' sales program finally reaped real rewards. With all the advertising, promotions, and product giveaways, farmers became less concerned about shoddy products and more anxious to use the miraculous new chemicals. Truth be told, they were pushed along by their bankers, the agricultural extension agents, the chemical salesperson, and the farm journals. A large factor contributing to the changeover was the use of heroic images, the propaganda surrounding the Green Revolution, and cash incentives from the government to modernize.

Effective advertising campaigns also convinced farmers in developing countries to use more chemicals in their search for cheaper, cleaner, and more efficient ways to farm. Large-scale U.S. farms were used as the model for farmers to copy in developing countries. The same war chemicals that chemical corporations sold to U.S. farmers became the centerpiece of pest and fertility management for Green Revolution projects all over the world and were held out as the saviors that would cure hunger and make farmers profitable.

Like the Green Revolution advocates before them, advertisements from the genetic manipulators tout their new solutions as the only hope to feed an exploding world population, and as the only hope to heal your incurably sick grandmother, or child, or wife. For more than 160 years we have been told that in order to survive, we have to use the newest breakthroughs from the chemical industry, whether or not they happen to be toxic or genetically polluting.

Many of us carry around a bucolic view of farming, ranching, and rural America. We think of farming as being toxic only after the introduction of DDT at the close of World War II. Such presumptions are wrong. Rural Americans began to be propagandized about the necessity of using deadly and dangerous pesticides and fertilizers well before the start of the Civil War and have been assaulted by several generations of chemical-corporation advertisements ever since.

Since many of the first pesticides had been used as medicines for hundreds of years before they were used to kill bugs, they were well known to farmers, and they were feared. In the late 1860s arsenic and sulfuric acid were first used widely as pesticides in the United States. This first generation of highly poisonous farm products opened the door to the use of thousands of toxic pesticides over the next 160 years. Unlike the first generation of pesticides, the second generation was completely unknown to the farmers since they were new poisons that had

Dr. Seuss also promoted pesticides for the war department, and as Capitan Geisel he helped create a demand for DDT with his cartoons about the anopheles mosquitoes that carried malaria, which DDT was used to control. From US War Department Office of Management and Budget.

This is *Ann*

she's dying to *meet you*.

Her full name is *Anopheles Mosquito* and her trade is dishing out *Malaria* She's at home in Africa, the Caribbean, India, the South and Southwest Pacific and other Hot Spots.

...she drinks BL**OO**d

And she stands on her head to get it.

She jabs that beak of hers in like a drill and sucks up the juice.

The head eats . . .
. . . the rest gets milked

CONSOLIDATED WORLD DAIRY — A. HITLER, Prop.

JUGO SLAVIA ROMANIA GREECE AUSTRIA

BELGIUM

HOLLAND

DENMARK

NORWAY

POLAND

CZECHO SLOVAKIA

FRANCE

Dr. Seuss

Another of Dr. Seuss's pro-intervention, anti-fascist cartoons, drawn for *PM*, the left-liberal magazine, in 1941. These cartoons had wide distribution and definitely impacted public opinion.

been synthesized in chemistry labs before and during World War II. Consequently, farmers were not as fearful of DDT or 2,4-D as they were of arsenic, lead, and mercury, which had much longer poisonous histories.

To those living in rural America, each successive generation of chemicals had terrible impacts, even though many of the products provided temporary pest control or fertilizer success. While rural environments and community health collectively suffered damage, each chemical in its turn injured specific sectors of the population or polluted different parts of the environment. Let's look at the major classes of chemicals that have been used in the last 160 years and who and what they have affected most.

Major Chemical Classes Introduced in the Last 160 Years and Their Impact

The high use of arsenic disproportionately injured and killed children, and it continues to do so today, but it also killed, sickened, and injured farmers, farmworkers, industrial workers, and careless householders. Arsenic products have been found to cause cancer (if swallowed or inhaled), serious damage to the lungs (if inhaled) or stomach (if minute amounts are swallowed), and serious debilitation if any arsenic product drifts onto an unprotected person or animal. Even in very small doses, arsenic is acutely toxic.

Arsenic's long persistence has left some soils toxic and even infertile for many plants and trees.[1] Because of arsenic's persistence, it is difficult to predict when its negative effects will end—on heavy clay soils it probably will take several hundred years. One thousand and sixty water systems in California exceed safe health limits for arsenic. Arsenic acid and several arsenic products, including cacadylic acid, MSMA, and MAMA, are still being used in the United States and throughout the world. For instance, arsenic is still added to baby chickens' water, and MSMA is being

used to control weeds resistant to the pesticide Roundup in Roundup Ready crop fields.

DDT and other chlorinated hydrocarbons killed bird populations and inflicted long-term environmental damage. As the early advertisements claimed, DDT was genocidal to certain species (and so were other chlorinated hydrocarbons). Chlorinated hydrocarbons are implicated in amphibian decline, female breast cancer, and reduced human sperm count.

DDT and many of its relatives will probably persist for more than a hundred years. No one knows all the future ramifications, but this class of chemicals seems to be a contributor to the degeneration of sex organs in amphibians and mammals, including humans. The California Environmental Protection Agency and the U.S. EPA have determined that most of the chlorinated hydrocarbons are probable human carcinogens and probably cause mutations, monstrous birth deformities, damage to the immune system, and poisoning of the fetus. Most of them cause anorexia, decreased sperm count, female breast cancer, hormonal changes, aplastic anemia, and toxic injury to the liver, kidneys, and central nervous system.

DDT-like chemicals also destroy bird and fish populations, are implicated in amphibian decline, and damage most other wildlife. In spite of their dangerous nature several chlorinated hydrocarbons are still being used in many parts of the world, including the United States. And DDT itself is still sold and used widely outside of the United States and the EU.

The bromines have greatly damaged aquifers, groundwater basins, soils, and, ultimately, the earth's atmospheric ozone layer. The bromines have increased incidences of cancer and birth defects and have caused serious environmental degradation—again, from aquifer to ozone. In hundreds of local disasters, farmers, citizens, and chemical workers have suffered high rates of injury, sterility, and death from EDB, methyl bromide, PCBs, and DBCP. Many communities in addition have suffered groundwater pollution, yet the manufacturers have contended in all their ads and promotions that bromines wouldn't accumulate in the water table.[2] Many bromines continue to cause damage, yet advertisers still aggressively defend and market them despite their dangers.

In one glaring example, agrichemical interests got an extension for the use of methyl bromide until the year 2005 by aggressively defending it and getting their state and federal representatives to defend its use. Horrendous long-term ozone damage resulted, because of this ten-year delay in suspending methyl bromide.

Methyl bromide also causes mutations, tumors, and monstrous birth defects. It is incredibly lethal in very

N-pHURIC is an example of how chemicals are combined in a cocktail that both provides fertility and controls weeds or other pests. From *California Farmer*, 1995.

small doses; consequently very few of its victims survive. Unlike the case for many other chemicals, pest resistance to methyl bromide has been low, with only a dozen or so organisms that have shown any tolerance to it after almost seventy years of continuous exposure. This lack of resistance is clearly due to the fact that the chemical kills almost all of the members of a population and leaves few if any resistant survivors.

In spite of such dangers the agrochemical industry continues to fight for its continued use, and they continue to win. Methyl bromide is protected by a small number of large-scale strawberry, grape, and fruit farmers who are powerful politically, especially in California and Florida. Strawberry farmers grow the same crops year after year on the same pieces of ground. Grape and fruit growers often replant trees and vines on the same pieces of ground that they previously used for grapes and fruit trees. So, in these systems, they need a fumigant to control pests that build up in the soil. There are only 7.7 metric tons left in existing stocks. But for special-exemption users, like the farmers listed above, the U.N. Environment Program has allowed the additional production and use of 11,000,000 pounds in 2007 and 10,000,000 pounds in 2008. It looks like the only thing that could stop farmers from using it is the price, which has more than doubled in the last four years.[3]

Methyl bromide is usually mixed with chloropicrin, another biocide that is more commonly known as tear gas. Recall that tear gas was one of the first antipersonnel weapons used in World War I, and it is still being used for crowd and riot control. It too is very poisonous and has few pests that have developed a resistance.

Carbamates and organophosphates especially attack, injure, and kill farmworkers, farmers, and chemical workers (as in the Bhopal disaster). While these populations suffer most severely, the health of everyone suffers from their continued excessive applications. Both types of chemicals are nerve poisons that are acutely toxic. Both are systemic poisons and thus their breakdown products remain in the crop, in our food, and in our bodies as low-level residues and metabolites.

Carbamate and organophosphate nerve poisons are related to the antipersonnel weapons that the world has criticized Saddam Hussein for using on the Kurds and Iranians. Yet we use them on our food, our clothing, our farm families, and our farmworkers. These two classes of chemicals cause about 80 percent of all the farmer and farmworker pesticide injuries. Both classes of chemicals cause severe nervous system, eye, lung, and internal organ damage. Exposure to prolonged low-level concentrations causes birth defects, cancer, circulatory-system damage, and heart attacks.

Nitrogen, phosphorus, and other toxic synthetic fertilizers especially injure infants, children, pregnant women, and the aquatic food chain. Their long-term effects on groundwater, aquifers, and rivers are just beginning to be felt. Already, however, groundwater basins in hundreds of rural communities are so contaminated with nitrate or phosphorous pollution that the water is unsafe for human consumption. While ignored by most environmentalists and advertised as safe by agrochemical corporations, these fertilizer materials are so toxic that they are used to kill pests, defoliate plants, and sterilize the soil as well as feed it. The combination of urea (a nitrogen fertilizer and a pesticide) and sulfuric acid (a pesticide and a fertilizer) sold as N-pHURIC is a currently popular and profitable example of this long history of dual use. Synthetic nitrogen fertilizer was used in 1946 as a defoliant for cotton to get rid of the leaves before mechanical harvesting of the cotton bolls and at the same time fertilized the next crop.

Synthetic nitrogen is now the most widely used fertilizer in America. In California, more than 1.1 billion pounds of nitrogen fertilizer and more than 500 million pounds of phosphorus and potash were used on farms every year from 1990 to 2004. California farmers applied about eight times as many pounds of fertilizer as pesticides over the same period, with pesticide use averaging slightly less than 200 million pounds per year.

California agriculture totals about 5 percent of all U.S. usage, so the amounts of chemical fertilizer applied nationwide are enormous. No watershed projects or cleanup of the Gulf of Mexico "dead zone" can hope to have long-term success unless synthetic nitrogen and phosphorus pollution is significantly reduced or eliminated.

Why Are These Chemicals Still on the Market?

One of the major obstructions to dealing with the pesticide and fertilizer poisonings and pollution since the turn of the twentieth century has been the demand made by the chemical-corporation lawyers that government regulators evaluate each chemical as a separate case. This tactic forces activists and the public to fight each chemical on its individual merits rather than as a family or group of chemicals, even when the whole family is known to be poisonous. Many elements in the fight against arsenic and lead seemed to set precedents that could be used in the fights against DDT, Agent Orange, or other poisons. And surely the fight over DDT should have set precedents that could be used against other closely related chlorinated hydrocarbon chemicals. But the corporations and their lawyers have thus far prevailed. No resolution of this problem is possible until the nation sets a goal of mandatory pesticide reduction, as some European countries have done. Perhaps more importantly, no resolution of this problem is possible until the regulators enforce existing laws and any new laws that are passed.

Another serious problem is the regulatory practice of not combining the residues from several different chemicals when tests are done for residues and toxicity. Based on data from California's mandatory pesticide use reports, compiled since 1970, and interviews with farmers and pesticide applicators, we found that farmers often apply multiple sprays on food crops. Many farmers spray cocktails of different pesticides to deal with the increasing problems of resistance or to control several pests and to eliminate multiple trips over the fields. Consequently, many crops have residues from multiple pesticide sprays and fertilizer applications, which on any one crop like lettuce or apples could include all the different classes of pesticides and fertilizers that we have reviewed in this book.

U.S. regulators know that farmers apply pesticides a number of times on the same crop, as well as in cocktails like the one advertised by the Olin Corporation, at right, and in the N-pHURIC ad on page 254. Yet the FDA and the EPA monitor and regulate poisons in food per insecticide, herbicide, miticide, fungicide, or fumigant without combining the accumulated amounts from all the pesticides used on the crop. The Food Quality Protection Act of 1996 was designed to analyze the cumulative use of toxins on food and fiber. But, once again, the chemical corporations and their legal staffs, abetted by willing government regulators, have roadblocked the act in the courts and in Congress.

Consequently, regulators in the EPA and FDA have failed to act and remain trapped in a morass of legislation and antiregulatory foot dragging. The public interest is the main victim of this corporate and regulatory stalling tactic. And as long as the stalling continues, the large-scale corporate farmers are permitted to use any cocktail of sprays the chemists can dream up.

Meanwhile, since multiresidue testing is in limbo, the corporations and the regulators continue to allege that most of our food and fiber does not have illegal residues. At this point only a small percentage of the pesticides used in California are analyzed on only 1 percent of the food produced and sold within that state. On the 1 percent of produce that is tested for pesticide residues, about 40 percent regularly shows up with illegal residues from a single pesticide. This means that 60 percent of the produce does not have illegal residues of any single pesticide (as defined by the EPA or FDA); however, this 60 percent of "safe" produce could still have highly toxic residue levels accumulated from several different sprays in several different pesticide cocktails.

Until FQPA is settled, the regulations do not require the regulators to count up the total residue load of pesticides tested, so they don't. Whether they account for them correctly or not, be assured that food and fiber crops are likely to have dangerous cumulative residues, regardless of the regulators' definition of what is legal and acceptable. These facts, which are becoming known to more consumers all the time, clearly underlie the rapidly growing market for safe organic food and prompted the European Union to create the REACH program.

As we reported, the REACH pesticide- and chemical-data review for 30,000 chemicals used in the EU was challenged by the U.S. chemical industry and the U.S. government. After much watering down of the bill, however, it finally passed in December 2006 and is now being implemented. Unfortunately, it feels flawed from the start. Its most important aims are to improve protection of human health and the environment from the risks of chemicals while enhancing the competitiveness of the EU chemicals industry. The seven objectives that needed to be balanced within the overall framework of sustainable development are:

1. Protection of human health and the environment;
2. Maintenance and enhancement of the competitiveness of the EU chemical industry;
3. Prevention of fragmentation of the internal market;
4. Increased transparency;
5. Integration with international effort;
6. Promotion of non-animal testing;

7. Conformity with EU international obligations under the WTO.

Objectives 1, 4, 5, and 6 seem to fit the goals of a registration and evaluation program. Unfortunately, the EU Competitiveness Council collaborated with the Environment Committee in the final draft. Consequently, objectives 2, 3, and 7 are designed to protect the chemical producers and downstream users, not evaluate chemicals.

More promising results of the REACH effort occurred when the U.S. Congress got involved and introduced its own version of the REACH bill, after politicians realized that the Toxic Substances Control Act, FIFRA, and FQPA were not sufficient to get rid of toxic poisons and that something more needed to be done. Stay tuned, we shall see what happens with this effort.

We saw earlier how the first state-supported farm research stations, which were designed to evaluate existing farm strategies and conduct research on new ideas, instead spent all their time analyzing bogus commercial fertilizer and pesticide products. But the researchers did nothing to stop these quack products from being sold and there were no laws at the time to protect consumers against fraudulent product claims.

In 1894, the U.S. Department of Agriculture published Farmers' Bulletin 127, *Important Insecticides: Directions for Their Preparation and Use*. More than 165,000 copies were distributed free for the asking by the department before 1900. By this time, as Farmers' Bulletin 127 shows, the USDA clearly favored poisons and discouraged the regulation of chemicals.

Propaganda such as Farmers' Bulletin 127 and equally clever ads, however, had to be continually fed to the farmers because the magazines knew that, while farmers were critical of chemicals, they were also avid readers. Educative and interesting pieces of literature, which had filled the earlier farm journals,

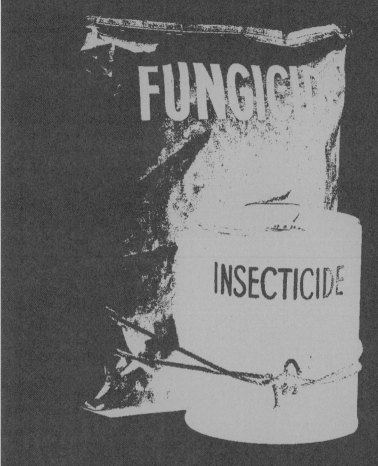

Cotton farmers:
Somebody found a way to put them together in one neat package.

From *Agrichemical West*. April 1967. Olin's promotion is for a mixture of very toxic insecticides and fungicides. This cocktail was a combination of pentachloronitrobenzene (PCNB), which is extremely toxic to fish, and Disyston, an organophosphate nerve poison that is extremely toxic to everything.

were replaced with propaganda about products, often thinly disguised as human-interest stories to entice the farmer to read them.

Protective laws and regulatory actions from the Pure Food and Drug Act in 1906 to the REACH law in 2006 were proposed to protect consumers from poisoned food, drugs, and fiber. In all of these confrontations, laws, and regulatory actions the public never won outright victories. Even in the victory over DDT, the triumph was only a partial one. DDT lost its registration in the United States, but it is still legally sold and used in many parts of the world and often comes back to poison us as an inert in another pesticide product or at breakfast, lunch, and dinner on imported food.

When the pesticide-control acts were proposed and passed, the public felt it had gained a real victory. Most of the protesters and the public relaxed, thinking they were protected with their new laws. The chemical corporations, however, did not relax. Instead, they immediately attacked the regulations and ultimately secured variances or watered down the laws. In many cases they just ignored the laws and kept doing "business as usual," since there was little or no regulatory pressure on them to obey the laws.

Instead of complying, companies tried to divert attention from the fact that farm chemicals had seriously damaged soil, water, and human health in most areas of rural America. Their lawyers tied the regulators up in legal knots. In order to control the regulators and the environmentalists, the chemical corporations hired the regulators. Monsanto even hired the environmental advocate Kate Wolf and for a time used the sustainable business guru Paul Hawken as a consultant. The corporations orchestrated their own version of "musical chairs," played out by regulators, scientists, and former high government officials who each took a turn sitting in a regulatory, corporate, science, environmental, or congressional chair.

It should come as no surprise then that, in spite of dangers to rural residents and urban consumers, the regulators consistently refused to cancel registrations on even the most deadly farm chemicals. To affect this filibuster, administrators and regulators have been hiding under the cover of one loophole or another ever since the passage of the Pesticide Act of 1910.

Even though they received almost constant support from the regulators, the chemical corporations and the large-scale farmers still cried foul in public and repeatedly tried to alarm both farmers and consumers about the pending loss of chemicals caused by "obsessive regulations" driven by "environmental loonies." Editors, scientists, and advertisers worried aloud that if farmers lost chemical tools, then food prices would certainly be higher and the crop losses to insects, weeds, and other pests would soar. Of course, this hasn't happened. Instead, crop prices are lower for farmers and pest losses are actually higher owing to increased resistance and the need for stronger chemicals.

Chemical corporations tried to alarm farmers about the impending loss of one pesticide after another, beginning with arsenic. Whenever the industry was trying to protect a dangerous old chemical or introduce a new one, this alarmist tactic worked. And it continues to work today! The chemical corporations and the magazines consistently described the fight over the regulation of chemicals as a struggle between the farmers and the regulators, or farmer struggles with labor, or the environmentalists, or the muckrakers, or schoolkids, or any other convenient straw man. Of course, the real fight was and is between the chemical corporations and all these complaining parties. The farmers are merely a pawn in the struggle between the toxic-chemical producers and their victims.

The chemical corporations tried to frame the debate so that it was perceived as the farmers fighting to protect their pesticide and fertilizer tools. But the chemical tools do not belong to the farmers; they are owned by giant corporations. The corporations put the farmers in the middle of this crisis because they were under fire from the public for chemical failure, chemical damage, health problems, deaths, and legitimate demands for regulation (just as with big tobacco). Bucolic-looking farms are always featured in the defensive propaganda of the chemical corporations, not the corporate factory farms that use their products, dominate U.S. agriculture, and are the opposite of pastoral and rural. Regular-looking farmers, versus slick corporate farmers, are trotted out to testify how invaluable the chemicals or the GMOs are.

Finally, many farmers have figured out this propaganda system and are beginning to read other published material. They are learning with their neighbors how to reduce or eliminate these poisons. Farmers and householders have begun to figure out a safer system of pest control and fertility management. At the same time the public continues to be editorially and graphically massaged to demand and protect the next generation of pesticides, plants, fertilizers, and genetically modified foods and fibers.

Today we find ourselves standing at a crossroads similar to the ones farmers faced at the beginning of the last two centuries. Farm populists during the Revolutionary period (1750–1830) witnessed precipitous losses in topsoil and declines in fertility, and populists after the Civil War (1865–1900) experienced even further soil and fertility loss. Farmers and farm advocates now are witness to the failure of many chemicals and the massive degradation of hundreds of millions of acres of farmland, grassland, wetlands, and forest caused by chemical-intensive farming. Ideas that were the legacy of heroic medicine and chemicals from the Industrial Revolution of the 1800s have dominated U.S. agriculture for more than one hundred years.

Farmers and consumers are in the process of deciding what kind of future they want, what kind of food they want. The stakes are high, perhaps higher than ever before because of the widespread damage to farmland and watersheds and the enormous corporate control of our food system. Our choice is between more ecological and biological systems, which are still used by most of the farmers in the world, versus a kind of chemical/genetic roulette that pays off in the short term for large-scale corporate farmers and the Wall Street investor class but is turning out to be a very unsound gamble for real farmers and consumers. No one has any idea of the long-term cost or potential for contamination from GMOs. No one has decided how or even whether the biotech companies will assume liability for the risks and consequences of genetic engineering, which could be permanent and irremediable.

Without addressing any of these concerns, the genetic-engineering corporations have already planted hundreds of millions of acres with genetically manipulated seed on farms in the United States, Canada, Argentina, China, and several other countries. Companies like Monsanto, Syngenta, DuPont, Hoechst, and Aventis are determined to plant hundreds of millions more acres as soon as possible.

Once the chemical corporations convinced the farmers that they should poison their farm pests, it was difficult to restore the balance between predator and pest on the farm for at least that crop season. Once farmers committed to chemicals for a couple of years, it was difficult to reestablish natural checks and balances. At that point, farmers became stuck on a chemical treadmill.

The strategy of the chemical corporations remains the same with their genetically manipulated products. If they get farmers hooked on genetically modified organisms it will be too late to go back, not just for those farmers, but for their entire community. Genetically manipulated crops drift everywhere, so no related crops in communities where GMOs are grown can be considered totally organic or GMO-free because of spillage in transport and pollen drift. For example, if GMO corn were grown within three or four miles of organic corn, it is unlikely that the organic corn would not be contaminated by GMO pollen drift or movement of bees and other pollinators.

We have looked back on 160 years of advertisements, promotions, and "agricultural science" that touted new technologies without ever considering precisely how they would fit into the ecological web of farming, rural communities, and the adjacent wilderness. Humans are fundamentally tool makers and risk takers, and farmers are gamblers—gambling on their crops, the weather, the market, and their skill. And yet, while agriculture entails manipulating the environment to a greater or lesser extent, many of America's farmers have also been naturalists, environmentalists, and willing stewards of the land. The best farmers learn their lessons from nature and from thoughtful, observant neighboring farmers, not from Liebig or Harvard or Yale or Monsanto.

Leonardo da Vinci once wrote that he would not allow some of his inventions, particularly underwater exploration devices, to fall into the public's hands because he feared that they would be used for violent or detrimental ends. Such, it seems, captures our present predicament with the chemical and genetic sciences. Granted, there could be times when GMO technologies might have to be used as a last resort. But as this survey of industrial agriculture has revealed, dangerous technologies will not be used as a last resort. Instead they have been and are being used as a first resort and will be broadcast irresponsibly across the planet, fueled by greed and pumped up with advertising campaigns that tell consumers only part of the story. Their hasty intrusion into the marketplace puts our farming future at risk and takes us on one dangerous leap after another into a world where it may not be possible to repair or reverse the damage.

Where Do We Go Now?

Just before the turn of the twentieth century about 75 percent of the American population farmed and resided in the bucolic countryside. While the chemical merchants enjoyed halcyon times through wars, panics, and depressions, bankruptcy has defined our farming and rural communities ever since the Civil War and the Panic of 1873.

When chemicals began to be widely used in American agriculture in the late 1880s, there were more than four million farmers who farmed about six hundred million acres. By the Great Depression, nearly seven million farmers farmed 1.2 billion acres.[4] Today there are only a million farmers left, but they still farm about 1.2 billion acres. Nearly 50 percent of U.S. farmland is rented or leased, and 75 percent of U.S. cotton land is rented. Farmers know that renters are likely to steward the land less carefully than owner-operators. In times of financial crisis, the way renters steward the land is almost always determined by price and market values, not by what is best for the long-term sustainability of soil, habitat, water, or other resources. Consequently, our soil, water, and rural communities are suffering under the weight of long-term land abuse and corporate control of most commodities.

We have been losing our farming base for the last one hundred years. The average age of today's farmer is fifty-nine. For the last three decades, many farm kids stopped going into farming because they knew they would have to work twelve-hour days while constantly facing bankruptcy and living in a poisoned environment. Those who wanted to farm felt frozen out of good markets or stuck with poor markets where the brokers made all the profit.

In order to help change the awful realities of our declining U.S. farm sector we must reinvest in the farms and we must do it now. U.S. consumers, who pay taxes for all the farm programs, must demand a shift in agricultural policy—one that will emphasize sustainability. This is a pivotal time to aggressively confront the chemical corporations and the factory-farm proponents because of a growing safe-food movement in the United States and a huge opposition to genetically modified foods and factory farms around the world. As a result, farm kids and first-time farmers are beginning to take the plunge into farming once again. We are in the midst of another farm-revitalization movement in the United States, but this time it is part of a global movement toward sustainable farming. This reaction is happening in spite of the national and international power of the food conglomerates and the chemical/seed giants. We must support this move back to farming and the land.

In the current movement, the focus is on local agriculture and is guided by the following mantras: "Buy your food as locally as you can." "Don't trade unless you can't produce it." "Support your local farmers and merchants." "Buy or grow organically grown food." "Eat food that doesn't rob your mind or bloat your body." "Preserve traditional cultures and their foods." This movement encourages consumers to develop and participate in sustainable communities and to support sustainable farms. This is a worldwide movement that rejects miracle rice, high-fructose corn syrup, junk food, toxic chemicals, genetic manipulation, food from factory farms, and the McDonald's fast-food culture.

With the defeat of big tobacco as a precedent, and a growing national and international safe-food movement, it finally seems possible to challenge the lies of the chemical giants and demand compensation for their knowingly poisoning the public for more than one hundred years. It is time to win outright victories.

This book alone cannot win those victories. But we hope that it will inspire consumers, researchers, investigative journalists, lawyers, and whistleblowers in corporations and government to act in their own interest and help win those victories. Perhaps then we can buy safe food—wherever we live, whatever our incomes—and the poisonous products will stop showing up as unwanted guests at our dining room tables. Obviously, the government regulators and corporations are not going to eliminate these terrible chemicals and stop producing genetically modified seed unless citizens stand up, oppose these dangerous products, and take matters into their own hands.

Our country is at an environmental and cultural crossroads. The growing awareness of global warming is prompting many to consider changing their lifestyle. Armed with the knowledge of toxicity in conventionally grown food, readers have a choice about the food they consume. The organic food industry has grown exponentially in the last fifteen years. The USDA Economic Research Service states that organic products are now available in nearly 20,000 natural food stores and almost three out of four conventional grocery stores. The number of organic farms and farmers' markets has also increased. The Local Harvest Web site (www.localharvest.org), a great place to find out where to source local organic food, states that, as of 2007, there are over 1,000 community-supported-agriculture farms across the country. Buying organic or growing your own food is a choice that people who care about the health of their families and the survival of the planet can and should make. It's a life choice that is gratifying and tasty!

Note: The appendices that follow contain sample analyses of pesticides on popular crops and some resources to help you fight agricultural pollution and find safe food.

Appendix: Analysis of Pesticide Use on Certain Popular Crops from California

The goal throughout this book has been to advise consumers to protect their health and safety by being fully informed about the foods they choose and the type of farming that produces their food. An analysis of residues on foods is a valuable guide to how many poisons are still lurking on your food when you eat it or cook it. Also valuable for anyone concerned with how our foods are grown is the amount and toxicity of the most used chemicals on the country's favorite foods.

The Environmental Working Group (EWG) Shopper's Guide to Pesticides in Produce ranked pesticide contamination on 44 popular fruits and vegetables based on an analysis of nearly 51,000 tests for pesticide residues on these foods. These analyses were conducted on sampling done from 2000 and 2005 by the USDA and the FDA. Contamination was measured in six different ways and crops were ranked based on a composite score from all categories.

As with previous Environmental Working Group residue studies, seven fruits were on the list of the twelve most contaminated fruits and vegetables.

Four fruits were among the top six. Peaches led the list, followed by apples, nectarines, and strawberries. Cherries, imported grapes, and pears were also in the top twelve. Among these seven fruits, nectarines had the highest percent of the products testing positive for pesticide residues (97.3), followed by peaches (96.6 percent) and apples (93.6 percent).

Peaches had the highest likelihood of multiple residues on a single sample (86.6 percent had two or more pesticide residues) followed by nectarines (85.3 percent) and apples (82.3 percent).

Peaches and apples had the most residues detected on a single sample, with nine pesticides each, followed by strawberries, with eight pesticides found on a single sample.

Apples had the most residues overall with some combination of up to fifty pesticides found on the samples tested, followed by peaches with forty-two pesticides and strawberries with thirty-eight.

Sweet bell pepper, celery, lettuce, spinach, and potato were the vegetables most likely to expose consumers to pesticides. Among these five vegetables, celery had the highest percentage of samples test positive for residues (94.1 percent), followed by sweet bell peppers (81.5 percent) and potatoes (81.0 percent).

Celery also had the highest likelihood of multiple pesticide residues on a single vegetable (79.8 percent of samples), followed by sweet bell peppers (62.2 percent) and lettuce (44.2 percent).

Sweet bell pepper was the vegetable with the most pesticides detected on a single sample (eleven found on one sample), followed by celery and lettuce (both with nine).

Sweet bell pepper was the vegetable with the most pesticides overall with sixty-four, followed by lettuce with fifty-seven and celery with thirty.

Source: Environmental Working Group Web site and residue analyses online, October 2007.

In addition to the EWG residue analyses, five popular crops grown in California in 2004 were analyzed for the total amount of pesticides used and their toxicity to illustrate current patterns of use. These analyses illustrate how farmers have come to depend on the most dangerous and deadly pesticides on the market. The data on amounts and toxicity came from the California Environmental Protection Agency (CalEPA), Department of Pesticide Regulation (DPR), Pesticide Use Reports, Priority Risk Lists, and Toxicology One-Liners and Conclusions. The pesticide use reports from farmers and applicators have been collected by the California Department of Food and Agriculture (CDFA) since 1970 and by CalEPA since 1990. The risk lists and toxicological analysis began in the 1990s. I obtained harvested acreage estimates from the California Department of Food and Agriculture Annual Crop Reports.

To derive an average of how many pesticides were used per acre I divided the total pounds of pesticides used on a crop by the total acreage harvested. For example, 11,135,407.85 pounds of pesticides were used on strawberries. There were an estimated 33,200 acres of chemically grown strawberries in 2004. I divided acreage estimates into total pesticides used, which gave an average of 335.40 pounds of pesticides per acre.

This type of analysis is an indicator of the amount of toxins in our food, on our soil, and in our water. I feel that the EWG residue findings are excellent guides, especially for shoppers. But residue testing and analysis does not tell the whole story, especially since very little produce is tested by government agencies.

I feel that a big part of that whole story is actual pesticide use.

Until recently, California was the only state to have mandatory pesticide use reporting. Finally, New York began to require farmers and applicators to report usage in 2005. The USDA and all other states depend on a volunteer survey sent to farmers and applicators asking them to report their usage. USDA surveys regularly underreport pesticide use by one-third to one-half. Because of this underreporting, mandatory use reporting should be required in all states and be transparently available to all citizens. Even though California is relatively exemplary in terms of reporting and data collection, it was still difficult to obtain and analyze all the information on California use, acreage, and toxicity that I present on the following pages.

Following are the analyses of most used pesticides on five favorite crops from 2004 and a separate analysis on peaches. I hope it supplements the EWG residue work, which tells how many pesticides could still be detected on the produce in the marketplace. Pesticide and fertilizer analyses of other vegetables, fruits, and fibers can be seen on the WarOnBugsBook Web site. As the following data show, buyer beware! Even if pesticides don't show up as residues, they still pollute the soil and water and injure farmworkers, farmers, and rural residents.

STRAWBERRIES

In 2004 California strawberry growers used 184 different pesticides. However, six accounted for 80.6 percent of use, nearly nine million pounds. Four were fumigants designed to kill all soil life and are among the most dangerous pesticides. Those four accounted for 74.1 percent of use, which averages out to about 249 pounds per acre. One, methyl bromide, is still on the market even though it has been scheduled to be banned for more than ten years. The Montreal Protocol on ozone-depleting substances determined that methyl bromide is the most destructive chemical reaching the ozone. In spite of its dangers, California and Florida strawberry growers are largely responsible for keeping it on the market for so long. An average of 335.40 pounds of pesticides were used on each acre to grow strawberries for our shortcakes.

2004 Strawberry Pesticides, Most Used

CHLOROPICRIN—tear gas, fumigant, biocide, causes birth defects, highly toxic to fish.	3,258,929.80
METHYL BROMIDE—fumigant, biocide, causes birth defects.	3,186,893.38
1,3-DICHLOROPROPENE—fumigant, causes cancer, biocide, groundwater contaminant, toxic to fish.	1,523,348.32
SULFUR—fungicide, causes birth defects.	492,322.40
METAM SODIUM—fumigant, nerve poison, causes birth defects.	274,472.00
CAPTAN—fungicide, insecticide, causes cancer and birth defects.	240,430.50
178 Other Pesticides Used on California Strawberries	2,159,011.45
Total Pounds of Pesticides Used on California Strawberries	**11,135,407.85**

Total Acreage of Chemical Strawberries Harvested in California	33,200.00
Average Pounds of Pesticide Used per Acre	**335.40**
Total California Organic Strawberry Acreage	1,382.19

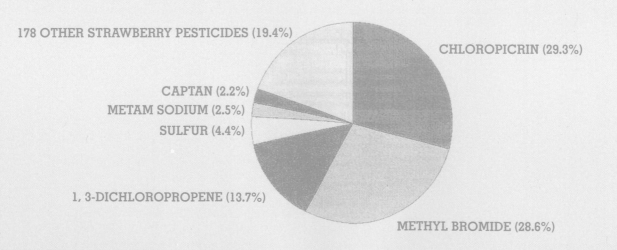

178 OTHER STRAWBERRY PESTICIDES (19.4%)

CHLOROPICRIN (29.3%)

CAPTAN (2.2%)
METAM SODIUM (2.5%)
SULFUR (4.4%)

1, 3-DICHLOROPROPENE (13.7%)

METHYL BROMIDE (28.6%)

CARROTS

We found that California carrot growers applied 145 different pesticides in 2004. Six pesticides accounted for 97.5 percent of use, or nearly 7 million pounds. But two soil fumigants, metam sodium and 1,3-dichloropropene, accounted for 89.7 percent of total use or 96 of every 108 pounds used on carrots . As with strawberries, these carrot fumigants were designed to kill most soil life.

2004 Carrot Pesticides, Most Used

METAM SODIUM—fumigant, nerve poison, causes multiple birth defects.	5,062,340.00
1,3-DICHLOROPROPENE—fumigant, carcinogen, biocide, groundwater contaminant, toxic to fish.	1,359,024.00
SULFUR—fungicide, causes birth defects.	324,141.00
POTASSIUM N-METHYLDITHIOCARBAMATE—fumigant, insecticide, carcinogen, nerve poison, causes birth defects, toxic to fish.	122,644.00
LINURON—herbicide, carcinogen, toxic to birds and fish.	55,628.00
CHLOROPICRIN—tear gas, fumigant, biocide, causes multiple birth defects, highly toxic to fish.	51,407.00
Other 139 Pesticides Used on California Carrots	182,127.00
Total Pounds of Pesticides Used on California Carrots	**7,157,311.00**
Total Acreage of Chemical Carrots Harvested in California	70,800.00
Average Pounds of Pesticides Used per Acre on Carrots	**108.00**
Total California Organic Carrots Acreage	4,681.00

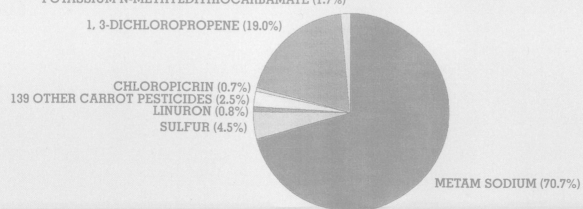

POTASSIUM N-METHYLDITHIOCARBAMATE (1.7%)

1, 3-DICHLOROPROPENE (19.0%)

CHLOROPICRIN (0.7%)
139 OTHER CARROT PESTICIDES (2.5%)
LINURON (0.8%)
SULFUR (4.5%)

METAM SODIUM (70.7%)

WATERMELON

One hundred fifty-eight pesticides were used on watermelons. But five highly toxic fumigants, designed to kill all soil life, accounted for 88.9 percent of use, or 65 of the 73 pounds used per acre. One chemical, metam sodium, accounted for more than half of all pesticides applied.

2004 Watermelon Pesticides, Most Used

METAM SODIUM—fumigant, nerve poison, causes multiple birth defects.	499,392.02
1,3-DICHLOROPROPENE— fumigant, carcinogen, biocide, groundwater contaminant, toxic to fish.	167,133.90
CHLOROPICRIN—tear gas, fumigant, biocide, causes multiple birth defects, highly toxic to fish.	80,867.51
METHYL BROMIDE—fumigant, biocide, ozone destroyer, causes multiple birth defects.	75,229.25
POTASSIUM N-METHYLDITHIOCARBAMATE—fumigant, insecticide, carcinogen, nerve poison, causes birth defects, toxic to fish.	57,661.26
SULFUR—fungicide, causes birth defects.	47,288.66
CYCLOATE—herbicide, cholinesterase inhibitor, possible groundwater contaminant, causes birth defects.	15,217.68
151 Other Pesticides Used on California Watermelons	47,242.86
Total Pounds of Pesticides Used on California Watermelons	**990,033.14**

Total Acreage of Chemical Watermelons Harvested in California	13,500.00
Average Pounds of Pesticides per Acre	**73.00**
Total California Organic Watermelon Acreage	539.70

POTASSIUM N-METHYLDITHIOCARBAMATE (5.8%)

153 OTHER WATERMELON PESTICIDES (11.1%)

1, 3-DICHLOROPROPENE (16.9%)

METAM SODIUM (50.5%)

METHYL BROMIDE (7.6%)

CHLOROPICRIN (8.0%)

SPINACH

One hundred nineteen pesticides were used on California spinach in 2004. Spinach was analyzed because it showed high residue samples in the EWG tests. It did not have an especially high number of average pounds applied per acre (only 13), but it did have some high percentages of certain fumigant chemicals, including metam sodium (53.7 percent),

potassium n-methyldithiocarbamate (15.2 percent), and 1,3-dichloropropene (6.0 percent). Again, the most used chemicals are the most toxic. Three poisonous fumigants accounted for 74.87 percent of use. Only seven accounted for 92.8 percent or more than 12 of every 13 pounds per acre of all pesticides used on spinach.

2004 Spinach Pesticides, Most Used

METAM SODIUM—fumigant, nerve poison, causes multiple birth defects.	256,699.00
POTASSIUM N-METHYLDITHIOCARBAMATE—fumigant, insecticide, carcinogen, nerve poison, causes birth defects, toxic to fish.	72,751.40
CYCLOATE—cholinesterase inhibitor, possible groundwater contaminant, causes birth defects.	35,344.00
1,3-DICHLOROPROPENE—fumigant, carcinogen, biocide, groundwater contaminant, toxic to fish.	28,766.58
DIAZINON—insecticide, cholinesterase inhibitor, causes birth defects, possible groundwater contaminant, toxic to wide variety of aquatic life and animals.	28,299.00
FOSETYL-AL—fungicide, possible groundwater contaminant, toxic to fish and variety of aquatic life.	12,705.00
MEFENOXAM—fungicide, toxic to fish.	9,641.00
112 Other Pesticides Used on California Spinach	34,215.00
Total Pounds of Pesticides Used on California Spinach	**478,421.00**

Total Acreage of Chemical Spinach Harvested in California	36,800.00
Average Pounds of Pesticides Used per Acre	**13.00**
Total California Organic Spinach Acreage	3,081.00

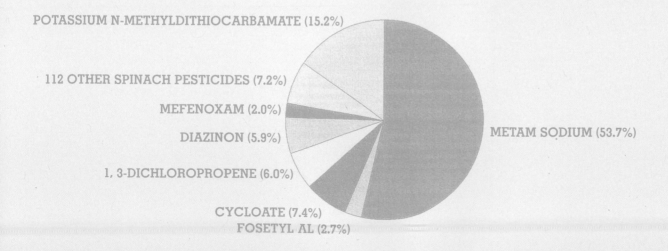

POTASSIUM N-METHYLDITHIOCARBAMATE (15.2%)

112 OTHER SPINACH PESTICIDES (7.2%)

MEFENOXAM (2.0%)

DIAZINON (5.9%)

1, 3-DICHLOROPROPENE (6.0%)

METAM SODIUM (53.7%)

CYCLOATE (7.4%)

FOSETYL AL (2.7%)

ONIONS

We analyzed the 169 pesticides used on onions and found that 54.5 percent of the pesticides were highly toxic fumigants. Only nine chemicals accounted for almost 80 percent of total pesticide use. As with the other crops analyzed in this brief presentation, the most used pesticides were among the most deadly and dangerous poisons. Again, the most used chemicals were for fumigation of the soil so that growers would be able to grow strawberries, carrots, watermelons, spinach, or onions year after year on the same pieces of ground.

2004 Onion Pesticides, Most Used

METAM SODIUM—fumigant, nerve poison, causes multiple birth defects.	342,862.68
1,3-DICHLOROPROPENE—fumigant, carcinogen, biocide, groundwater contaminant, toxic to fish.	120,159.64
CHLOROPICRIN—tear gas, fumigant, biocide, causes multiple birth defects, highly toxic to fish.	65,995.88
MANCOZEB—fungicide, reproductive toxin, toxic to birds and fish.	51,633.85
CHLOROTHALONIL—fungicide, carcinogen, potential groundwater contaminant, toxic to broad range of terrestrial and aquatic life.	45,527.82
CHLORTHAL-DIMETHYL—herbicide, possible carcinogen, groundwater contaminant, toxic to wide variety of animal and aquatic life.	41,718.86
SULFUR—fungicide, causes birth defects.	41,553.90
MANEB—fungicide, noted reproductive effects at high levels of exposure. Organs affected by Maneb include the thyroid, kidneys, and heart. Highly toxic to fish and aquatic species.	33,193.09
MALEIC HYDRAZIDE, POTASSIUM SALT—growth regulator, herbicide.	29,992.58
160 Other Pesticides Used on California Onions	197749.19
Total Pounds of Pesticides Used on California Onions	**970,387.49**

Total Acreage of Chemical Onions Harvested in California	44,000.00
Average Pounds of Pesticides Used per Acre	**22.05**
Total California Organic Onions Acreage	606.00

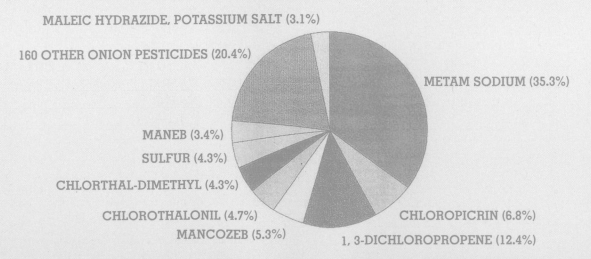

MALEIC HYDRAZIDE, POTASSIUM SALT (3.1%)
160 OTHER ONION PESTICIDES (20.4%)
METAM SODIUM (35.3%)
MANEB (3.4%)
SULFUR (4.3%)
CHLORTHAL-DIMETHYL (4.3%)
CHLOROTHALONIL (4.7%)
MANCOZEB (5.3%)
CHLOROPICRIN (6.8%)
1, 3-DICHLOROPROPENE (12.4%)

PEACHES

As an added illustration we are presenting peaches in both a graphic and a total pesticide use format. This is different from the other five analyses but instructive about the gross use of poisons on our food. The top twelve peach pesticides accounted for 74.3 percent of use. More than 25.6 percent of these top twelve were herbicides (simazine, paraquat, glyphosate, oxyfluororfen, oxyzalin), 27.9 percent were insecticides (chlorpyrifos, diazinon, phosmet, and carbaryl), another 14.9 percent were fungicides (copper oxide, iprodione), and only 5.8 percent were fumigants (methyl bromide). Unlike the case for the other five crops analyzed, the top pesticides in peaches were not fumigants but herbicides, insecticides, and fungicides. Only when trees are replaced do high levels of fumigants show up.

The following chart contains all the pesticides used on peaches in California in 2004 amounting to more than 1 pound in total. Peach cobbler, anyone?

Chemical	Pounds of chemicals
PHOSMET	62,348
GLYPHOSATE, ISOPROPYLAMINE SALT	60,412
COPPER OXIDE (OUS)	55,989
CHLORPYRIFOS	32,233
METHYL BROMIDE	27,390
DIAZINON	25,456
PARAQUAT DICHLORIDE	16,680
IPRODIONE	14,938
ORYZALIN	14,808
SIMAZINE	14,178
CARBARYL	12,457
OXYFLUORFEN	10,164
2,4-D, DIMETHYLAMINE SALT	7,552
CYPRODINIL	7,220
CHLOROPICRIN	6,281
GLYPHOSATE	5,576
FENBUTATIN-OXIDE	5,348
PENDIMETHALIN	4,432
XYLENE RANGE AROMATIC SOLVENT	4,348
METHIDATHION	4,314
COPPER SULFATE (PENTAHYDRATE)	4,170
SODIUM TETRATHIOCARBONATE	3,719
CHLOROTHALONIL	3,656
COPPER OXYCHLORIDE	3,487
ESFENVALERATE	3,477
PERMETHRIN	3,389
PETROLEUM DISTILLATES	3,310
NORFLURAZON	3,267
CAPTAN	3,176
BIFENAZATE	3,127
PROPICONAZOLE	2,954
GLYPHOSATE, DIAMMONIUM SALT	2,818
TEBUCONAZOLE	2,730
LIME-SULFUR	2,193
HEXYTHIAZOX	2,116
BACILLUS THURINGIENSIS, SUBSP. KURSTAKI, STRAIN ABTS-351, FERMENTATION SOLIDS AND SOLUBLES	2,032
AZINPHOS-METHYL	1,994
PETROLEUM OIL, PARAFFIN BASED	1,832
GLYPHOSATE-TRIMESIUM	1,815
CRYOLITE	1,773
KAOLIN	1,758
CLOFENTEZINE	1,736
METHOXYFENOZIDE	1,294
Z-8-DODECENYL ACETATE	1,185
GLYPHOSATE, MONOAMMONIUM SALT	1,154
BOSCALID	1,088
PROPARGITE	1,033
NAPROPAMIDE	914
THIOPHANATE-METHYL	824
GLYPHOSATE, POTASSIUM SALT	776
SPINOSAD	729
METAM-SODIUM	635
PYRACLOSTROBIN	553
FENAMIPHOS	530
MYCLOBUTANIL	514
FORMETANATE HYDROCHLORIDE	495
METHOMYL	462
POTASSIUM BICARBONATE	448
LAMBDA-CYHALOTHRIN	400
CLARIFIED HYDROPHOBIC EXTRACT OF NEEM OIL	383
FENBUCONAZOLE	288
BACILLUS THURINGIENSIS, SUBSP. KURSTAKI, STRAIN HD-1	279
AZOXYSTROBIN	225
TRIFLURALIN	218
BACILLUS THURINGIENSIS (BERLINER), SUBSP. KURSTAKI, SEROTYPE 3A,3B	199
XYLENE	149
PYRIDABEN	147
BENOMYL	110
DIFLUBENZURON	93
DIURON	90
BACILLUS THURINGIENSIS (BERLINER), SUBSP. KURSTAKI STRAIN SA-12	89
E-8-DODECENYL ACETATE	78
CAPTAN, OTHER RELATED	71

MALATHION	66	
DICOFOL	55	
ENDOSULFAN	51	
FLUAZIFOP-P-BUTYL	48	
BACILLUS THURINGIENSIS (BERLINER), SUBSP. *KURSTAKI*, STRAIN SA-11	47	
SETHOXYDIM	46	
MANEB	45	
PYRIPROXYFEN	45	
ISOXABEN	43	
MYROTHECIUM VERRUCARIA, DRIED FERMENTATION SOLIDS & SOLUBLES, STRAIN AARC-0255	33	
BACILLUS THURINGIENSIS (BERLINER), SUBSP. *ISRAELENSIS*, SEROTYPE H-14	29	
1,2-DICHLOROPROPANE, 1,3-DICHLOROPROPENE, AND RELATED C3 COMPOUNDS	22	
NALED	22	
BACILLUS THURINGIENSIS, SUBSP. *AIZAWAI*, STRAIN ABTS-1857	20	
VINCLOZOLIN	15	

Z-8-DODECENOL	14	
ENCAPSULATED DELTA ENDOTOXIN OF *BACILLUS THURINGIENSIS* VAR. *KURSTAKI* IN KILLED *PSEUDOMONAS FLUORESCENS*	13	
E,E-8,10-DODECADIEN-1-OL	12	
(E)-5-DECENYL ACETATE	10	
TRIFLOXYSTROBIN	9	
BACILLUS THURINGIENSIS (BERLINER), SUBSP. *AIZAWAI*, GC-91 PROTEIN	8	
STRYCHNINE	7	
PETROLEUM DISTILLATES, AROMATIC	6	
ALUMINUM PHOSPHIDE	5	
PIPERONYL BUTOXIDE	3	
PYRETHRINS	3	
COPPER AMMONIUM COMPLEX	3	
ACETAMIPRID	2	
COPPER OXYCHLORIDE SULFATE	2	
DIATOMACEOUS EARTH	2	
(E)-5-DECENOL	2	
GIBBERELLINS	2	

E-11-TETRADECEN-1-YL ACETATE	2	
BACILLUS THURINGIENSIS SUBSP. *KURSTAKI*, GENETICALLY ENGINEERED STRAIN EG7841 LEPIDOPTERAN ACTIVE TOXIN	2	
LAURYL ALCOHOL	1	
AZADIRACHTIN	1	
FENHEXAMID	1	
IMIDACLOPRID	1	
PIPERONYL BUTOXIDE, OTHER RELATED	1	
BACILLUS THURINGIENSIS (BERLINER)	1	

Total Use on Peaches   ***468,804***

For additional information on pesticide use and abuse contact our Web site, as well as the EWG, Organic Consumers Association, and Center for Food Safety Web sites.

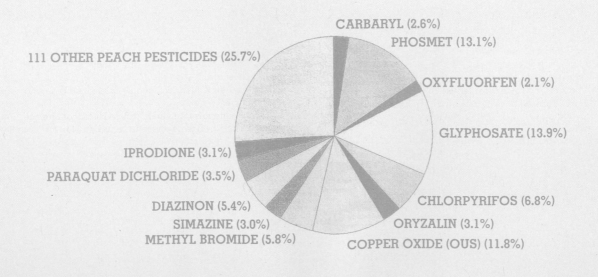

- CARBARYL (2.6%)
- PHOSMET (13.1%)
- 111 OTHER PEACH PESTICIDES (25.7%)
- OXYFLUORFEN (2.1%)
- GLYPHOSATE (13.9%)
- IPRODIONE (3.1%)
- PARAQUAT DICHLORIDE (3.5%)
- DIAZINON (5.4%)
- SIMAZINE (3.0%)
- METHYL BROMIDE (5.8%)
- CHLORPYRIFOS (6.8%)
- ORYZALIN (3.1%)
- COPPER OXIDE (OUS) (11.8%)

Notes

Full citations for works mentioned here can be found in the bibliography.

Preface

1. Summary of Toxicology Data: *Toxicology One-liners and Conclusions*, from *Methyl Bromide Inhalation Study on Dogs*. California EPA, July 24, 1995.
2. Both of these Cyrus H. K. Curtiss quotes are from Frank Rowsome, *They Laughed When I Sat Down: An Informal History of Advertising in Words and Pictures* (1959).

Introduction

1. The California Department of Food and Agriculture (CDFA) Department of Pesticide Regulation collected and tabulated the pesticide use reports from 1970 to 1990. In the 1990s the California Environmental Protection Agency (CalEPA) took over the administration of the Department of Pesticide Regulation and issued the permits and collected the pesticide use reports from 1991 to the present.

   The data from CDFA is not totally comparable on an exact pound-for-pound basis to the CalEPA data, because of different ways in which the quantity of pesticides were tallied. But in determining current use and trends, the data is irreplaceable. The California data is the best data from any of the states. In order to buy or use pesticides legally in California, farmers or applicators must get a permit and provide dates and rates of use. California also taxes pesticide sales so there is a backup record of every purchase.

   Almost all the pesticide data from other states is based on voluntary surveys of use. These surveys are almost totally useless and seem to be designed to underreport the use of some of the most toxic poisons on the planet. Federal surveys for Califorma are at least 50 percent under what the state pesticide use reports tally. Clearly, not all pesticide use is legal or reported, even in California. Many tons of stolen pesticides are on the black market and are high on the lists of stolen goods in the rural United States. And when stolen pesticides are bought, no permits are required to use them and no survey or reporting system is going to detect them.
2. J. W. Turrentine, *Potash*, 1.
3. Including the infamous formulation Zyklon-B of the Nazi concentration camps.

Chapter 1

1. Anthony Bimba, *The History of the American Working Class*, 12.
2. W. B. Bizzell, *The Green Rising: An Historical Survey of Agrarianism, with Special Reference to the Organized Efforts of the Farmers of the United States to Improve Their Economic and Social Status*.
3. Gustavus Myers, *The History of the Great American Fortunes*, vol. 1, 11–14, 22.
4. James O'Neal, *The Workers in American History: From Conspiracy to Collective Bargaining*, 34–35.
5. Bimba, *The History of the American Working Class*, 35.
6. See especially James E. Rogers, *History of Agriculture and Prices*; Karl Marx, *Capital*, 788–848; and Del Sweeney, *Agriculture in the Middle Ages*.
7. John Stow, *Annales of England* (London, 1605).
8. By this time, Spain, France, and Portugal had already spent more than a hundred years pillaging South America, Mexico, and the Caribbean.
9. Ironically, because of the need for cheap labor in the European factories, several countries restricted their workers from migrating. Industrial and garment factories expanded so rapidly and so many men had left or had been deported that the industrialists began the practice of exploiting women and children.
10. Jared Diamond, *Guns, Germs, and Steel*.
11. See Malcolm Margolin, *The Ohlone Way: Indian Life in the San Francisco–Monterey Bay Area* and *The Way We Lived: California Indian Stories, Songs, and Reminiscences*; Jack Weatherford, *Indian Givers*; Julian Steward, *Handbook of South American Indians*; and Charles Mann, *New Revelations of the Americas before Columbus* for more extensive discussions of these issues.
12. Daniel Quinn, *The Story of B*.
13. William Allen, *Some Ecological Factors Influencing the Prehistoric Population Density in the Alto Pachitea Region of Peru* and *A Ceramic Sequence from the Alto Pachitea, Peru: Some Implications for the Development of Tropical Forest Cultures in South America*; William Allen and Judy Holshouser Tizon, *Land-Use Patterns in the Alto Pachitea Region of Peru*; and Jack Weatherford, *Native Roots*.
14. Diamond, *Guns, Germs, and Steel*, 210–14.
15. *Las Visitas de Monterey, Santa Barbara, y La Purisima Conception*: reports from central coast California missionaries and the Guardia Civil to government and church officials in Monterey, Mexico. I read some of them that were in the archives of Santa Barbara Mission and Santa Ynez Mission in 1962 and 1963. Also see Diamond, *Guns, Germs, and Steel*
16. Informational materials from the Fort Ross Historical Museum, Fort Ross, California. Fort Ross is a reconstructed Russian fort near the Russian River on the northern California coast. It was an outpost of the Russian American Company until about 1840.
17. See Jack Weatherford's two books, *Indian Givers* and *Native Roots*, in which he asserts that about 60 percent of the food products eaten in the world today came from crops developed by North and South American tribal farmers.
18. O'Neal, *Workers in American History*, 34–35; Bimba, *The History of the American Working Class*.
19. O'Neal, *Workers in American History*, 61.
20. Catherine McNichols Stock, *Rural Radicals*.

21. See David P. Szatmary, *Shays's Rebellion: The Making of an Agrarian Insurrection*; Steven R. Boyd, *The Whiskey Rebellion: Past and Present Perspectives*; Wayne E. Lee, *Crowds and Soldiers in Revolutionary North Carolina: The Culture of Violence in Riot and War*; and Leonard L. Richards, *Shays's Rebellion: The American Revolution's Final Battle*.
22. See Lee, *Crowds and Soldiers in Revolutionary North Carolina*, and Stock, *Rural Radicals*.
23. Lewis Cecil Gray, *History of Agriculture in the Southern United States to 1860*, vol. 2, 612, 779.

Chapter 2

1. Eric Sloane, *Our Vanishing Landscape*, 14.
2. John Taylor, *Arator*, 134.
3. Ibid., 1, 82, 84.
4. A. L. Demaree, *The American Agricultural Press*.
5. Sweeney, *Agriculture in the Middle Ages*.
6. We now know that beans, or, more broadly, legumes, have the capability of gathering ammonia (nitrogen) from the atmosphere, which is 78 percent nitrogen. By 1888 scientists knew that bacterial inoculants increased bean plants' ability to extract atmospheric nitrogen. We know that mustards (which belong to the botanical family Brassicaceae) accumulate available phosphorus and that grasses (the Graminaceae) increase soil organic matter and accumulate minerals and other essential plant-root nutrients. We also know that members of the onion genus (*Allium*) and several other families of plants enhance the production of mycorrhizae, a beneficial soil fungi. Farmers then and now also knew that cover crops improve the texture and workability of the soil.

In the nineteenth century specific cover crops for nitrogen and nutrient accumulation were endorsed by scientists. Even E. B. Voorhees—director of the New Jersey agricultural extension service and professor of agriculture at Rutgers University—in his classic book *Fertilizer* (first published in 1898) continued to recommend cover crops. As Voorhees's books and the farm-journal editorials, ads, and letters illustrate, the knowledge of fertility crops was very advanced in both the scientific and the farming communities well before the twentieth century. The main point here, however, is that the farmers knew that these crops and rotation strategies worked to improve their soil and produce better yields long before the scientists figured out *why* various crops and rotation strategies worked.

Chapter 3

1. Frank Presbrey, *The History and Development of Advertising*, 121.
2. Ibid.
3. Quoted in Sloane, *Our Vanishing Landscape*, 55.
4. *The Cultivator* (January 1840); *South-Western Farmer* (January 5, 1844); see also Demaree, *American Agricultural Press*.

Chapter 4

1. Jimmy M. Skaggs, *The Great Guano Rush*, pages 160–61.
2. Demaree, *The American Agricultural Press*, 143
3. Richard Wines, *Fertilizer in America*.
4. *Country Gentleman*, May 26, 1853.
5. See Wines, *Fertilizer in America*; Margaret Rossiter, *The Emergence of Agricultural Science*; and Skaggs, *The Great Guano Rush*, for more complete discussions of guano and post-guano fertilizer history.

Chapter 5

1. Richard Sasuly, *I.G. Farben*.
2. Lewis Henry Morgan, *Ancient Society*.
3. Sir Albert Howard, *An Agricultural Testament*.
4. See L. F. Haber, *The Chemical Industry, 1900 to 1930*, for a more thorough discussion.
5. Justus von Liebig, *Organic Chemistry in Its Applications to Agriculture and Physiology*.
6. Rossiter, *Emergence of Agricultural Science*.
7. Ibid.
8. Ibid.
9. David Wells, *American Journal of Science*, July 1852, and Rossiter, *Emergence of Agricultural Science*. See also Forest Ray Moulton, ed., *Liebig and after Liebig: A Century of Progress in Agricultural Chemistry*.
10. Rossiter, *Emergence of Agricultural Science*, 123.

Chapter 6

1. Diamond, *Guns, Germs, and Steel*, 247.
2. *Scientific American*, August 7, 1847: 362.
3. *Scientific American*, December 21, 1895: 388.
4. Turrentine, *Potash*, 1.

Chapter 7

1. James Whorton, *Before Silent Spring*.
2. See James Harvey Young, *The Toadstool Millionaires: A Social History of Patent Medicines in America before Federal Regulation* and *The Medical Messiahs*. Both are excellent analyses of patent-medicine swindlers and false advertising.
3. From *International Directory of Company Histories*, various years 1985–2004.
4. Barbara Griggs, *Green Pharmacy*.
5. Whorton, *Before Silent Spring*
6. Salvarsan was known as Compound 606; it was developed by Hoechst (the largest German chemical company) after 605 tries. Compound 606 was about 50 percent effective against syphilis. See Haber, *The Chemical Industry, 1900 to 1930*, and *The Chemical Industry in the Nineteenth Century*.

Chapter 8

1. Robert Snetsinger, *The Ratcatcher's Child*.
2. Ibid.
3. "Please, God, stop these aphids" and "Please, God, eliminate these boll weevils" are still common prayers in parts of the world. Asking for divine intervention isn't outside the realm of hope, as many pray to God that the ants will leave or the mosquitoes or ticks they encounter will not transmit Lyme disease or West Nile virus!
4. See Sam Alewitz, *Filthy Dirty: A History of Unsanitary Philadelphia in the Late Nineteenth Century*, and Cathy Murillo, "Rats: Are There More Around This Year?"

Chapter 9

1. Edward F. Treadwell, *The Cattle King*, and Ellen Liebman, *California Farmland: A History of Large Agricultural Land Holdings*.
2. Lawrence Goodwyn, *The Populist Moment*.
3. See *Pacific Rural Press*, which began publication in 1870.

4. See especially Goodwyn, *The Populist Moment*, and N. A. Dunning, ed., *The Farmers' Alliance History and Agricultural Digest*. For what happened to the Grange in the Depression years see Clarke A. Chambers, *California Farm Organizations: A Historical Study of the Grange, the Farm Bureau, and the Associated Farmers 1921–1941*.
5. The Caucasus is a Eurasian mountain region located between the Black and Caspian seas.
6. Nonetheless, farmers had applied tens of millions of pounds of pyrethrum powder since 1880, but a large percent of that was imported. Pyrethrum is currently being aggressively marketed once again, especially for control of animal pests—such as lice—and as an organic insecticide.
7. "History of Biological Control in California" in *California Cultivator*, October 25, 1947, 720, 729, 754, 755; and Carson, *Silent Spring*. In another instance of total irony, Rachel Carson reported in *Silent Spring* that in the 1940s, "citrus growers began to experiment with glamorous new chemicals against other insects. With the widespread use of DDT and the even more toxic chemicals to follow, the populations of the vedalia in many sections of California were wiped out. Its importation, for $2,000 had saved the fruit growers several millions of dollars a year, but in a moment of heedlessness, the benefit was quickly canceled out. Infestations of the scale insect quickly reappeared and damage exceeded anything that had been seen in fifty years."

Chapter 10
1. *Country Gentleman*, July 19, 1877.
2. Whorton, *Before Silent Spring*, 20–22. Chemically, London purple and Paris purple poisons were always some form of calcium arsenite.
3. Ibid., 22.
4. Ibid., 24. See also *Country Gentleman, Pacific Rural Press, Farm Journal, Practical Farmer, California Cultivator*, and *New England Farmer* from 1870 until 1910 for editorials, promotionals, testimonials, and ads for arsenic. Widespread farmer resistance and regulatory uncertainty plagued arsenic before the beginning of the twentieth century, so it was difficult to sell it to farmers, and repetitious advertising and promotions were used to break down this resistance. A farmer campaign against shoddy formulations and an antipesticide campaign led the fight against arsenic from the early 1900s. The public's regulatory fight continued through the Depression until well after World War II. It wasn't until the 1970s that lead arsenate and most other arsenic pesticide formulations had their registrations suspended. Until the 1970s there was always a vigorous anti-arsenic campaign in the United States and Europe. However, in spite of the suspensions, many arsenic formulations are still registered and widely used on American farms today.
5. See Whorton, *Before Silent Spring*, for a wider discussion of medical and household use of arsenic, especially pages 36–65.
6. See Whorton, *Before Silent Spring*, for a more complete description of this hoax, especially pages 53–57, 78.
7. Arsenic turned out to be *the* most deadly pesticide to children in the more than one hundred years of its use as a pesticide. See *Agricultural Chemicals*, 1964: 72 and 132.
8. See L. F. Haber, *The Chemical Industry during the Nineteenth Century* and *The Chemical Industry, 1900 to 1930*.
9. California Environmental Protection Agency, Department of Pesticide Regulation, *Toxicology One-liners and Conclusions*, 1996.
10. See the *2003 Farm Chemicals Handbook*, by Meister Publishing Corporation for a more complete description of both carbon bisulfide (carbon disulfide) and carbon tetrachloride.

Chapter 11
1. See Whorton, *Before Silent Spring*, for a thorough discussion of the struggle for enforcement of this tolerance in the United States.
2. Upton Sinclair, *The Jungle*
3. The U.S. Department of Agriculture (USDA) has never been a good regulatory or public-health agency, because its mandate is to promote agriculture and consumption, not to protect consumers. The department serves its corporate masters, first, last, and always. This is as true today as it has ever been—as can be seen when one looks at how the agency defends pesticides and toxic fertilizers, the problems with genetically manipulated organisms, the assault on the National Organic Program, or even the squabbles over nutritional information with the food pyramid.

Chapter 12
1. Richard Wines, *Fertilizer in America*; Turrentine, *Potash*; and Haber, The *Chemical Industry: 1900–1930*.
2. Patented by Adolph Frank and Nicodem Caro, the production of cyanamide (ammonium cyanide) is referred to as the Frank-Caro process. Cyanidgesellschaft, a German corporation, owned and franchised the Frank-Caro process around the world.
3. Alvin Cox, "The History of Fertilizer in California," *California Cultivator*, October 25, 1947; Haber, *The Chemical Industry: 1900–1930*; and Fred Aftalion, *A History of the International Chemical Industry*.
4. See Borkin, *The Crime and Punishmnet of I. G. Farben*, 19; Haber, *The Chemical Industry: 1900–1930*; and Aftalion, *A History of the International Chemical Industry*, for a more thorough discussion of these early war-hero chemicals.

Chapter 13
1. Carey McWilliams, *Factories in the Field*.
2. F. H. King, *Farmers of Forty Centuries*.
3. William Albrecht, *The Albrecht Papers 1918–1947*. These two volumes summarize Albrecht's work and celebrate his extensive contributions to the study of soil fertility. Note: Every agricultural extension advisory on beans, alfalfa, peas, vetch, or clover recommends the use of appropriate bacterial inoculants when planting legumes.
4. Rudolf Steiner, *Agriculture*. Note: Ruminants ingest enormous quantities of forage (hay or green leaves), which is very high in carbon. That carbon material is mixed in the cud and in the stomach system with microorganisms that help break down that carbon as the manure composts. Steiner added additional biodynamic mixtures to his composts and fertilizer teas. The result is a highly available source of humic acid for plant roots and organic matter for certain nutrients and water to cling to (absorb). The compost also provides for the soil microorganism inoculants, that help break down the carbon in organic matter.

Chapter 14
1. Roland Marchand, *Advertising the American Dream: Making Way for Modernity, 1920–1940*, 162, 210; Presbrey, *History and Development of Advertising*, 616.

2. The rural population remained large until well after World War II, still accounting for more than 40 percent of the citizenry at the start of World War II.

3. Chemically, Flit was a petroleum fraction—a dimethyl phthalate.

4. Adelynne Hiller Whitaker, *A History of Federal Pesticide Regulation in the U.S. to 1947*. PhD dissertation, Emory University, 1974.

Chapter 15

1. Presbrey, *History and Development of Advertising*.

2. Whorton, *Before Silent Spring*, 228–35. Letter from Cannon to Department of Agriculture, Sept. 8, 1938, and letter from Cannon to H. L. Brown, Sept. 12, 1938 (Records Office, Secretary of Agriculture).

3. Whorton, *Before Silent Spring*, 174–75. Statement from the St. Louis FDA bureau chief is reprinted in the *Congressional Record*, 69, 9, 524–529, 531 (1928).

4. Isadore Kallett and F. J. Schlink, *100,000,000 Guinea Pigs*; F. J. Schlink, *Eat, Drink, and Be Wary*; and Rachel Palmer and Isadore Alpher, *40,000,000 Guinea Pig Children*.

5. "Resistance in Pesticides" in *California Cultivator*, November 9, 1935: 476. The toxicity of the soil was especially acute in the Washington apple region, according to Whorton, *Before Silent Spring*, 217.

6. George P. Weldon, "California Insects Develop Resistance," *Pacific Rural Press*, November 1935: 476.

7. Donald L. Kieffer, "Condemn Airplane Dusting," *Pacific Rural Press*, November 1935: 562.

8. *Farm Chemicals Handbook*, 2003.

9. CalEPA, Priority Risk List, US EPA.

Chapter 16

1. The use of and aggressive advertising for fluorine-based cryolite continues today.

2. D. B. Mackie, "Shipping Bartlett Pears under Fumigation," *Pacific Rural Press*, May 23, 1940: 218.

3. Editorial in *Pacific Rural Press*, February 23, 1935: 199.

4. BASF began this review in the 1920s under the direction of Fritz Haber in the Haber Bureau studying war raw materials; but by 1933, Haber no longer directed this project, since, being a Jew, he had been removed for racist reasons and finally went into exile for fear of his life.

5. Germany's chemical industry lost only about 15 percent of its productive capacity to Allied bombing in World War II, with BASF suffering the most structural damage. So at the end of the war its industrial capacity remained largely intact. From a business perspective, however, I. G. Farben's losses were enormous, since after the war all of Germany's patents for products were eliminated at Nuremberg. Patents were given to the French, English, Swiss, and American chemical corporations.

Chapter 17

1. Much of the material for this chapter was obtained from Richard Wines's seminal history *Fertilizer in America* and Jesse W. Markham's *The Fertilizer Industry: Study of an Imperfect Market*. The data on antitrust infractions from 1906 until 1958 are documented by Markham on pages 21–22. Keep in mind that the total production of synthetic nitrogen in all of America was a mere 390,300 tons in 1940, much less than California's average tonnage used per year in the 1990s, which was 560,000 tons.

2. Later on the United States used its increased nitrogen capacity to bomb North Korea, Vietnam, Cuba, Cambodia, Laos, Thailand, Nicaragua, East Timor, Afghanistan, Panama, Grenada, Guatemala, Haiti, Iraq, Peru, Colombia, Bosnia, Venezuela, Afghanistan, and Iraq (again). Sales of nitrogen to the government proved to be profitable to U.S. corporations during these conflicts, but their sales and capacity especially soared during World War II, the Korean War, and the Vietnam War.

3. World War II casualties totaled 407,316 dead and 670,846 wounded. Estimates are that between 60 and 70 percent of the enlistees were rural men and women, so the estimates used here for rural casualties are probably conservative.

4. Markham, *The Fertilizer Industry*, 103.

5. An example of the great disparity in use can be seen in California. There, the tonnage of nitrogen fertilizer alone has been more than five times as high as pesticide use for any year recorded since the 1970s. California farmers in 2000 applied about 200 million pounds of pesticides but used 1.1 billion pounds of nitrogen fertilizer and more than 300 million pounds of phosphorus and potash. The disparity between fertilizer and pesticide use throughout the United States is roughly similar to the use differences in California. Source: *Pesticide Use Reports*, 1970–1995, California Department of Food and Agriculture and California EPA, *Fertilizer Use Reports*, 1979–1995, California Department of Food and Agriculture.

6. Miracle seeds required miracle chemicals, miracle pesticides, and regular water, from either irrigation or a regular rainfall. Many farmers could afford only the seeds. Some bought fertilizer only occasionally. Since 2000, more than a million African cotton farmers who were growing for export went bankrupt. Since 1999 in India more than 40,000 cotton farmers who were growing for export committed suicide. See Vandana Shiva, A. H. Jafri, Ashok Emani, and Manish Pande, *Seeds of Suicide*, for a discussion of the Green Revolution and the attempted globalization of chemical genetic agriculture.

Chapter 18

1. Editorial, *California Cultivator*, 1943. We now know that DDT is not good for us. It is a cumulative carcinogen, a mutagen, a fetotoxin, an embryotoxin, and an immunotoxin and causes hormone changes, decreased fertility, aplastic anemia, and liver damage. In spite of these horrific human dangers, the toxicology experts contend that DDT's greatest damage was to the environment—to birds, fish, and wildlife habitats. Clearly, DDT wasn't safe then or now, yet its use continues. The 1943 editorial also lauds DD for its soil-fumigation properties. Chemically, DD was 1,3-dichloropropene (later called Telone) and 1,2-dichloropropane. Since this early endorsement DD/Telone has become the most widely used soil fumigant in the world. Tests by the manufacturers have shown that Telone II is carcinogenic, oncogenic, and acutely toxic. This chemical is also a carcinogenic air pollutant. In 1990 the California EPA determined that Telone II was detected (from adjacent agricultural applications) over a Merced Grammar School at concentrations 885 times higher than levels shown to cause cancer after prolonged exposure (*California Farmer*, June 1990). Because of these dangers Telone II's registration was suspended in 1990 in California and use of the leftover stocks dropped to 2,122 pounds by 1994.

In spite of all these dangers, a major effort was mounted by the manufacturer to prove that the problems with Telone II could be mitigated. Of course, the California EPA Department of Pesticide Regulation agreed, and, once again, Telone II can be used in California—its registration is only restricted, not canceled (*California Farmer*, January 1995). Though it is supposedly restricted in California, enormous volumes continue to be sold in the state (9,355,205 pounds in 2003) and in other regions of the United States and the rest of the world, where it is not restricted.

2. For additional information on DDT, see Kenneth Mellanby, *The DDT Story*, 53–54.

3. Basel AG was reestablished in 1997 as Novartis and then combined with AstraZenica in 2001 to form Syngenta, the largest chemical corporation in the world and the third-largest seed corporation.

4. Chemically, 2,4-dichlorophenoxyacetic acid and initially called Tributon.

5. Chemically 2,4,6-D dinitrophenyl.

6. World Wildlife Fund, *Environmental Report*, 159–161.

7. The California Environmental Protection Agency lists toxaphene as a chlorinated camphene with an average of 69 percent chlorine. California's Environmental Protection Agency lists aldrin as a cumulative carcinogen and a suspect teratogen that also causes major liver and kidney damage. Dieldrin is listed as a suspect cumulative carcinogen, a suspect teratogen, and an immunotoxin that causes abnormal brain waves and behavior changes. Endrin was shown to be a suspect carcinogen, a teratogen, and a suspected neurotoxin. All of the drins were very toxic to birds and fish. Endosulfan is listed as sulfuric acid chlorinated esters. Endosulfan is a central nervous system stimulant for which there is no antidote. It produces convulsions and is toxic to fish and birds. Sources: California EPA pesticide priority risk lists, and *Toxicology One-Liners and Conclusions*.

8. Both have been proven by the California Environmental Protection Agency to be probable human carcinogens, probable mutagens, teratogens, immunotoxins, and suspect fetotoxins. Both cause anorexia and toxic injury of the liver, kidney, and the central nervous system. These are nerve poisons that kill and debilitate *rapidly*. See also "Prevention and Treatment of Pesticide Exposure," in *Agrichemical West*, June 1965: 8, 9, 10, 30.

9. Ron Kroese, "Agriculture's War Against Nature" in *Fatal Harvest*, ed. Andrew Kimbrell.

10. Interested readers should avail themselves of the information available from the California EPA. Its priority risk lists, the Proposition 65 cancer data, and *Toxicology One-Liners and Conclusions* contain valuable information on most pesticides in use today and on many chemicals used in the past. However, since this data is generally not available on either the farmer's or the consumer's radar screen, many people remain unaware of the real dangers from exposure to these chemicals or their use on food products.

Chapter 19

1. Orville Schell, *Modern Meat: Antibiotics, Hormones and the Pharmaceutical Farm*, 23.

2. Ibid., and Stephen Hill, "The Food on Your Plate," *Philadelphia Trumpet*, June 2001, http://www.thetrumpet.com/index.php?page/magazine&q=33.

3. Schell, *Modern Meat*, 24–27.

4. Ibid., 41.

5. Ibid., 68.

6. Ibid., 190.

7. Ibid., 187, 190, 192.

8. Ibid., 301.

9. Ibid., 333.

10. Rick Weiss, "New Cattle Antibiotic Nears OK; Groups Warn of Danger to People" *Washington Post*, March 2007.

Chapter 20

1. *Saturday Evening Post*, January 6, 1945.

2. *California Cultivator*, July 20, 1946: 459.

3. J. K. Terres, *New Republic*, 114 (1946): 415.

4. Mellanby, *The DDT Story*, 54–55.

5. Chemically, 2,2,2-trichloro-1,1-bis (chlorophenol) ethanol. Kelthane is oncogenetic (causes tumors) and causes reproductive harm. Its observable toxic effect level is low, though it is probably cancer-causing and is very damaging to both bird and fish populations.

Chapter 21

1. W. H. Lange, "The Artichoke Plume Moth and Other Pests Injurious to the Globe Artichoke," *California Agricultural Experiment Station Bulletin*, 1941: 653.

2. Personal communication: Will Allen was hired by the California Energy Commission to work with the Knolls to help design and monitor the artichoke study.

3. This was possible because Sir Albert and Gabriella Howard always conducted their experiments in India on farm-sized plots, not on ten-by-ten-foot test plots, which are common among American and European university researchers.

4. Rotenone is derived from at least two varieties of derris roots and was, and still is, used as a fish poison in the tropical forests of South America and Southeast Asia.

5. Imhoff, interview with Deke Dietrick, *Farmer to Farmer*.

6. Ibid.

7. Ibid.

8. Perkins, *Insects, Experts, and the Insecticide Crisis*, 61–96.

9. Peter Rosset, *Small Farm Agriculture: Meeting Future Food Needs Sustainably*.

Chapter 22

1. *Seed World*, September 1981: 6.

2. Barry Commoner, "Unravelling the DNA Myth: The Spurious Foundation of Genetic Engineering," *Harpers*, February 2002.

3. Jeffrey Smith, *Seeds of Deception*.

4. By 2002, they had lost that bet to canola and then soy, but 86 percent of U.S. cotton is GMO, so though they lost, GMOs dominate U.S. cotton.

5. Allen R. Meyerson, "Seeds of Discontent," *New York Times*, November 19, 1997: Section D,1–2.

6. *Crop Life*, April 2003: 50.

7. Mindy Laff, "The Process of Biotech Communication," *Seed World*, November 1997: 4.

8. Charlotte Sine, *Farm Chemicals*, 2000 and later.

9. Charlotte Sine, *Farm Chemicals*, October 1999: 276.

10. *Cotton Grower*, Fall 1996: 11.

11. Monsanto promoted the belief in Roundup's safety and effectiveness. Consumers dump millions of pounds of Roundup per year on fields and yards, believing it to be safer than other herbicides. As a result, Monsanto's Roundup is the world's most widely used weed killer.

Roundup is not universally effective, however, and many weeds have begun to develop resistance. Roundup is also not an environmentally benign chemical, as the advertisers claim. Tests by both independent and chemical-corporation labs found that Roundup caused liver damage and tumors in test animals. Recent reports allege that it causes non-Hodgkins lymphoma, sterilizes soil, and damages fish and wildlife habitats.

12. Monsanto needed farmers to sign contracts that required the use of their brand of glysophate (Roundup Ultra) because their exclusive patent ran out in the year 2000. But if farmers violated the contract in any way, or if Monsanto perceived that they were violators, then Monsanto took them to court.
13. Meyerson, "Seeds of Discontent."
14. Ibid.
15. Center for Food Safety, *Monsanto vs. U.S. Farmers*. A 2004 report on the impact on farmers of Monsanto's aggressive patent protection schemes.
16. In 2001 GMO cottonseed in California shattered when ginned and spread seed-coat fragments on cleaned fiber. This required additional cleaning, which the farmers paid for. After three years of use the amount of Roundup needed for weed control greatly increased. After a decade, five major weeds were resistant to Roundup. By 2007, the number of resistant weeds had reached twelve.
17. "Fighting Marestail," *Cotton Grower*, January 2003: 24–25; "Resistant Horseweed," *Cotton Grower*, April 2004: 28. Weed species showing serious resistance include marestail, rigid ryeweed, waterhemp, velvetleaf, and giant ragweed. Ron Swoboda, "Weed Worries," *Wallaces Farmer*, March 2003: 19–20. Pam Henderson and Wayne Wenzell, "Glyphosate Is Under Attack and More Weeds Are Winning," *Farm Journal*, February 2007: 12.
18. "EPA Toughens Bt Refuge Rules," *Wallaces Farmer*, January 2004: 18.
19. *Farm Chemicals*, November 1999.
20. *California Farmer* editorial; November, 1999.
21. Pew Initiative on Food and Biotechnology, "US vs. EU: An Examination of the Trade Issues Surrounding Genetically Modified Food," August 2003. Reported in the 2004 Center for Food Safety report *Monsanto vs. U.S. Farmers*, 279.
22. Archer Daniels Midland and Continental Grain September 1, 1999 directive to grain elevators to not accept GMO crops.
23. *CropLife*, January 2003: 46.
24. These debacles include FlavrSavr tomatoes and Frostban in the early 1990s, Roundup Ready canola (rapeseed) crossing with a wild mustard (*Brassica* sp.) in 1996, and pollen from Bt corn drifting onto milkweed where the monarch butterflies were feeding in 1999. In 2000, the seed giant Aventis (a merger of Hoechst and Rhone-Poulenc) released Starlink corn, which was supposed to be fed only to animals. Aventis failed to inform many farmers and grain elevators that this corn could not be made into human food. When environmentalists found Starlink in taco shells, the disaster hit Aventis and made thousands of consumers ill from allergic reactions to Starlink. Starlink and the butterflies are the tip of a potentially very large iceberg of trouble for the chemical corporations. In 2001, genetically engineered corn was found to have contaminated the wild races of corn in the heartland of Mexico.

Other potentially disastrous possibilities abound. If an escaped salmon, altered with human and poultry genes to increase its size, mates with a native fish and DNA is transferred from one species to the other, there will be no way to trace or reverse that process. (In fact, such fish already exist, and their "parent" corporation is awaiting government permission to market them.) If a cotton plant engineered with Roundup Ready genes mates with one of its close relatives in the Mallow family and makes that weed resistant to herbicides, farmers will need to use even more and stronger herbicides than they now use. If an unexpected mutation, such as a susceptibility to some pathogen, appears in the third or fourth generation as an unforeseen result of genetically altering a plant or animal and is let loose on the commercial marketplace, it will suddenly be too late; the damage will already have been done.

25. Galen P. Dively et al., "Effects on Monarch Butterfly Larvae (*Lepidoptera Danaidae*) after Continuous Exposure to Cry1Ab-Expressing Corn during Anthesis," *Environmental Entomology*, August 2004: 1116–25.
26. Stanley Ewan and Arpod Pusztai, "Effects of Diets containing Genetically Modified Potatoes Expressing *Galanthus Nivalis* Lectin on Rat Small Intestines," *Lancet* 354 (1999): 1353.
27. *Ecologist*, December/January 2006.
28. *Center for Food Safety Newsletter*, December 2005.
29. Ibid., 27–29.
30. "GMO Crops: Who Is Responsible for the Consequences of Pollen Drift?" *Farmer's Digest*, 2004.

Chapter 23

1. See especially Roland Marchand, *Advertising the American Dream*, and Rowsome, *They Laughed When I Sat Down*.
2. Gene Logsdon, "What Goes a Roundup Comes a Roundup: Even If Herbicides Are Safe, Wouldn't We Be Better Off without Them?" *The New Farm*, March/April 1992: 43.
3. The Haber Bureau consisted of Haber and other Nobel Prize winners: Walther Nernst, Emil Fischer, and Richard Willstaetter. They informed the government that the German dyestuff industry was the source of poisons such as bromine, chlorine, and phosgene, which could easily be converted to weapons of mass asphyxiation or deadly pesticides. The German government and scientists conducted their experiments without Haber, after 1933, however, because being a Jew, he was forced to escape into exile. He never returned to Germany. See Borkin, *The Crime and Punishment of IG Farben*. The USDA also focused most of its research energy during World War II on an evaluation of the wartime potential of pesticides and other toxic chemicals. See John H. Perkins, *Insects, Experts, and the Insecticide Crisis*, 3–10.
4. California EPA Department of Pesticide Regulation, *Pesticide Use Reports*, 1970–96; California EPA Department of Pesticide Registration, *Toxicology One-Liners and Conclusions*, 1998, 2002. See especially Borkin, *The Crime and Punishment of I.G. Farben*.
5. Acute toxicity means that it doesn't take very much of these poisons to kill or seriously injure. Farmworkers and children are the most common victims of organophosphates and carbamates, because of the frequency of exposure of the former and generally smaller body size of the latter. Even though the most injured sectors of the population are kids and smaller-bodied farmworkers, the thresholds for danger were established based on exposure for white adult males.
6. Department of Commerce, *EU White Paper: Strategy for a Future Chemicals Policy* (undated, but probably January or February 2002).
7. Department of Commerce, *Chemicals White Paper* (undated, but probably February 2002).

8. World Wildlife Fund, "How the US Can Benefit from Chemicals Policy Reform in Europe," presentation to National Foreign Affairs Training Center, June 24, 2003.

9. Cable from J. J. Foster, deputy chief of mission, U.S. mission to the EU, to Secretary Powell, January 9, 2003.

10. E-mail from U.S. Trade Representative Barbara Norton to certain trade associations and companies, April 4, 2003.

11. U.S. House of Representatives, Committee on Government Reform, *The Chemical Industry, the Bush Administration, and European Efforts to Regulate Chemicals: A Special Interest Case Study* (April 1, 2004). Much of our analysis for REACH comes from this revealing report prepared for Congressman Henry Waxman.

12. Cable from Secretary of State Colin Powell to posts in the EU member states and the mission to the EU, April 29, 2003. Copied to EU candidate states.

13. Letter from Tony Blair, prime minister of the UK, Jacques Chirac, prime minister of France, and Gerhard Schroeder, chancellor of Germany, to Professor Romano Prodi, president of the European Commission, September 20, 2003.

14. *EEB Position on Commission Proposal for a Regulation on REACH*, December 10, 2003. The European Environmental Bureau (EEB) is a federation of 143 NGO environmental groups from thirty-one European countries.

Chapter 24

1. Arsenic significantly contaminated tree- and fruit-growing regions of Washington, Oregon, California, and New York and the cotton-growing regions of the South.

2. The groundwater basins on the east side of the San Joaquin Valley are contaminated with DBCP—dibromochloropropene—though the chemical corporations promised it would not get in the groundwater.

3. Terry Kelley, "Grower Guidelines: Finding Methyl Bromide Alternatives." Editorial in *American Vegetable Grower*, July, 2007, Meister Publishing: 35. Will Allen, *Out of the Frying Pan, Avoiding the Fire: Ending the Use of Methyl Bromide*.

4. The 1954 *U.S. Census General Report*, Table 20: "Number of Farms, All Land in Farms, and Value of Farms 1850–1954," 46–47.

Bibliography

Aftalion, Fred. *A History of the International Chemical Industry.* Philadelphia: University of Pennsylvania Press, 1991.

Albrecht, William. *The Albrecht Papers (1918–1947).* Kansas City, MS: Acres, USA Publishing Co., 1975.

Alewitz, Sam. *Filthy Dirty: A History of Unsanitary Philadelphia in the Late Nineteenth Century.* New York: Garland Publishing Inc., 1989.

Allen, Will, and Eric Sotelo. *A Comparison of the National Agricultural Statistical Service Surveys and the California Environmental Protection Agency, Department of Pesticide Regulation Use Reports for Cotton Pesticides: USDA's Smokescreen of Pesticide Abuse.* 2004. Unpublished manuscript.

Allen, William. *A Ceramic Sequence from the Alto Pachitea, Peru: Some Implications for the Development of Tropical Forest Cultures in South America.* PhD diss., University of Illinois, 1968.

———. *Out of the Frying Pan, Avoiding the Fire: Ending the Use of Methyl Bromide.* Ed. John Pasacantando. Washington: Ozone Action, 1995.

———. *Some Ecological Factors Influencing the Prehistoric Population Density in the Alto Pachitea Region of Peru.* Paper read at the 1967 Annual Meeting of the Society for American Archaeology. Milwaukee.

Allen, William, and Judy Holshouser Tizon. *Land-Use Patterns in the Alto Pachitea Region of Peru.* In *Variation in Anthropology: Essays in Honor of John C. McGregor.* Urbana: Illinois Archaeological Survey Publication, 1971.

Bimba, Anthony. *The History of the American Working Class.* New York: International Publishers Company, 1927.

Bizzell, W. B. *The Green Rising: An Historical Survey of Agrarianism, with Special Reference to the Organized Efforts of the Farmers of the United States to Improve Their Economic and Social Status.* New York: The Macmillan Co., 1926.

Borkin, Joseph. *The Crime and Punishment of I. G. Farben.* New York: The Free Press, 1978.

Boyd, Steven R. *The Whiskey Rebellion: Past and Present Perspectives.* New York: Greenwood Press, 1985.

Carson, Rachel. *Silent Spring.* Greenwich, CT: Fawcett Publications, 1962.

Chambers, Clarke A. *California Farm Organizations: A Historical Study of the Grange, the Farm Bureau and the Associated Farmers, 1921–1941.* Berkeley: University of California Press, 1952.

Commoner, Barry. "Unravelling the DNA Myth: The Spurious Foundation of Genetic Engineering." *Harpers* (February 2002).

Demaree, A. L. *The American Agricultural Press.* New York: Columbia University Press, 1940.

Diamond, Jared. *Guns, Germs, and Steel.* New York: W.W. Norton and Co., 1997.

Dunning, N. A., ed. *The Farmers' Alliance History and Agricultural Digest.* Washington, DC: The Alliance Publishing Company, 1891.

Ewan, Stanley, and Arpod Pusztai. "Effects of Diets Containing Genetically Modified Potatoes Expressing *Galanthus Nivalis Lectin* on Rat Small Intestine." *Lancet* 354 (1999): 1353.

Goodwyn, Lawrence. *The Populist Moment.* New York: Oxford University Press, 1978.

Gray, Lewis Cecil. *History of Agriculture in the Southern United States to 1860.* 2 vols. Carnegie Institution of Washington, publication no. 430. New York: Peter Smith Publishers, 1941.

Griggs, Barbara. *Green Pharmacy.* New York: Viking Press, 1981.

Haber, L. F. *The Chemical Industry in the Nineteenth Century.* London: Oxford University Press, 1958.

———. *The Chemical Industry, 1900 to 1930.* London: Oxford University Press, 1971.

Herbst, Arthur, et al. *Comprehensive Gynecology.* St. Louis: Mosby-Yearbook, 1992.

Hindly, Diana and Geoffrey. *Advertising in Victorian England, 1837–1901.* London: Wayland Publishers Ltd., 1972.

Howard, Sir Albert. *An Agricultural Testament.* New York: Oxford University Press, 1943.

Imhoff, Dan. *Farmer to Farmer.* Davis, California: Alliance of Family Farmers (CAFF), 1997.

Kallett, Isadore, and F. J. Schlink. *100,000,000 Guinea Pigs*. New York: Grosset and Dunlap, 1933.

Karolevitz, Robert F. *Old Time Agriculture in the Ads* (Aberdeen, SD: North Plains Press, 1970).

Kimbrell, Andrew, ed. *Fatal Harvest*. Washington: Island Press, 2001.

———. *Monsanto vs. U.S. Farmers*. Washington: Center for Food Safety, 2004.

King, F. H. *Farmers of Forty Centuries*. 1911. Reprint, Emmaus, PA: Rodale Press, 1949.

Kroese, Ron. "Agriculture's War against Nature." In *Fatal Harvest*, ed. Andrew Kimbrell. Washington: Island Press, 2001.

Lange, W. H. "The Artichoke Plume Moth and Other Pests Injurious to the Globe Artichoke." *California Agricultural Experiment Station Bulletin*, 1941, 653.

Lee, Wayne E. *Crowds and Soldiers in Revolutionary North Carolina: The Culture of Violence in Riot and War*. Gainesville, FL: University Press of Florida, 2001.

Liebig, Justus von. *Organic Chemistry and Its Applications to Agriculture and Chemistry*. Edited from author's manuscript by Lyon Playfair. Taylor and Walton, 1840.

Liebman, Ellen. *California Farmland: A History of Large Agricultural Land Holdings*. Totowa, NJ: Rowman and Allenhead, 1983.

Malthus, *Thomas Robert. An Essay on the Principle of Population*. Reprinted with a four-volume set. London: Routledge, 1797.

Mann, Charles C. *New Revelations of the Americas before Columbus*. New York: Knopf, 2005.

Marchand, Roland. *Advertising the American Dream*. Berkeley: University of California Press, 1985.

Margolin, Malcolm. *The Ohlone Way: Indian Life in the San Francisco–Monterey Bay Area*. Berkeley, California: Heyday Books, 1978.

———. *The Way We Lived: California Indian Stories, Songs and Reminiscences*. Berkeley, California: Heyday Books, 1993.

Markham, Jesse W. *The Fertilizer Industry: Study of an Imperfect Market*. Nashville, TN: Vanderbilt University Press, 1958.

Marschall, Richard. *The Tough Coughs as He Plows the Dough*. New York: William Morrow and Co., 1987.

Marx, Karl. *Capital*. The Modern Library. New York: Charles Kerr and Co., 1906.

McWilliams, Carey. *Factories in the Field: The Story of Migratory Farm Labor in California*. 1935. Reprint, Santa Barbara and Salt Lake City: Peregrine Smith Inc., 1971.

Mellanby, Kenneth. *The DDT Story*. London: The British Crop Protection Council, 1992.

Minear, Richard H. *Dr. Seuss Goes to War*. New York: The New Press, 1999.

Moulton, Forest Ray, ed. *Liebig and after Liebig: A Century of Progress in Agricultural Chemistry*. Washington, DC: American Association for the Advancement of Science, 1942.

Morgan, Judith, and Neil Morgan. *Dr. Seuss and Mr. Geisel*. New York: De Capo Press, 1995.

Morgan, Lewis Henry. *Ancient Society*. London: Macmillan and Co., 1877.

Murillo, Cathy. "RATS: Are There More Around This Year?" *Santa Barbara Independent*, July 22, 1999.

Myers, Gustavas. *The History of the Great American Fortunes*. Vol. 1. The Modern Library. New York: Charles Kerr and Co., 1936.

Nevett, T. R. *Advertising in Britain*. London: Heineman Publishing, 1982.

O'Neal, James. *The Workers in American History: From Conspiracy to Collective Bargaining*. 1912. Reprint, New York: Arno and the New York Times, 1971.

Palmer, Rachael, and Isadore Alpher. *40,000,000 Guinea Pig Children*. New York: Vangaurd Press, 1937.

Pendergrast, Mark. *For God, Country, and Coca Cola*. New York: Orion Books, 1993.

Perkins, John H. *Insects, Experts and the Insecticide Crisis*. New York: Plenum Press, 1982.

Presbrey, Frank. *The History and Development of Advertising*. Garden City, NY: Doubleday, Doran and Company, Inc., 1929.

Quinn, Daniel. *The Story of B*. New York: Bantam Books, 1997.

Richards, Leonard L. *Shays's Rebellion: The American Revolution's Final Battle*. Philadelphia: University of Pennsylvania Press, 2002.

Rogers, James E. *History of Agriculture and Prices*. London: Oxford Press, 1866.

Rosset, Peter. *Small Farm Agriculture: Meeting Future Food Needs Sustainably*. Proceedings of the Second Asia Pacific Conference on Sustainable Agriculture. October 1999, Phitsanulok, Thailand.

Rossiter, Margaret. *The Emergence of Agricultural Science*. New Haven: Yale University Press, 1975.

Rowsome, Frank. *They Laughed When I Sat Down: An Informal History of Advertising in Words and Pictures*. New York: Bonanza Publishing, 1959.

Sasuly, Richard. *I. G. Farben*. New York: Boni and Gaer Publishers, 1947.

Schell, Orville. *Modern Meat: Antibiotics, Hormones and the Pharmaceutical Farm*. New York: Random House, 1984.

Schlink, F. J. *Eat, Drink, and Be Wary*. New York: Covici, Friede Publishers, 1935.

Seuss, Dr. *The Lorax*. New York: Random House, 1971.

Shiva, Vandana, A. H. Jafri, A. Emani, and M. Pande. *Seeds of Suicide: The Ecological and Human Costs of Globalisation of Agriculture*. Delhi: Research Foundation for Science, Technology, and Ecology, 2002

Sinclair, Upton. *The Jungle*. New York: Doubleday Press, 1906. Reprint, Heritage edition, Doubleday Press, 1965.

Skaggs, Jimmy M. *The Great Guano Rush*. New York: St. Martin's Press, 1944.

Sloane, Eric. *Our Vanishing Landscape*. New York: Wilfred Funk Inc., 1955.

Smith, Jeffrey. *Seeds of Deception*. Fairfield, IA: Yes Books, 2003.

Snetsinger, Robert. *The Ratcatcher's Child*. Cleveland: Franzak and Foster, 1983.

Steiner, Rudolf. *Agriculture*. London: Biodynamic Agricultural Association, 1924.

Steward, Julian. *Handbook of South American Indians*. Washington, DC: Bulletin of the Bureau of American Ethnology, no. 143, 1949.

Stock, Catherine McNichols. *Rural Radicals*. New York: Penguin Books, 1996.

Sweeney, Del, ed. *Agriculture in the Middle Ages*. Philadelphia: University of Pennsylvania Press, 1995.

Szatmary, David P. *Shays's Rebellion: The Making of an Agrarian Insurrection*. Boston: University of Massachusetts Press, 1980.

Taylor, John. *Arator*. 1813. Reprint, Indianapolis: The Liberty Fund, 1977.

Treadwell, Edward F. *Cattle King*. New York: The Macmillan Company, 1931

Turrentine, J. W. *Potash*. New York: John Wiley and Sons, 1926.

Voorhees, E. B. *Fertilizers*. New York: The Macmillan Co., 1916.

Weatherford, Jack. *Indian Givers*. New York: Ballantine Books, 1988.

————. *Native Roots*. New York: Ballantine Books, 1991.

Whitaker, Adelynne Hiller. *A History of Federal Pesticide Regulation in the U.S. to 1947*. PhD diss., Emory University, 1974.

Whorton, James. *Before Silent Spring: Pesticides and Public Health in Pre-DDT America*. Princeton: Princeton University Press, 1974.

Wines, Richard. *Fertilizer in America*. Philadelphia: Temple University Press, 1985.

Young, James Harvey. *The Medical Messiahs*. Princeton: Princeton University Press, 1967.

————. *The Toadstool Millionaires: A Social History of Patent Medicines in America before Federal Regulation*. Princeton: Princeton University Press, 1961.

Index

Italicized page references refer to illustrations.

Gutenberg Bible, *17*
gypsy moths, 79–80

H

Haber, Fritz, 94, 217
herbal medicine, in colonial America, 52
herbicides
    advertisements, *121, 148, 153, 154*
    and GMO crops, 200–201
    released after World War II, 150, 152–154
heroic medicine, 49–50, 52
Hertz, Roy, 168
Hester, John, 202
Hippocrates, 49
historical markers
    1865-1945, *60–63*
    1946-2007, 134–137
    through 1870, *xxviii–xxix*
Hitler, Adolph, 96
hogs, confinement management of, 165–166
holistic medicine, origins of, 49
hormones
    food additives, risks of, 167–169
    use in livestock farming, 166–167
Horsford, Eben Norton, 37
Howard, Sir Albert, 181, 183, 184
Huffaker Project, 186
Human Genome Project, 192
hunger, world, actual reasons for, 228
Hunt, T. Steery, 38

I

Indian tribes in America, destruction of, 4, 5
industrial disasters
    Bhopal, India, 157
    Union Carbide, *xxvii*
infectious drug resistance and antibiotics, 164–165
inoculants, bean, *104*
insecticides
    advertisements for, *23, 237*
    biological, 184
    labels, early, *x, xiii, xiv, xv*
    warnings, *x*
        See also specific insecticides
insects, beneficial, 70, 172, *185*
Integrated Pest Management, 186
international markets, and GMO products, 203–205
IPM (Integrated Pest Management), 186

J

Johnson, Samuel, 37
journalists
    investigative, and the need for activism, 241
    muckraking, and toxic food, 120
journals
        See farm journals
Jukes, Thomas, 163–164

K

Kallett, Arthur, 120–121
Kelthane, 174–175, *176*
King, F. H., 102
Koebele, Albert, 70
Kroese, Ron, 161

L

labeling
    chemical, consumer advocacy for, 225
    for GMO products, 205, 208
laboratories, chemical, *39*
laborers, for the new U.S. colonies, 3
Laff, Mindy, 194–195
land grabs
    the enclosure movement in Europe, 3–4
    Miller and Lux, 65–66
    near roads, canals, and railroads, 22, 66
    in North America, 3–4, 5
Lange, H.D., 181
Lavoisier, Antoine, 93
Law of the Minimum, *34*, 36
lawyers, and the need for activism, 241
Le Chaterlier, Henri, 94
lead
    on apples in Washington State, 175–176
    and arsenic, 80
    editorial supporting, 126
    poisoning, 120
legislation, requiring chemical testing, 217–218
lice, advertisement for pest control products, *59–60*
Liebig, Justus von
    and the industrialists, 35–39
    Law of the Minimum, the, *34*
    misleading of farmers, 229
    and the populist press, 72
livestock, confinement operations, 163–169
Local Harvest web site, 241
London purple, 78, 79, *81*
The Lorax, 226
Lux, Charles, 66

M

machinery, advertisements for, *64*
magazines, farm
        See farm journals
mail service, 21
malathion, 155–156, *157*, 158
Manifest Destiny, 6
manure
    as fertilizer, propaganda against, 93
    recycling programs, 57
    vs. guano, 28
Mapes, Robert, *29*, 31, 72
marker-assisted genetic research, 210
marketing of GMO products, 203–205
materia medica, 49
meat, tainted, medical problems from, 167